A FIRST COURSE IN FUNCTIONAL ANALYSIS

PURE AND APPLIED MATHEMATICS

A Wiley-Interscience Series of Texts, Monographs, and Tracts

A complete list of the titles in this series appears at the end of this volume.

A FIRST COURSE IN FUNCTIONAL ANALYSIS

S. DAVID PROMISLOW
Department of Mathematics and Statistics
York University
Toronto, Ontario, Canada

WILEY-INTERSCIENCE
A JOHN WILEY & SONS, INC., PUBLICATION

Published by John Wiley & Sons, Inc., Hoboken, New Jersey
Published simultaneously in Canada

Library of Congress Cataloging-in-Publication Data:

Promislow, S. David.
A first course in functional analysis / S. David Promislow.
p. cm.
Includes bibliographical references and index.
ISBN 978-0-470-14619-4 (cloth)
1. Functional analysis. I. Title.
QA320. P765 2008
515'.7–dc22 2007042712

10 9 8 7 6 5 4 3 2 1

To Ashley, William, Rachel, and Jessica

CONTENTS

PREFACE

This book is intended to present the main ideas of functional analysis to beginning graduate or advanced undergraduate students, assuming a minimal background. The impetus for this project came from a one-semester course in functional analysis that I taught for many years. It was a compulsory course for several types of graduate mathematics students, including those in pure mathematics, applied mathematics, and statistics. There was a diversity of backgrounds, and many students had limited prior exposure to such topics as measure theory or general topology. Although there are many excellent books on functional analysis, I found it difficult to select one that was entirely appropriate for this particular course. Many of the standard texts are too encyclopedic and contain much more material than could be covered in one semester. Some require more background than most of the students have, and many proceed at a faster pace than I deemed desirable. Others are too elementary and do not cover topics that I thought essential to include, such as L^p spaces or the elements of topological vector spaces. Accordingly, I designed my own lecture notes, which were well received by the students and formed the basis for this book. Hopefully, it will provide a useful resource for similar classes of students, including mathematics students, as well as those who require a knowledge of functional analysis for applications to other fields, such as the physical sciences, probability theory, economics, or finance.

The main prerequisite is a knowledge of basic real analysis, including such topics as sequences and series, continuity, uniform convergence of functions, topology of the real line, and elements of metric spaces. Readers should also be familiar with topics from elementary linear algebra, such as linear independence, bases, and matrix manipulation.

The book does not assume a prior knowledge of measure theory. This topic is covered in the book in two ways. In the body of the text, the basic facts about measurable functions and the integral are summarized without detailed proofs. This should allow all readers to follow the material on L^p spaces. For those interested in more details, an appendix on measure and integration fills in the gaps by proving most of the assertions stated in the body of the text.

Moreover, the book does not assume any knowledge of general topology. Most of the book deals with normed spaces, for which metric topology is sufficient.

General topology is introduced in a natural way later in the book, after the topic of weak convergence is covered, and it is shown that something more than sequences is necessary in order to properly define a concept of closure associated with this type of convergence. In Chapter 7, concepts of general topology are developed from the beginning and applied to topological linear spaces.

As well as the appendix on measure and integration, appendices are included on a few other topics which could be considered more as background material rather than part of the main subject matter. These are Zorn's lemma, the Stone–Weierstrass theorem, Tychonoff's theorem on product spaces, and upper and lower limit points of sequences.

Throughout the book, I have tried to provide a great deal of motivation. A common complaint that students have of many texts, and of functional analysis books in particular, is that they appear to be a listing of one theorem after another, without much indication of how these theorems are connected or why one is interested in the conclusions. I try, whenever possible, to fully explain the concepts and to provide a sense of unity in the development. The motivation extends to an analysis of proofs. Rather than simply writing down proofs in the most direct and shortest form possible, I frequently outline for the reader the logic behind the steps in an attempt to help them see how the argument develops.

The book for the most part covers standard topics in functional analysis. These include the study of Banach spaces, Hilbert spaces, bounded linear functionals and operators, and an introduction to topological vector spaces. The emphasis is on basic principles of the subject as opposed to particular applications. Nonetheless, I do try to allude briefly to various applications, when they arise, and to provide references for those who wish to pursue them further. The one place where applications are dealt with in more detail is Chapter 10, which covers differential and integral equations.

There are frequent references to the finite-dimensional case, since it is assumed that most readers will be at least somewhat familiar with this area. Many topics in functional analysis can be viewed as generalizations of facts in linear algebra or geometry. However, students are not always aware of this from standard presentations. The book points out these connections in various places, including the discussion of such topics as adjoint operators, spectral theory, functional calculus, and convexity. The chapter on the Hahn–Banach theorem includes a proof of the particular finite-dimensional statement, which is not usually found in functional analysis books.

Although this book is intended to be suitable for those who wish a general knowledge of the subject and are not necessarily going further or specializing in functional analysis, it should also provide a good first book for those planning to study more advanced work in this field. The student who masters the material here will be well prepared to undertake the study of such topics as spectral theory, unbounded operators, additional topics in topological vector spaces, and operator algebras.

The chapters are intended more or less to follow each other logically, although some flexibility in the order of studying the material is possible. Section 4.6 on

Hilbert space operators is not needed for Chapters 5, 6, or 7 and could be covered after those chapters if deemed desirable. That section is however essential for Chapters 8 through 11. Chapter 7, dealing with topology and topological vector spaces, is not needed for the chapters that follow.

While the book was based on the lecture notes for a one-semester course, there is somewhat more material here than can normally be covered in that time. A suitable one-semester course might consist of most of the first nine chapters. For courses in which there is more of an emphasis on applied mathematics, instructors would want to include Chapter 10, adjusting some of the other topics accordingly. Chapter 11 deals with material that might be considered more advanced and could be described as the first chapter in a second course in functional analysis.

Exercises appear at the end of each chapter. They range from routine, which test knowledge of the basic ideas and definitions, to some that are quite challenging.

This book is supported by the following website:
ftp://ftp.wiley.com/public/sci_tech_med/functional_analysis

ACKNOWLEDGMENTS

I want to acknowledge, with appreciation, the influence of Donald Bures, who introduced me to a serious study of analysis, and exposed me to several of the topics covered in this work. Many thanks to Virginia Young for a truly significant contribution. She read the entire manuscript, worked many of the exercises, corrected numerous misprints, and provided valuable comments. Finally, I want to thank my wife Shirley, for her never failing encouragement and understanding during the writing of this book.

S. David Promislow

Toronto, Ontario, Canada
February 2008

1

LINEAR SPACES AND OPERATORS

1.1 INTRODUCTION

A major object of study, which runs through all of mathematics, is that of a *function*. The origins of the subject of *functional analysis* can be traced back to a time, near the beginning of the twentieth century,when mathematicians began to deal with a whole collection or space of functions, rather than looking at one function at a time. This idea proved particularly fruitful to those studying differential and integral equations motivated from problems in physics. It gradually became clear that the problems were remarkably similar to familiar ones of solving algebraic systems of equations, but rather than looking for unknown numbers or vectors, they were looking for unknown elements in a space of functions.

In order to carry out the analysis, they needed notions of distance or length in spaces of functions and were able to formulate this by imitating the geometry of finite-dimensional Euclidean spaces using the Lebesgue integral in place of sums. The type of structure that emerged was later termed *Hilbert space*, named after the German mathematician David Hilbert, one of the early pioneers in this endeavor. It soon became clear that there were other applications, such as Weierstrass's theorem on the approximation of continuous functions with polynomials, which used different types of distances and length. In 1920, the Polish school of mathematicians, led by Stefan Banach, formulated axioms for a general abstract structure that applied to various collections of functions and methods of measuring distance. This is known today as a *Banach space*.

Mathematicians did not just conceive of a set of functions as a static object but rather looked at associated dynamic aspects. One often starts with a function and assigns another function to it. Familiar examples abound in elementary calculus through the operations of differentiation or integration. In other cases one takes a

A First Course in Functional Analysis. By S. David Promislow

function and assigns a number, or a constant function, such as taking a limit of a sequence or summing an infinite series. In many applications of mathematics dealing with quantities changing over time, one has a function at one point of time and is interested in the resulting function at a future point of time. Another familiar example, as indicated above, is a system of linear equations. This can be viewed as a transformation on vectors, which are functions on a finite set. In many cases, the correspondence is linear (defined below), and the assignment of one function to another can be expressed through the abstract concept of a linear operator, which is a particular kind of function defined on one space of functions and taking values in another.

Through this abstraction, one then had a powerful approach that combined aspects of analysis, algebra, and geometry. The same techniques could be brought to bear in diverse settings. Subsequently, applications were found in fields outside of mathematics and physics. These include probability theory, statistics, economics and finance, engineering, and many others. As we go along, we will allude to some of these developments.

Although the subject has evolved considerably, we believe that the main goal of a first course is to investigate this original abstract scheme, which can be compactly described as the study of *bounded linear operators on normed linear spaces*. In this brief introductory chapter we will summarize the basic facts about linear spaces and linear operators, somewhat quickly, as we assume that the typical reader has some knowledge of linear algebra. Most examples will be left to the exercises. The concepts of boundedness and norms are introduced in Chapter 2.

1.2 LINEAR SPACES

1.2.1 Basic Definitions

Throughout the book, we will have occasion to refer to the field of real numbers or the field of complex numbers. In most cases, our statements will hold for both, and we will use the symbol $\mathcal{F}$ to denote either field. For the occasions when we definitely need one or the other, we will use $\mathcal{R}$ for the reals and $\mathcal{C}$ for the complex numbers.

We use standard notation for elements of $\mathcal{C}$. Given a complex number $z = x + iy$ with x, y real, we denote x by Re z, y by Im z, and we let $\bar{z} = x - iy$ denote the *complex conjugate* of z. The absolute value $|z|$ is given by $(x^2 + y^2)^{1/2}$. We can write $z = |z|e^{i\Theta}$, where $e^{i\Theta} = \cos(\Theta) + i\, sin(\Theta)$ has absolute value 1. This is known as the *polar* form of z.

A *linear space* over the field $\mathcal{F}$ is a set X equipped with operations of addition and scalar multiplication subject to some natural rules. For completeness, we state these axioms.

Given any x and y in X and α in $\mathcal{F}$, there are elements $x + y$ and αx in X satisfying the following:

Rules of addition:

(i) $x + y = y + x$;
(ii) $(x + y) + z = x + (y + z)$;
(iii) there is an element 0 in X such that $x + 0 = x$ for all $x \in X$;
(iv) for each $x \in X$, there is an element $-x$ such that $x + (-x) = 0$.

Rules of scalar multiplication:

(i) $1x = x$;
(ii) $\alpha(\beta x) = (\alpha\beta)x$.

Rules connecting the two operations:

(i) $(\alpha + \beta)x = \alpha x + \beta x$;
(ii) $\alpha(x + y) = \alpha x + \alpha y$.

We will use the notation $x - y$ to denote $x + (-y)$.

Addition and scalar multiplication carry over to operations on sets. Given subsets A and B in a linear space and a scalar α, $A + B$ is the set of all elements that can be written as $a + b$ for some $a \in A$ and $b \in B$ and αA is the set of all elements of the form αa for some $\alpha \in \mathcal{F}$ and $a \in A$. In particular, $-1A$ is denoted by just $-A$ and $A + (-B)$ is denoted by $A - B$.

Two trivial examples of linear spaces are $\mathcal{F}$ itself and the set consisting of the single element 0. We discuss more pertinent examples in the next section. Of course, associated with any linear space are several others, namely its linear subspaces defined as follows.

Definition. Let Y be a nonempty subset of the linear space X that is closed under addition and scalar multiplication. That is, for all $x, y \in Y$ and $\alpha \in \mathcal{F}$, the elements $x + y$ and αx are in Y. Then, Y becomes a linear space in its own right under the same operations, and it is called a *linear subspace* of X. (We will normally drop the word *linear* and just use *subspace*.)

Linear spaces are also commonly known as *vector spaces*. Accordingly, we will frequently refer to elements of a linear space as vectors.

We next summarize the concepts of basis and dimension.

Definitions. Given a set B in a linear space, a *finite linear combination of elements of B* is an element x of the form

$$x = \sum_{i=1}^{n} \alpha_i b_i,$$

where for each i we have $\alpha_i \in \mathcal{F}$ and $b_i \in B$. We refer to the set $\{\alpha_i\}$ as the *coefficients*.

We let span(B) denote the set of all finite linear combinations of elements of B. It is the smallest subspace containing the set B, in the sense that if Y is a subspace containing B, then span$(B) \subseteq Y$.

Definition. A set B is said to be a *basis* for the linear space X if every element in X can be written uniquely as a finite linear combination of elements of B with *nonzero* coefficients. (We need nonzero here to ensure uniqueness, for without it we could add an unused element with a zero coefficient.)

Definitions. A linear space is *finite-dimensional* if it has a basis with a finite number of elements. It is a basic theorem of linear algebra that in a finite-dimensional space all bases have the same number of elements, which is known as the *dimension* of the space.

Definition. A set B in a linear space is said to be *linearly independent* if no element in B can be written as a finite linear combination of the others. An equivalent formulation is that if a finite linear combination of elements of $B = 0$, then all coefficients must be zero.

Many authors give the definition of a basis in an alternate form, namely, that it is a linearly independent set B such that span$(B) = X$. That is, the linear independence replaces the uniqueness criterion.We invite the reader to show the equivalence of these two definitions. We prefer to state the definition in the above form, as it is the way in which we most frequently use the concept of basis; additionally, it ties in better with the definition of a Schauder basis, which we present in Chapter 2.

1.2.2 Spaces of Functions

The main reason why linear spaces are important in analysis is that the interesting classes of functions form such an object. To fix the ideas, let S be any set, and let $F(S)$ denote the set of all functions from S to $\mathcal{F}$. By defining $f + g$ and αf in the obvious way as

$$(f + g)(s) = f(s) + g(s), \qquad (\alpha f)(s) = \alpha f(s)$$

for all $s \in S$, it is easy to verify that all of the above rules hold, and $F(S)$ becomes a linear space. The 0 element of this space is just the function that assigns 0 to all points.

When S consists of n elements for some finite n, a function on S can be represented as an ordered n-tuple, and $F(S)$ is nothing more than the familiar n-dimensional space consisting of n-dimensional vectors. Here, $F(S)$ is denoted by $\mathcal{R}^n$ for real scalars or $\mathcal{C}^n$ for complex scalars.

When S is infinite, $F(S)$ is too large to be of much interest, and our focus will be on certain subspaces of $F(S)$. Elementary calculus and analysis courses are filled with examples showing that a certain class of functions forms a subspace of $F(S)$ for various S. When S is an interval of the real line, these include bounded functions, continuous functions, differentiable functions, and polynomial functions, and the list goes on. When S is the set $\mathcal{N}$ of positive integers, our functions are sequences. Interesting subspaces include bounded sequences, convergent sequences, sequences converging to 0, summable sequences, and others. We will investigate many of these subspaces in more detail in the remaining chapters.

Rarely will we again have occasion to verify the rules above in order to demonstrate that we have a linear space. Almost all of the examples we discuss will be subspaces of $F(S)$ for some S, with a couple of variations, which we point out as they arise.

Remark on Notation. To emphasize the functional aspect, we will sometimes write vectors or sequences with arguments in brackets, as opposed to the more traditional use of subscripts. So, for example, when talking about several vectors or sequences, we will often write $x = (x(1), x(2), \ldots, x(n))$ and reserve subscripts to distinguish between different vectors or sequences.

Given $s \in S$, we will let $\delta_s \in F(S)$ denote the function that takes the value 1 at the point s and 0 elsewhere. For a general subset $A \subseteq S$, we will use 1_A to denote the function that takes the value of 1 on A and 0 on the complement of A. (This is often referred to as *the characteristic function* of A. Note that this term is used in probability theory with a completely different meaning.)

1.3 LINEAR OPERATORS

1.3.1 Basic Definitions

If X and Z are two linear spaces, a *linear operator* from X to Z is a rule T that associates to each $x \in X$ a unique element $Tx \in Z$ such that, for all $x, y \in X$ and $\alpha \in \mathcal{F}$, we have

$$T(x+y) = Tx + Ty, \qquad T(\alpha x) = \alpha Tx.$$

Linear operators are also referred to as *linear transformations* or *linear mappings*.

We will use the familiar notation $T : X \to Z$ to denote a mapping from X to Z, and unless other specified, this notation will imply that the mapping is linear.

For most readers, one of the most familiar examples of a linear operator will likely be the operation on $\mathcal{R}^n$ of multiplying a vector by a matrix. Indeed, all linear operators on $\mathcal{R}^n$ can be so expressed, as we show in Section 1.3.4. Various examples of operators on sequence spaces appear in Exercises 1.1, 1.3, and 1.4.

1.3.2 Null Spaces and Ranges

To every $T : X \to Z$ we associate two important subspaces. The *null space* (also known as the *kernel*), denoted by $N(T)$, consists of all $x \in X$ such that $Tx = 0$.

The *range* of T (also known as *the image*), denoted by $R(T)$, consists of all $z \in Z$ such that $Tx = z$ for some $x \in X$. These are easily seen to be subspaces of X and Z, respectively.

By definition, a linear operator T is *surjective* (onto) if and only if $R(T) = Z$. We next verify that a linear operator T is *injective* (one-to-one) if and only if $N(T) = 0$. It is easily seen that we always have $T(0) = 0$, so if x is a nonzero vector in $N(T)$, then $Tx = T(0)$ and the map is not injective. Conversely, if $N(T) = 0$, then by invoking linearity, we have $Tx = Ty$ if and only if $T(x - y) = 0$ if and only if $x - y = 0$ if and only if $x = y$.

1.3.3 Equations and Inverses

Many problems that arise in mathematics are of the following type. We are given $T : X \to Z$, an element $z \in Z$, and we want to find an element $x \in X$ satisfying the equation

$$Tx = z.$$

Systems of linear equations can be put into this form. Indeed, the beginning of functional analysis, to which we referred to in the introduction, involved problems of this type with T an integral operator on continuous functions or a differential operator on differentiable functions.

The first questions one usually asks about such an equation is whether a solution exists and, if so, whether it is unique. Answers can be easily formulated in terms of the subspaces of the previous section. A solution exists for a given z if and only if $z \in R(T)$; therefore, one exists for all z if and only if $R(T) = Z$. Similarly, if one exists, it is unique if and only if $N(T) = 0$ since $Tx = Ty$ if and only if $x - y \in N(T)$.

If $N(T) = 0$, we can define the *inverse* transformation $T^{-1} : R(T) \to X$ by setting $T^{-1}z$ equal to the unique x such that $Tx = z$. It is straightforward to show that T^{-1} is linear. In this case, the solution to the above equation is given for all $z \in R(T)$ by $x = T^{-1}(z)$. This shows the importance of inverses, and we will deal much more with this concept in the chapters to follow.

1.3.4 Space of Linear Operators

We now observe one of the variations to which we alluded in Section 1.2 when we said that nearly all of the linear spaces which we would encounter would be linear subspaces of $F(S)$. In defining the linear space of functions on S we can allow the functions to take values in any linear space Z, rather than be restricted to $\mathcal{F}$. The definitions of addition and scalar multiplication of functions are the same as given, using the corresponding operations of Z. Additionally, suppose that our domain S is itself a linear space X. Then, the subset of the space of functions from X to Z consisting of the linear operators is easily seen to be a subspace of all the functions on X with values in Z.

To summarize, given any two linear spaces X and Z, we get a third linear space, namely, the linear operators from X to Z, with the usual pointwise addition and scalar multiplication inherited from Z. We will denote this space by $\mathcal{L}(X, Z)$. When $X = Z$, we shorten this to $\mathcal{L}(X)$.

Most readers will have previously encountered a version of $\mathcal{L}(X, Z)$ in the finite-dimensional case as a space of matrices. Given spaces X with a basis $\{b_1, b_2, \ldots, b_n\}$ and Z with a basis $\{c_1, c_2, \ldots, c_m\}$, any $T \in \mathcal{L}(X, Z)$ is determined by its value on the b_j's. This follows since, for $x = \sum_{j=1}^{n} \alpha_j b_j$, the linearity of T (which extends from two to any finite number of terms by induction) shows that

$$Tx = \sum_{j=1}^{n} \alpha_j \, Tb_j.$$

Now, for each j, we can write

$$Tb_j = \sum_{i=1}^{m} a_{i,j} \, c_i$$

for some unique scalars $a_{i,j}$. All the information we need about the operator is encoded by the matrix A, whose entry in the ith row and jth column is $a_{i,j}$. In fact, multiplying the column vector $(\alpha_1, \alpha_2, \ldots, \alpha_n)$ on the left by the matrix A gives us, as a column vector, the coefficients of Tx with respect to the given basis of Z. Moreover, the matrix associated with the sum of two operators is the usual pointwise sum of two matrices, and the matrix associated with αT is that of T multiplied by the scalar α. So, after choosing the two bases, we can identify $\mathcal{L}(X, Y)$ as the linear space of all $m \times n$ matrices.

1.4 PASSAGE FROM FINITE- TO INFINITE-DIMENSIONAL SPACES

An example of an n-dimensional function space is $F(S)$ in which $S = \{1, 2, \ldots, n\}$. We can take $\{\delta_i : i = 1, 2, \ldots, n\}$ as a basis since for any $f \in F(S)$ we can write f uniquely as

$$f = \sum_{i=1}^{n} f(i)\delta_i.$$

This is a special case, and nearly all of the interesting function spaces are infinite-dimensional in the sense that they do not possess a finite basis. When one passes to the infinite-dimensional case, several of the conclusions and properties of finite-dimensional spaces no longer hold. However, many of these results can be recovered if one deals with *closed* linear subspaces and *continuous* linear operators, in place of arbitrary linear subspaces and operators, provided that we take suitable definitions for these terms. Such notions are often not needed in problems involving finite-dimensional spaces, since the linear structure dominates in that case. In finite dimensions, linear subspaces are automatically closed and linear operators are automatically continuous for any reasonable definitions of these concepts, as we will show later. Therefore, in studying infinite-dimensional spaces, we must go beyond linear algebra and incorporate topological notions that allow us to speak of such things as convergence, continuity, and closure. We will begin to do this in the next chapter.

EXERCISES

In Exercises 1.1 through 1.4, let s denote the linear space consisting of all sequences.

1.1 Let c_{00} denote the set of all sequences that have only finitely many nonzero entries.

(a) Show that c_{00} is an infinite-dimensional subspace of s. Find a basis.

(b) Show that the mapping $T : c_{00} \to \mathcal{F}$, given by $Tx = \sum_{i=1}^{\infty} x(i)$, is linear.

1.2 For any subset A of positive integers, let s_A denote the set of all sequences x such that $x(n) = 0$ for all $n \in A$. Show that s_A is a subspace of s.

1.3 Given any $y \in s$, define a function $M_y : s \to s$ by $M_y(x)(n) = y(n)x(n)$.

(a) Show that M_y is a linear operator.

(b) Describe the spaces $N(M_y)$ and $R(M_y)$.

1.4 Define functions S and T from s to itself by

$$S(x(1), x(2), x(3), \ldots) = (0, x(1), x(2), \ldots),$$
$$T(x(1), x(2), x(3), \ldots) = (x(2), x(3), \ldots).$$

(a) Show that S and T are linear operators.

(b) For both S and T, find the null space and range.

(c) Compute the compositions ST and TS.

1.5 Verify the assertion made in Section 1.4 that $\mathcal{L}(X, Z)$ is a subspace of the space of all functions from X to Z.

1.6 Consider the space $\mathcal{L}(X, Z)$, where X and Z are infinite-dimensional. Which of the following, if any, are subspaces?

(a) $\{T : N(T) \text{ is finite-dimensional }\}$;

(b) $\{T : R(T) \text{ is finite-dimensional }\}$.

1.7 Suppose that for each λ in some index set Λ we are given a subspace Y_λ of a linear space X such that for all $\lambda, \mu \in \Lambda$ either $Y_\lambda \subset Y_\mu$ or $Y_\mu \subset Y_\lambda$. Show that $\bigcup_{\lambda \in \Lambda} Y_\lambda$ is a subspace of X.

1.8 Let S be an infinite set, and let X be a linear space of functions that includes δ_s for all $s \in S$. Show that X is infinite-dimensional.

1.9 Show that the two definitions of a basis given in this chapter are equivalent.

2

NORMED LINEAR SPACES: THE BASICS

2.1 METRIC SPACES

The concept of a *metric space* (X, d), extends notions such as convergence, closure, and continuity from the real numbers to more general settings. We assume that most readers are somewhat familiar with these ideas and will summarize the main points quickly. Here, X is any set, and a *metric* d is a rule that associates with any two points x and y in X, a nonnegative number $d(x, y)$, which we think of as giving the distance between x and y. The motivating example is the standard metric on the reals given by $d(x, y) = |x - y|$.

A metric d must satisfy the following axioms: For all x, y, z in X:

(i) $d(x, y) \leq d(x, z) + d(z, y)$;
(ii) $d(x, y) = d(y, x)$;
(iii) $d(x, y) = 0$ if and only if $x = y$.

The most frequently exploited axiom is (i), which tells us that if we want to show that the distance between x and y is "small," we can do so by showing that for some intermediate point z the distance from x to z and the distance from z to y are both small since the distance from x to y is less than the sum of these two distances. We can, of course, introduce more intermediate points and iterate the procedure.

We also want to define the distance from a point x to a subset $A \subseteq X$. In terms of the metric d, this distance is given by

$$d(x, A) = \inf\{d(x, a) : a \in A\}.$$

(Definitions of sup and inf appear in Appendix C.)

A sequence (x_n) in a metric space (X, d) *converges* to a point $x \in X$ if the sequence eventually gets and stays arbitrarily close to x. Precisely, we require

A First Course in Functional Analysis. By S. David Promislow

that the sequence of numbers $d(x_n, x)$ converges to 0. We call x the *limit* of the sequence and write $x = \lim x_n$ or $x_n \to x$.

Given two metric spaces (X_1, d_1) and (X_2, d_2), a function f from X_1 to X_2 is continuous at a point $p \in X_1$ if, intuitively, it maps points close to p into points close to $f(p)$. There are various ways of making this precise. A standard definition is that, given any $\epsilon > 0$, there is a $\delta > 0$ such that $d_1(p, q) < \delta$ implies that $d_2(f(p), f(q)) < \epsilon$. Another formulation requires that for any sequence (p_n) converging to p the sequence $(f(p_n))$ converges to $f(p)$. We leave it to the reader to show the equivalence of these two definitions.

Definition. For any subset A of a metric space, we define the *diameter* of A by

$$\text{diam}(A) = \sup\{d(x, y) : x, y \in A\}.$$

The set A is said to be *bounded* if the diameter of A is less than ∞.

We will not dwell further here on the concept of general metric spaces. When dealing with linear spaces, we normally want not just any metric but rather one that is compatible with linear operations. We make this precise in the following section.

2.2 NORMS

2.2.1 Definitions and Consequences

A *norm* on a linear space X is a rule that assigns to each $x \in X$ a nonnegative number $\| x \|$ such that for all x, $y \in X$ and $\alpha \in \mathcal{F}$:

(i) $\| x + y \| \leq \| x \| + \| y \|$ (triangle inequality);
(ii) $\| \alpha x \| = |\alpha| \, \| x \|$ (scalar property);
(iii) $\| x \| = 0$ implies that $x = 0$.

If $\| \cdot \|$ is a norm on a linear space X, then $(X, \| \cdot \|)$ is called a *normed linear space*. (We will usually just refer to X as a normed linear space.) If axiom (iii) is omitted, we call $\| \cdot \|$ a *seminorm*.

A norm immediately induces a metric d via

$$d(x, y) = \| x - y \| .$$

The norm then gives us the distance of an element from 0, and this determines all other distances since it is immediate from the definition that

$$d(x + z, y + z) = d(x, y),$$

a property known as *translation invariance*.

Example 2.1. The most familiar example of a norm to most readers is likely the Euclidean norm on $\mathcal{R}^n$ or $\mathcal{C}^n$. Given a vector $x = (x(1), x(2), \ldots x(n))$, this norm is defined by

$$\| x \| = \left(\sum_{i=1}^{n} x(n)^2 \right)^{1/2} .$$

The induced metric corresponds to physical distance in two- or three-dimensional space. Moreover, the use of this norm is prevalent throughout applied mathematics and statistics through the idea of least squares approximation. It is not obvious that the triangle inequality holds, but a proof can be found in almost all books on linear algebra or multivariate calculus. We will provide a proof later in a more general context in Section 2.9. Note that, when $n = 1$, this norm reduces to the familiar absolute value.

Example 2.2 Another familiar method of measuring distance in $\mathcal{R}^n$ or $\mathcal{C}^n$ is by adding the absolute value of all the deviations between the coordinates. That is, we define

$$\| x \| = \sum_{i=1}^{n} |x(n)|,$$

which also reduces to the absolute value when $n = 1$. In this case, all the norm axioms, including the triangle inequality, are readily verified.

Other examples of norms will follow later.

There are various basic facts that follow immediately from the norm axioms. By taking $\alpha = 0$ in axiom (ii), we see that $\| 0 \| = 0$. By taking $\alpha = -1$, we see that $\| -x \| = \| x \|$.

Another observation is that, by induction, axiom (i) extends from two to any finite number of elements. That is, given $x_1, x_2, \ldots, x_n \in X$, we get a general triangle inequality,

$$\left\| \sum_{i=1}^{n} x_i \right\| \leq \sum_{i=1}^{n} \| x_i \| .$$

Finally, we get the *backwards triangle inequality*

$$| \| x \| - \| y \| | \leq \| x - y \|$$

by noting that $\| x \| = \| x - y + y \| \leq \| x - y \| + \| y \|$. By subtracting $\| y \|$ from each side and then interchanging x and y, we obtain the given inequality.

When we speak of convergence with respect to a norm, we are referring to convergence in the induced metric. That is, a sequence (x_n) converges to x if $\| x_n - x \|$ converges to 0. Similarly, we can speak of continuity with respect to a norm.

The fact that the metric induced from a norm is connected with the linear operations is shown most vividly by the following theorem, which we will use repeatedly.

Theorem 2.3 (Continuity of Operations). *Suppose that (x_n) and (y_n) are sequences in a normed linear space and (α_n) is a sequence in $\mathcal{F}$ such that* $\lim x_n = x$, $\lim y_n = y$, *and* $\lim \alpha_n = \alpha$. *Then:*

(*a*) $\lim(x_n + y_n) = x + y$;
(*b*) $\lim \alpha_n x_n = \alpha x$;
(*c*) $\lim \| x_n \| = \| x \|$.

Proof. We leave the proof of (a) and (c) to the reader. (Use the backwards triangle inequality for the latter.) We will demonstrate (b) by the standard adding-and-subtracting technique used with two-variable functions. Since (α_n) is convergent, it is bounded and we can find K such that $|\alpha_n| \leq K$ for all n. Then,

$$\begin{aligned} \| \alpha_n x_n - \alpha x \| &= \| \alpha_n x_n - \alpha_n x + \alpha_n x - \alpha x \| \\ &\leq |\alpha_n| \, \| x_n - x \| + |\alpha_n - \alpha| \, \| x \| \\ &\leq K \| x_n - x \| + |\alpha_n - \alpha| \, \| x \| . \end{aligned}$$

Since both sequences on the right converge to 0, so does the sequence on the left. □

Another important property of convergence is given in the following proposition.

Proposition 2.4 (Uniqueness of Limits). *If a sequence (x_n) in a normed linear space converges to both x and y, then $x = y$.*

We leave the proof to the reader. Note that this proposition does not hold in a seminormed space. Indeed, if (x_n) converges to x and $\| x - y \| = 0$, one can easily verify that (x_n) also converges to y.

2.2.2 Open and Closed Sets

Many ideas involving convergence and continuity are more easily dealt with through the concept of open and closed sets. Given a normed linear space X, we define for all $x \in X$ and $r > 0$

$$\begin{aligned} B(x; r) &= \{y \in X : \| x - y \| < r\} \quad \text{(open } r\text{-ball around } x); \\ \bar{B}(x; r) &= \{y \in X : \| x - y \| \leq r\} \quad \text{(closed } r\text{-ball around } x). \end{aligned}$$

In both cases we refer to r as the *radius* of the ball. If the first argument x is omitted, it is taken to be 0. That is, $B(r)$ consists of all elements with norm $< r$. It is clear that

$$B(x; r) = x + B(r).$$

Definitions. Given any set $A \subseteq X$ and $x \in X$, we say that x is

(i) an *interior* point of A if $B(x; r) \subseteq A$ for some $r > 0$;
(ii) an *exterior* point to A if it is an interior point of $A^c = \{x \in X : x \notin A\}$, the complement of A;
(iii) a *boundary point* of A if it is neither interior or exterior. So, any open ball around a boundary point x will contain both points in A and points not in A.

The set A is said to be *open* if all points in A are interior, that is, A contains none of its boundary points; A is said to be *closed* if it contains all of its boundary points.

Clearly, A is open if and only if A^c is closed since a set and its complement have the same boundary points.

The well-known combinatorial properties of these types of sets follow easily. Specifically, arbitrary unions of open sets are open, and arbitrary intersections of closed sets are closed. Finite intersections of open sets are open, and finite unions of closed sets are closed.

We can define convergence and continuity in terms of open balls. A sequence (x_n) converges to x if, given any open ball B around x, the elements x_n are in B for sufficiently large n. Similarly, a function f from one normed linear space X to another normed linear space Y is continuous at $x \in X$ if, given any open ball B' around $f(x)$, there is an open ball B around x such that $f(B) \subseteq B'$.

Definition. We say that f is *continuous* if it is continuous at all points of X.

From the above definition of continuity, it is easy to verify that f is continuous if and only if, for any open set V in Y, the inverse image $f^{-1}(V)$ is an open subset of X. By taking complements, the same statement holds for closed sets in place of open sets.

Definition. The function f is said to be *uniformly continuous*, if given $\epsilon > 0$, there is a $\delta > 0$, depending only on ϵ, such that, for all $x \in X, f(B(x; \delta)) \subseteq B(f(x); \epsilon)$.

This is a stronger condition than just being continuous since the same δ must work for all points in X.

Given any set A, the *closure* of A, denoted by $\bar{A}$, consists of all points that are in a sense arbitrarily close to A. Points in A are automatically arbitrarily close to A; thus, the closure contains A itself. We can give a precise definition of closure in various equivalent ways, each of which is useful in certain contexts. We leave it to the reader to verify that all of the following define the same set.

Definition. The *closure* of A is defined by any of the following equivalent statements:

(i) The intersection of all closed sets containing A. This shows that $\bar{A}$ is closed and equal to A if A is closed to begin with.
(ii) The union of A and its boundary points.

(iii) The set of all $x \in X$ such that, for all $r > 0$, $B(x; r)$ contains a point in A.
(iv) The set of all $x \in X$ for which there is a sequence (a_n) of points in A converging to x.
(v) The set of all $x \in X$ such that $d(x, A) = 0$, in which d is the metric induced by $\| \cdot \|$.

Remark. When mathematicians choose a word to denote a concept, they generally attempt to relate its usual meaning to the particular mathematical idea involved. In this, they are sometimes successful and sometimes not. A closed set is, indeed, "closed" in the sense that you cannot get out of it. A sequence in the closure cannot converge to anything outside. An open set, however, at least in the opinion of this author, is not particularly well described by the usual meaning of this word. Indeed, an open set is also closed, but now in the sense that you cannot get into it. One can view an open set as protecting each of its points from an encroaching sequence from the outside trying to converge to it. The choice of the word was, no doubt, motivated by the fact that an open set is the complement of a closed set.

The following result was anticipated by our terminology and notation.

Proposition 2.5

(a) $B(x; r)$ is open.
(b) The closure of $B(x; r)$ is $\bar{B}(x; r)$.

Proof.

(a) Given $y \in B(x; r)$, choose a positive number $s < r - \| y - x \|$. If $\| z - y \| < s$, then $\| z - x \| \leq \| z - y \| + \| y - x \| \leq s + (r - s) = r$. So, $B(y; s) \subseteq B(x; r)$ showing that y is interior to $B(x; r)$. See Figure 2.1.
(b) If (z_n) is a sequence in $\bar{B}(x; r)$ that converges to z, then $(z_n - x)$ converges to $z - x$, and continuity of the norm shows that $\| z - x \| \leq r$. This shows that $\bar{B}(x; r)$ is closed. Therefore, $\bar{B}(x; r)$ contains the closure of $B(x; r)$, as shown by statement (i) in the definition of closure.

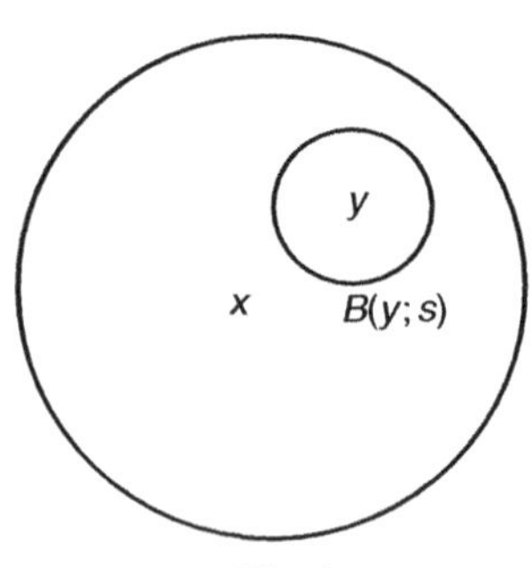

Figure 2.1. Graphical derivation of Proposition 2.5(a).

Conversely, if $\| y - x \| = r$, define a sequence by $y_n = x + (1 - 1/n)(y - x)$. Then, $\| y_n - x \| < \| y - x \| = r$, so $y_n \in B(x; r)$ (as we can easily visualize, say, in the plane, where y_n lies on the line segment joining x and y). Moreover, by the continuity of addition and scalar multiplication, (y_n) converges to y showing that y is in the closure of $B(x; r)$, using statement (iv) in the definition of closure. □

Remark. All of the definitions of this section can be taken over to a general metric space, and the conclusions that we have stated above hold, except for part (b) of Proposition 2.5, which fails drastically. For example, take X to be any set with the metric $d(x, y) = 1$ for all points $x \neq y$. Then, $B(x; 1) = x$, and it is clear that its closure is also equal to x. However, $\bar{B}(x; 1)$ is the entire space X. The scalar multiplication property of a norm is the fact that makes Proposition 2.5(b) true for a normed linear space.

We close this section with a definition and remark that we will use frequently in later material.

Definition. A subset Y of a normed linear space X is said to be *dense in X if* $\bar{Y} = X$. We leave it to the reader to verify that a set is dense if and only if it intersects any nonempty open set.

It is useful to identify such sets since a continuous function on X is determined completely by its values on any dense set Y. Indeed, for all $x \in X$, there is a sequence (y_n) in Y converging to x, and $f(x) = \lim f(y_n)$.

Remark. A fact that we will use frequently is that in a normed linear space a set is bounded if and only if it is contained in $B(r)$ for some $r > 0$. This follows since, given a nonempty bounded set A and $x \in A$, it is clear that A is contained in the ball of radius $\| x \| + \text{diam}(A)$ around 0.

2.2.3 Compact Sets

One of the most important concepts of analysis is that of *compactness*. Compactness is a generalization of the properties of closed and bounded subsets of the real line. There are various equivalent forms of the definition, of which we now give two.

A subset K of a normed linear space is said to be *compact* if, given any collection of open sets with union containing K, there is a finite subcollection of these sets with union containing K. (We often speak of such a collection of sets as an *open cover* of K.)

Finite sets are obviously compact, and this form of the definition suggests that compact sets are in some sense "almost" finite. (We elaborate on this statement in Chapter 9.) Certain statements that are obvious for finite sets carry over to compact ones. For example, a real-valued continuous function assumes a minimum and maximum value on any compact set.

A second form of the definition is as follows: A set K is compact if and only if any sequence $(x_n) \in K$ has a subsequence that converges to a point in K.

See Royden (1988, Theorem 7.21) for a proof of the equivalence of these two definitions. This second form of the definition indicates the usefulness of the concept because analysts are continually looking for convergent sequences. This formulation shows that there are plenty of them when dealing with a compact set.

This second definition also makes it easy to prove the main properties of compact sets. For example, a compact set must be bounded, for if not, we could choose a sequence of elements with norms getting arbitrarily large, and there would not be any convergent subsequence. Also, a compact set K is closed. Indeed, take a point x in the closure of K, and choose a sequence (x_n) in K converging to x. There is a subsequence that converges to a point $y \in K$, but it must also converge to x. By uniqueness of limits, $x = y \in K$. (It follows from this derivation that a compact set in a seminormed space need not be closed.)

We have just shown that compact sets are closed and bounded, and it is a well-known result that the converse is true in a finite-dimensional normed linear space. However, the converse is not true in an infinite-dimensional normed linear space, as we will show later.

Two other useful properties of compact sets that follow easily from the sequential form of the definition are that a closed subset of a compact set is compact and that if f is a continuous function and K is compact, then $f(K)$ is compact.

2.3 SPACE OF BOUNDED FUNCTIONS

It is time to give an example of an infinite-dimensional normed linear space. This is not merely an illustrative example but a crucial one, for it embodies the notion of uniform approximation and convergence, which was behind several early results in analysis. The polynomial approximation result of Weierstrass, which we mentioned in Chapter 1, involves this type of distance measurement. Another major theorem of analysis is the well-known result that continuity is preserved under uniform convergence. The norm we are about to define is also a very flexible one, for it applies to functions on an arbitrary set S. No special structure on the set need be imposed.

For an arbitrary set S and any $f \in F(S)$, define

$$\| f \| = \sup\{|f(s)| : s \in S\}.$$

Despite our notation, this cannot be a norm when S is infinite since it could take a value of ∞. Suppose, however, we restrict ourselves to the subset $B(S)$ of *bounded* functions, which are those functions f such that there exists $K > 0$ (which depends on f), with $|f(s)| \leq K$ for all $s \in S$. For any such f, $\| f \| \leq K < \infty$. Given f, $g \in B(S)$, we have that

$$|(f + g)(s)| = |f(s) + g(s)| \leq |f(s)| + |g(s)| \leq \| f \| + \| g \| < \infty$$

for all $s \in S$. This shows that $f + g \in B(S)$ and, moreover, that the triangle inequality holds for $\| \cdot \|$.

The scalar property follows from the fact that $|\alpha f| = |\alpha||f|$ as well as from the fact that if you multiply a set of reals by a positive scalar, the supremum gets multiplied by the same scalar.

Finally, if $\| f \| = 0$, then it is clear that f is the zero function. We have verified that $B(S)$ is a linear space and $\| \cdot \|$ is a norm on $B(S)$. This norm is often referred to as the *sup norm* and denoted by $\| \cdot \|_{\text{sup}}$. The space is infinite-dimensional, except when S is finite; see Exercise 1.8.

What does convergence mean with respect to the sup norm? To say that (f_n) converges to f means that for any $\epsilon > 0$ we can find a positive integer N such that for all $n > N$ we have $\| f_n - f \| < \epsilon$, which implies that $|f_n(s) - f(s)| < \epsilon$ for all $s \in S$. This is none other than the familiar *uniform* convergence of functions, as opposed to *pointwise* convergence, which is less stringent, requiring only that $(f_n(s))$ converge to $f(s)$ for all $s \in S$.

2.4 BOUNDED LINEAR OPERATORS

We are about to give a key definition, and some remarks on the language used are in order. What should it mean for a linear operator to be bounded? Since a linear operator is a function, one might think that we simply mean what we did in the previous section when we defined a bounded function. Using this, $T : X \to Y$ would be bounded if there were some $K > 0$ such that $\| Tx \| \leq K$ for all $x \in X$. A moment's reflection shows that such a definition does not make much sense. Suppose $\| Tx \| = 1$ for some $x \in X$; then, for any positive integer n, $\| T(nx) \| = \| nTx \| = n \| Tx \| = n$. Since we can choose n as large as we like, the only possible operator satisfying this definition of boundedness would be the 0 operator.

The key to the proper definition is to note that a linear operator is completely determined by its values on the so-called *unit vectors*, that is, those elements with norm equal to 1. Indeed, for any $x \neq 0$, the element $x' = x / \| x \|$ is a unit vector and $Tx = \| x \| Tx'$. We can therefore view a linear operator as a function on the unit vectors. (Of course, not all functions on the unit vectors arise in this way.) With this point of view, we easily recast Section 2.3 to apply to linear operators.

Given normed linear spaces X and Y, we define for all $T \in \mathcal{L}(X, Y)$

$$\| T \| = \sup\{\| Tx \| : \| x \| = 1\}.$$

Note that in the above we are using the symbol $\| \cdot \|$ in three different ways. The symbol $\| x \|$ refers to the norm on X, $\| Tx \|$ refers to the norm on Y, and $\| T \|$ refers to the defined function on $\mathcal{L}(X, Y)$. We will continue this practice of just using the single symbol $\| \cdot \|$ whenever the context makes it clear which norm is intended.

Definition. We say that $T \in \mathcal{L}(X, Y)$ is *bounded* if $\| T \| < \infty$.

Note that for all $x \neq 0$, with x' as above,

$$\| Tx \| = \| x \| \| Tx' \| \leq \| T \| \| x \| .$$

This inequality gives us another characterization of boundedness. Namely, T is bounded if and only if for all bounded subsets A of X the image $T(A)$ is bounded in Y. (The "if" statement follows immediately from our original definition since the set of unit vectors is clearly bounded.)

A particular class of bounded operators, which we will consider in more detail later, is as follows.

Definition. A linear operator T that preserves length is known as an *isometry*. Specifically, $\| Tx \| = \| x \|$ for all $x \in X$.

Clearly, an isometry has norm equal to 1.

Let $\mathcal{B}(X, Y)$ denote the set of bounded linear operators from X to Y, which we shorten to $\mathcal{B}(X)$ when $Y = X$. By arguing exactly as in Section 2.3, we see that $\mathcal{B}(X, Y)$ is a subspace of $\mathcal{L}(X, Y)$, and $\| \cdot \|$ is a norm on $\mathcal{B}(X, Y)$.

What does convergence mean with respect to this norm? (This is a question we want to ask about any norm.) We leave it to the reader to verify that (T_n) converges to T with respect to the norm on $\mathcal{B}(X, Y)$ if and only if convergence is uniform on bounded sets. That is, given any bounded $B \subseteq X$ and any $\epsilon > 0$, there is a positive integer N such that for all $x \in B$ and all $n \geq N$ we have $\| T_n x - Tx \| \leq \epsilon$.

Since we are dealing with normed spaces, we are naturally interested in *continuous* linear operators. Remarkably, it turns out that these coincide exactly with the bounded ones. The following is our first major theorem on linear operators.

Theorem 2.6. *Given $T \in \mathcal{L}(X, Y)$, the following are equivalent*:

(i) T is uniformly continuous;

(ii) T is continuous at some point;

(iii) T is bounded.

Proof. (i) $\Rightarrow$ (ii). This is trivial.

(ii) $\Rightarrow$ (iii). Suppose that T is continuous at the point x. Choose $\delta > 0$ such that $T(B(x; 2\delta)) \subseteq B(Tx; 1)$. We will show that $\| T \| \leq 1/\delta$. For any unit vector z, the vector $x + \delta z$ is in $B(x, 2\delta)$ so that $Tx + \delta\, Tz \in B(Tx; 1)$. This gives

$$\delta \| Tz \| = \| Tx + \delta Tz - Tx \| < 1,$$

thereby proving the assertion.

(iii) $\Rightarrow$ (i). If $\| T \| = K$, take $\delta = \epsilon / K$ in the definition of uniform continuity. Clearly,

$$\| y - x \| < \delta \quad \text{implies} \quad \| Ty - Tx \| \leq K \| y - x \| < \epsilon,$$

thereby verifying uniform continuity of T. □

Let us again look at $B(S)$ with the sup norm. In addition to adding functions, we can multiply them. This leads to the following example; compare Exercise 1.3.

Example 2.7. Fix some $g \in B(S)$. [We always assume, if not mentioned otherwise, that $\|\cdot\|$ denotes the sup norm on $B(S)$.] Define a linear operator M_g on $B(S)$ by

$$M_g f(s) = g(s) f(s).$$

Here, M_g is known as a *multiplication operator.* What is $\| M_g \|$? For any unit vector f, we have $|M_g f(s)| = |g(s)|\,|f(s)| \leq \| g \|$, which shows that $\| M_g \| \leq \| g \|$. Conversely, for any $s \in S$,

$$|g(s)| = |(M_g \delta_s)(s)| \leq \| M_g \delta_s \| \leq \| M_g \|,$$

and by taking sups we have $\| g \| \leq \| M_g \|$. It follows that $\| g \| = \| M_g \|$. We can now say that the mapping that sends g to M_g, a mapping from $B(S)$ to the linear operators from $B(S)$ to itself, is an isometry.

Another operation on operators that we will deal with frequently is that of composition. Given bounded linear operators $T : X \to Y$ and $S : Y \to Z$, we can form the composition $ST : X \to Z$, in which ST is given by $STx = S(Tx)$. It is well known that ST is continuous because it is the composition of continuous functions. We can, however, make a more precise statement.

Proposition 2.8 *ST is linear and $\| ST \| \leq \| S \| \| T \|$.*

Proof. The linearity is clear. Moreover, for any unit vector x,

$$\| STx \| \leq \| S \| \| Tx \| < \| S \| \| T \| . \qquad \square$$

2.5 COMPLETENESS

2.5.1 Basic Definitions

Suppose we want to prove that a certain sequence (x_n) converges. To do so directly from the definition without further information, we have to know the limit. However, we often want to define an element in a normed linear space by saying it is the limit of a certain sequence. [As an example in the real numbers, we define the number $e = \lim(1 + 1/n)^n$.] To do this, we have to know beforehand that the sequence converges. This leads us to the notion of *completeness*, which parallels the familiar situation on the real line.

Definition. A sequence (x_n) in a normed linear space is said to be a *Cauchy sequence* if given any $\epsilon > 0$, there is a positive integer N such that for all $m, n \geq N$ we have $\| x_m - x_n \| < \epsilon$.

Definition. A normed linear space is said to be *complete* if every Cauchy sequence converges. Such a space is known as a *Banach space.*

Thus, from the definition, in a Banach space we have a way to verify convergence without knowing the limit. To investigate this idea further, we list some basic properties of Cauchy sequences.

Proposition 2.9. *In any normed linear space:*

(*a*) *A convergent sequence is Cauchy.*
(*b*) *A Cauchy sequence* (x_n) *is bounded. That is, there is a* $k > 0$ *such that* $\| x_n \| \leq k$ *for all* n;
(*c*) *All subsequences of a Cauchy sequence are Cauchy;*
(*d*) *If* (x_n) *is Cauchy and some subsequence converges to* x, *then* (x_n) *converges to* (x).

Proof. The proofs of (a), (b), and (c) are straightforward, so we leave them to the reader. We prove (d). This statement is intuitively clear since the Cauchy property means that the sequence entries are all eventually close to one another, and therefore they must be close to the subsequence entries, which should imply that the sequence converges to the same limit as the subsequence. Formally, given $\epsilon > 0$, choose N to satisfy the definition of a Cauchy sequence for $\epsilon/2$. Next, choose $J \geq N$ so that $\| x_J - x \| < \epsilon/2$, which we can do by the fact that there is a subsequence converging to x. Then, for $n \geq N$,

$$\| x_n - x \| \leq \| x_n - x_J \| + \| x_J - x \| < \epsilon. \qquad \square$$

The following definition simplifies the verification of completeness in many cases.

Definition. A sequence (x_n) is called a *fast Cauchy sequence* if $\| x_{n+1} - x_n \| \leq 1/2^n$ for $n = 1, 2, \ldots$. (In place of $1/2^n$, we could use any convergent series.)

Proposition 2.10

(*a*) *A fast Cauchy sequence is a Cauchy sequence.*
(*b*) *Any Cauchy sequence contains a fast Cauchy subsequence.*

Proof.

(a) Given a fast Cauchy sequence (x_n) and $\epsilon > 0$, choose a positive integer N such that $1/2^{N-1} < \epsilon$. Then, for $n > m \geq N$, the general triangle inequality gives

$$\| x_n - x_m \| \leq \sum_{k=m}^{n-1} \| x_{k+1} - x_k \| \leq \sum_{k=m}^{n-1} \frac{1}{2^k} \leq \frac{1}{2^{N-1}} < \epsilon.$$

(b) Given a Cauchy sequence (x_n), choose inductively a sequence of integers (N_k) as follows: $N_0 = 0$, and for $k \geq 1$, $N_k > N_{k-1}$ is such that for $m, n \geq N_k$ we have $\| x_m - x_n \| \leq 1/2^k$. Then, (x_{N_k}) for $k = 1, 2, \ldots$ is a fast Cauchy subsequence. □

It follows from Proposition 2.10(b) and Proposition 2.9(d) that, in order to show a space is complete, we need only show that fast Cauchy sequences converge.

2.5.2 Series

A statement that is phrased in terms of limits of sequences can always be rephrased in terms of series. In this section, we characterize completeness via series, which is useful in many cases.

We notice first that in any normed linear space we can both add and take limits, so the concept of an infinite series makes sense. Given (x_i) in X, we say that

$$\sum_{i=1}^{\infty} x_i = x$$

if the partial sums (s_n) converge to x, where $s_n = \sum_{x=1}^{n} x_i$. This is precisely the same definition as for a series of real numbers.

We are easily able to extend the triangle inequality to infinite series.

Proposition 2.11. *If* $x = \sum_{i=1}^{\infty} x_i$ *exists, then*

$$\| x \| \leq \sum_{i=1}^{\infty} \| x_i \| .$$

Proof. By using the fact that $\| \cdot \|$ is continuous, we have

$$\| x \| = \| \lim s_n \| = \lim \| \sum_{i=1}^{n} x_i \| \leq \lim \sum_{i=1}^{n} \| x_i \| = \sum_{i=1}^{\infty} \| x_i \|,$$

where we allow ∞ as a value for the last sum. □

The inequality of the last proposition is useful only when the sum on the right is finite. This brings us to the next definition.

Definition. The series $\sum_{i=1}^{\infty} x_i$ is said to be *absolutely convergent* if the series of nonnegative reals $\sum_{i=1}^{\infty} \| x_i \|$ is convergent.

Showing that a series converges directly from the definition could be a horrendous task. It may be impossible to compute what the partial sums are in a complicated space. If the space is complete, however, the task is reduced to showing convergence of a nonnegative valued series, as the following result indicates.

Theorem 2.12. *A normed linear space X is complete if and only if every absolutely convergent series is convergent.*

Proof. Suppose that X is complete and that $\sum_{x=1}^{\infty} \| x_i \| < \infty$. Then, for $m < n$, the triangle inequality gives

$$\| s_n - s_m \| \leq \sum_{i=m+1}^{n} \| x_i \| .$$

The term on the right tends to 0 as m gets large, which shows that (s_n) is a Cauchy sequence and therefore convergent.

Conversely, suppose that every absolutely convergent sequence is convergent. Completeness follows if we show that any fast Cauchy sequence converges. Given such a sequence (x_n), the series $\sum_{i=1}^{\infty}(x_{i+1} - x_i)$ is by definition absolutely convergent; therefore, it converges to some element x. For this series, the partial sum s_n is just $x_{n+1} - x_1$, so the original series converges to $x + x_1$. □

2.5.3 Completeness Examples

It is time for some examples, and we will demonstrate the completeness of some of the spaces that were introduced above. We first discuss a general strategy for showing completeness in spaces of functions. Given a Cauchy sequence of functions (f_n) on a set S, the first step is to find a candidate for a limit. A natural choice for this is the *pointwise limit* f given by $f(s) = \lim f_n(s)$ for each $s \in S$. We must first show that this f exists, but this still leaves a major task. We then must show that f belongs to the space in question and that (f_n) converges to f with respect to a particular norm. This is not at all obvious since it can be easier for a sequence of functions to converge pointwise than in some other mode. This is certainly true of the sup norm, which, as we indicated, implies uniform convergence. The following lemma can facilitate the procedure.

Lemma 2.13. *Suppose that X is a subspace of $F(S)$ and that $\| \cdot \|$ is a norm on X for which the closed unit ball $\bar{B}(1)$ is closed under pointwise limits.* [*That is, given $f_n \in \bar{B}(1)$ such that (f_n) converges pointwise to f, it follows that $f \in X$ and has norm ≤ 1.*] *If a sequence (f_n) in X is Cauchy and converges pointwise to f, then $f \in X$, and (f_n) converges to f with respect to $\| \cdot \|$.*

Proof. Let (f_n) be a sequence in X with the given properties and suppose $\epsilon > 0$ is given. Choose N such that $n, m \geq N$ implies that $\| f_n - f_m \| \leq \epsilon$. By using the scalar multiplication properties for the space and the norm, the hypothesized condition implies that $\bar{B}(\epsilon)$ is also closed under pointwise limits. Given $n_0 \geq N$, the sequence $(f_n - f_{n_0})$ lies in $\bar{B}(\epsilon)$ and converges pointwise to $f - f_{n_0}$. It follows that $f - f_{n_0} \in X$, so that $f = f_{n_0} + (f - f_{n_0}) \in X$ and $\| f - f_{n_0} \| \leq \epsilon$, thereby verifying the required convergence. □

We can now present our first example of an infinite-dimensional Banach space. We take it as an established fact that the field $\mathcal{F}$ with the absolute value norm is complete.

Theorem 2.14. *$B(S)$ with the sup norm is complete.*

Proof. Let (f_n) be a Cauchy sequence in $B(S)$. For any $s \in S$,

$$|f_m(s) - f_n(s)| \leq \| f_m - f_n \|,$$

showing that $(f_n(s))$ is a Cauchy sequence in $\mathcal{F}$. Therefore, $(f_n(s))$ converges to some point in $\mathcal{F}$, which we call $f(s)$, so that the pointwise limit f exists. Moreover, if $\| f_n \| \leq 1$ for all n, then, for any $s \in S$, $|f_n(s)| \leq \| f_n \| \leq 1$, so that $|f(s)| = \lim |f_n(s)| \leq 1$, showing that $f \in \bar{B}(1)$. The condition of Lemma 2.13 holds, and we conclude that (f_n) converges to f with respect to the sup norm. □

We next investigate the completeness of $\mathcal{B}(X, Y)$. One might guess that if both X and Y are complete, then this space is as well. Actually, one can do better. We need only the target space to be complete.

Theorem 2.15. *If Y is complete, then so is $\mathcal{B}(X, Y)$.*

Proof. By viewing elements of this space as functions on the unit vectors of X, we can simply imitate the proof of Theorem 2.14 while replacing $\mathcal{F}$ with Y. There is, however, one more point to verify. Our candidate for a limit is a function T such that for some sequence (T_n) of linear operators, $(T_n x)$ converges (with respect to the norm on Y) to Tx for all unit vectors x. As such, T is defined only on unit vectors, but we extend it in the obvious way to a mapping on all of X by $Tx = \| x \| \, T(x')$, in which $x' = x / \| x \|$. We must show that T defined in this way is linear. To this end, by the scalar multiplication properties, notice that convergence at unit vectors implies that $(T_n x)$ converges to Tx for all $x \in X$ and then simply use the continuity of the operations from Theorem 2.3. □

2.5.4 Subspaces of Banach Spaces

Given a Banach space X, which of its subspaces are Banach spaces? The answer is surprisingly easy.

Theorem 2.16. *A subspace Y of a Banach space X is itself a Banach space if and only if Y is closed.*

Proof. Suppose Y is closed. Given a Cauchy sequence (y_n) in Y, the sequence converges to some $x \in X$, and necessarily $x \in \bar{Y} = Y$.

Conversely, if there is a point $x \in \bar{Y}$ that is not in Y, choose a sequence (y_n) in Y converging to x. Then (y_n) is a Cauchy sequence, which by uniqueness of limits cannot converge to anything in Y, so that Y is not complete. □

Example 2.17. As an application of this result, let $[a, b]$ be a closed interval of the reals. A linear subspace of $B[a, b]$ that is of considerable importance in applications is $C[a, b]$, the continuous functions. We know that this is closed by a basic theorem of real analysis that states that a uniform limit of a sequence of continuous functions is continuous (Rudin, 1976, Theorem 7.13). Therefore, we know that $C[a, b]$ with the sup norm is complete.

Following any definition of a mathematical property, there should be—if at all possible—an example of something that does *not* satisfy the property. After all, if no such examples exist, there is little need for the definition in the first place. So what is an example of an incomplete space? The result above provides a method for finding such. We can simply take any nonclosed subspace of a Banach space, but this still leaves open the question of whether any such object exists. This is not obvious since, as we indicated in Chapter 1 and will illustrate later, all subspaces of a finite-dimensional space are closed.

Example 2.18. The easiest example of a nonclosed subspace is c_{00}, the subspace of the bounded sequences consisting of those sequences that have only finitely many nonzero entries; see Exercise 1.1. The closure of c_{00} with respect to the sup norm is c_0, those sequences converging to 0; see Exercise 2.7. Because c_0 is strictly bigger than c_{00}, the latter is not closed and therefore not complete. The space c_{00} will figure prominently in the exercises to provide counterexamples to results depending on completeness.

Example 2.19. For an example similar to the one above, suppose we wish to consider the continuous functions on the whole real line (or on any unbounded subset). We no longer have a subspace of the bounded functions since continuous functions on a noncompact set can be unbounded. There are certain subspaces of the bounded continuous functions that have been extensively studied. Let $C_{00}(\mathcal{R})$ be the space of all continuous functions f on $\mathcal{R}$ *with compact support.* This means that for some compact K we have $f(x) = 0$ for all $x \notin K$. Equipped with the sup norm, this is a normed linear space, but it is not closed and therefore not complete. Its closure, as the reader can verify, is the space $C_0(\mathcal{R})$ of all continuous functions vanishing at infinity; see Exercise 2.8. This means that given any $\epsilon > 0$ there is a compact set K (which could depend on ϵ) such that, for all $x \notin K$, $|f(x)| < \epsilon$. The space $C_0(\mathcal{R})$ is closed and therefore another example of a Banach space.

2.5.5 Extension Theorem

Extension is at the heart of analysis. In numerous situations, we have a function defined on a certain set, and we want to extend it to a larger set while preserving many of the properties of the function on the smaller set. The following is one such extension theorem which illustrates the importance of completeness. We will use it frequently in the material to follow.

Theorem 2.20. *Suppose that X_0 is a dense subspace of the normed linear space X and that $T_0 \in \mathcal{B}(X_0, Z)$, where Z is a Banach space. Then, T_0 has a unique extension to an operator $T \in \mathcal{B}(X, Z)$. Moreover, $\| T \| = \| T_0 \|$, and if T_0 is an isometry, so is T.*

Proof. Uniqueness follows easily from the facts that X_0 is dense and T is bounded. The main task is to show existence. Given any $x \in X$, we can find a sequence (x_n) in X_0 converging to x. Because the sequence (x_n) is Cauchy and $\| T_0 x_m - T_0 x_n \| \leq \| T_0 \| \| x_m - x_n \|$, it follows that $(T_0 x_n)$ is Cauchy. By completeness of Y, the sequence $(T_0 x_n)$ converges to an element of Y, which we call Tx. There are several things to verify.

(i) T is well-defined: As a basic principle of mathematics, we must show when making a definition that, when something depends on a particular choice, any other choice leads to the same result. So, suppose we choose a different sequence (y_n) converging to x. Then,

$$\| T_0 x_n - T_0 y_n \| \leq \| T_0 \| \| x_n - y_n \| .$$

The term on the right is easily seen to approach 0, so $(T_0 y_n)$ has the same limit as $(T_0 x_n)$.

(ii) T extends T_0: For any $x \in X_0$, we can take the sequence that consists only of the term x, and clearly $Tx = T_0 x$.

(iii) T is linear: For any $x, y \in X$, we can choose sequences (x_n) and (y_n) in X_0 converging to x and y, respectively. Then, $(x_n + y_n)$ converges to $x + y$, so

$$T(x+y) = \lim T_0(x_n + y_n) = \lim (T_0 x_n + T_0 y_n) = \lim T_0 x_n + \lim T_0 y_n = Tx + Ty.$$

The proof for scalar multiplication is similar.

(iv) $\| T \| = \| T_0 \|$: As an extension, T clearly has a bigger norm since we are taking the supremum over a bigger set. Conversely, for any unit vector $x \in X$, choose (x_n) in X_0 converging to x. Then, by using the continuity of the norm,

$$\| Tx \| = \| \lim T_0 x_n \| = \lim \| T_0 x_n \| \leq \lim \| T_0 \| \| x_n \| = \| T_0 \| \| x \| = \| T_0 \|,$$

thereby showing that $\| T \| \leq \| T_0 \|$.

(v) Suppose T_0 is an isometry. Arguing as in (iv) shows that $\| Tx \| = \| x \|$ for all $x \in X$. □

Example 2.21. As a major application of this theorem, we give a very simple definition of the integral that handles all continuous functions.

We begin by considering the space $B[a, b]$ with the sup norm, where $[a, b]$ is a closed and bounded interval in $\mathcal{R}$. A function f on $[a, b]$ is called a *step function* if there exists a sequence of points $a = t_0 < t_1 < t_2 < \cdots < t_n = b$ such that f is a constant c_i on each of the open intervals (t_i, t_{i+1}), $i = 0, 1, \ldots, n-1$. Let X_0 denote the set of step functions. Define a mapping I_0 from X_0 to $\mathcal{F}$ by

$$I_0(f) = \sum_{i=0}^{n-1} (t_{i+1} - t_i)c_i.$$

Note that $I_0(f)$ is simply the usual integral of f.

One can verify that X_0 is a subspace of $B[a, b]$ and that I_0 is linear, with $\| I_0 \| \leq b - a$. (In fact, $\| I_0 \|$ is equal to $b - a$.) The extension I of I_0 gives us a definition of the integral on the space $\bar{X}_0$ that is easily seen to contain the continuous functions $C[a, b]$. Note also that, by replacing $\mathcal{F}$ with any Banach space, we have a straightforward definition of a vector-valued integral.

2.5.6 Completion

Mathematics is full of examples in which we are given a set that does not satisfy a desirable property, but we can achieve a set with this property by adding points. This is true when we have a normed linear space that is not complete.

Definition. Given any normed linear space X, a *completion* of X consists of a Banach space $\tilde{X}$ and an isometry $J : X \to \tilde{X}$ such that $J(X)$ is dense in $\tilde{X}$.

This definition serves as an introduction to some general discussion as to what it means for two Banach spaces to be the same. This is a problem that arises in all branches of mathematics. Intuitively speaking, we regard two mathematical objects as being the same from an abstract point of view if there is a *bijective* mapping (that is, both injective and surjective) from one to the other that preserves all of the mathematical structure and such that the inverse mapping also preserves the mathematical structure. Any such mapping is called an *isomorphism*, and we say the two objects are *isomorphic*. Of course, we must define precisely in each case what we mean by the underlying mathematical structure.

To illustrate, suppose we are given two normed linear spaces X and Y. We consider these the same as *linear spaces* if there is a bijective linear operator from one to another. The inverse will automatically be linear. However, if we want to consider them the same as *normed linear spaces*, we need to preserve more structure, and there are different possibilities. For some applications, two normed linear spaces are considered isomorphic if there is a bijective linear map between them that is bounded and has a bounded inverse. A stronger requirement is that this mapping be an isometry. (See Exercise 2.30 for the connection between these two types of isomorphisms.) In this book, whenever we say that two normed linear spaces are the same, we will be referring to this stronger criterion. That is, we will mean that there is a surjective linear isometry between them, in which case the two spaces are said to be *isometric*.

Note that the definition of completion above identifies the given space with an isometric copy in a larger space, which ties in with our statement that we enlarge the space to get the completion.

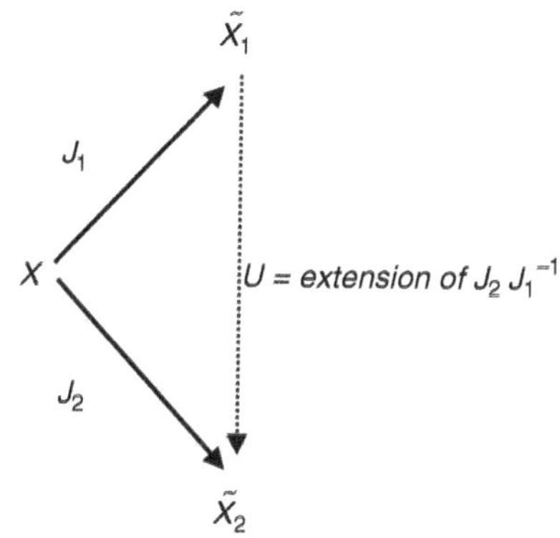

Figure 2.2. Uniqueness of a completion.

Theorem 2.22. *For any normed linear space X, a completion exists. Moreover, it is unique in the following sense: If $(\tilde{X}_i, J_i)$ for $i = 1, 2$ are two completions, there is a surjective isometry $U : \tilde{X}_1 \to \tilde{X}_2$ such that $UJ_1 = J_2$.*

Proof. Uniqueness: The mapping $U_0 = J_2 J_1^{-1}$ is an isometry from $J_1(X)$ to $\tilde{X}_2$, which from Theorem 2.20 extends to an isometry U from $\tilde{X}_1$ to $\tilde{X}_2$. (See Figure 2.2.) Since $\tilde{X}_1$ is complete, the isometric image $U(\tilde{X}_1)$ is easily seen to be complete and therefore closed in $\tilde{X}_2$ by Theorem 2.16. Since $U(\tilde{X}_1)$ contains $J_2(X)$, it is dense as well as closed; therefore, $U(\tilde{X}_1)$ can only be the whole space $\tilde{X}_2$.

Existence: Choose a Banach space Y containing X as a subspace. Let $\tilde{X}$ be the closure of X in Y and define $Jx = x$. □

The reader is invited to criticize the above existence proof, which for the pun aficionados we may characterize as shockingly *incomplete*. The problem comes from the word "choose," which is fraught with danger and the source of many errors in mathematical proofs. Whenever we use this word, we have to ensure that the object we want to choose actually exists. How do we know in the present case that there is a Banach space containing a given normed linear space? This will follow quite easily from a result in Chapter 6, and this will fix the proof, which the reader will have to take on faith for the time being.

For an alternate proof of existence, one can directly construct the Banach space by imitating the procedure for obtaining the real numbers from the rationals. This is reasonably involved, however, and we only sketch the details. Since a Cauchy sequence needs a limit, we will simply enlarge our space by adding each Cauchy sequence as a point. This point then serves as the limit for that sequence. In fact, we just consider the space of all Cauchy sequences and note that this set contains a copy of the original space by identifying each element x with the corresponding constant sequence. This space is too big, however, since different Cauchy sequences can converge to this same point. We therefore consider a equivalence relation on this space of Cauchy sequences in which two such sequences are equivalent if their difference tends to zero. We must show that the natural pointwise operations are well defined on the set of equivalence classes and give us a linear space. We next show that we can norm an equivalence class by taking the limit of norms of elements in a representative sequence. Finally, we show that this gives a Banach space and that the constant sequences are dense. For complete details, see Kadison and Ringrose (1983, Theorem 1.5.1).

2.6 COMPARISON OF NORMS

There are situations in which we want to consider several norms on the same linear space X, and there is a natural way to compare them.

Definition. Given two norms, we say that $\|\cdot\|_1$ is *weaker* than $\|\cdot\|_2$ if any $\|\cdot\|_2$ convergent sequence is also $\|\cdot\|_1$ convergent. (We sometimes describe this by saying that $\|\cdot\|_2$ is *stronger* than $\|\cdot\|_1$.)

The use of "weaker" is another example where it is not clear that the usual meaning of the English word adequately describes the idea. It is best to just remember that in this context weaker means *easier to converge.* It may help to go back to the metaphor of attacking sequences given in Section 2.2.2. In the weaker norm, any point is in a sense weaker, since it is more vulnerable to attack by a sequence trying to converge to it.

Following are some useful characterizations of this concept.

Proposition 2.23. *$\|\cdot\|_1$ is weaker than $\|\cdot\|_2$ if and only if there is a $K > 0$ such that for all x*

$$\| x \|_1 \leq K \| x \|_2 .$$

Proof. The existence of such a K clearly proves that $\|\cdot\|_1$ is weaker. Conversely, suppose that $\|\cdot\|_1$ is weaker. Then, the identity transformation that sends each x to itself is a continuous mapping from X equipped with $\|\cdot\|_2$ to X equipped with $\|\cdot\|_1$. From Theorem 2.6, this map is bounded and has a finite norm K, which implies the inequality. □

Proposition 2.24. *$\|\cdot\|_1$ is weaker than $\|\cdot\|_2$ if and only if every $\|\cdot\|_1$ open ball contains a $\|\cdot\|_2$ open ball.*

Proof. For the "only if" direction, choose K as in Proposition 2.23. Then, $B(x; r)$ for $\|\cdot\|_1$ contains $B(x; r/K)$ for $\|\cdot\|_2$. We leave the converse as an exercise for the reader. □

This proposition is intuitively clear since the larger open balls in the weaker norm make it easier for a sequence eventually to get inside and remain in the ball, making it easier to converge.

As a immediate consequence of Proposition 2.24, we have the following.

Corollary 2.25. *$\|\cdot\|_1$ is weaker than $\|\cdot\|_2$ if and only if every $\|\cdot\|_1$ open set is $\|\cdot\|_2$ open.*

The intuition for this corollary is that if a set protects its points in the weaker norm, it will certainly protect them in the case that convergence is harder.

To summarize, the space with the weaker norm has more convergent sequences, has bigger closures, has fewer open sets, has fewer closed sets, admits fewer continuous functions from the space, and admits more continuous functions to the space.

Definition. We say that two norms are equivalent if each is weaker than the other.

The results above show that two equivalent norms have the same convergent sequences and the same open and closed sets and admit the same continuous functions to and from the space. They also show that equivalent norms have the same bounded sets, a fact that is not true in a general metric space. Another important fact, which we now prove, is that equivalency preserves completeness, which is also not true in general metric spaces; see Exercise 2.21.

Theorem 2.26. *If X is a Banach space with respect to a norm $\|\cdot\|_1$, it is also a Banach space with respect to any equivalent norm.*

Proof. Suppose that (x_n) is a Cauchy sequence with respect to the norm $\|\cdot\|_2$, which is equivalent to $\|\cdot\|_1$. We know that, for some $K > 0$, $\| x \|_1 \leq K \| x \|_2$ for all x, which easily implies that (x_n) is $\|\cdot\|_1$ Cauchy and therefore $\|\cdot\|_1$ convergent. By equivalence, it is also $\|\cdot\|_2$ convergent. □

2.7 QUOTIENT SPACES

The concept of a quotient space is useful in many instances. We first review the basic algebraic facts.

2.7.1 Algebraic Aspects

Let N be a subspace of a linear space X. We define an equivalence relation on X by $x \sim y$ if $(x - y) \in N$. For any $x \in X$, let $\bar{x}$ denote its equivalence class, so $\bar{x} = x + N$, a translation of N. These equivalence classes are often known as *linear manifolds.* (In general algebraic structures, they are referred to as *cosets.*) As an example, if N is a one-dimensional subspace in the plane, the equivalence classes consist of parallel lines.

The set of all equivalence classes is denoted by X/N. We define operations of addition and scalar multiplication on X/N by

$$\bar{x} + \bar{y} = \overline{x + y}, \qquad \alpha\bar{x} = \overline{\alpha x}.$$

One must verify that these are well defined, since the operations are given by choosing a particular representative in each class. This follows easily from the fact that N is a subspace. Under these operations X/N is a linear space, known as a *quotient space* of X. We leave the verification of these statements to the reader.

The mapping π_N that sends each element into its equivalence class is known as the *canonical projection* with respect to N. It is a surjective linear operator.

For our purposes, the main application of quotient spaces is to the factorization of mappings. Given $T : X \to Y$ with the null space $N(T) = N$, we define a mapping $\tilde{T}$ from X/N to Y by

$$\tilde{T}\bar{x} = Tx.$$

The key point to notice is that, since $N(T) = N$, $Tx = Ty$ if and only if $x \sim y$. The *if* part of this statement shows that $\tilde{T}$ is well defined, and the *only if* part shows that $\tilde{T}$ is injective. It is clear that $\tilde{T}$ is linear. We then have the factorization

$$T = \tilde{T}\pi_N$$

as a composition of a surjective and an injective mapping. Later material will indicate various applications of this factorization.

2.7.2 Norms on Quotient Spaces

If X is a normed linear space, we would like to put a norm on a given quotient space of X. Whenever we define a linear space using other linear spaces, we usually want to choose a norm so that the pertinent mappings from or to the new space are bounded. In the present case, we would like this to be true for the canonical projection.

We proceed as follows. We are given an element of X/N, which consists of a whole equivalence class of elements in X. How do we assign a norm to this? Obviously, we will want to relate this norm to the norms of the individual elements. One way to ensure that π_N is bounded is to choose the smallest such norm or, more precisely, the infimum of all the norms. Accordingly, we make the formal definition

$$\| \bar{x} \| = \inf\{\| x - z \| : z \in N\} = d(x, N).$$

This is clearly well defined.

Theorem 2.27. *Let N be a closed subspace of the normed linear space X.*

(*a*) *The above definition defines a norm on X/N.*

(*b*) *If X is a Banach space, then X/N is a Banach space.*

(*c*) $\| \pi_N \| = 1$.

(*d*) *In the case that $N = N(T)$ for some bounded linear transformation T, the associated map $\tilde{T}$ on X/N has the same norm as T.*

Proof. We apply the definition of the quotient space norm in the following manner. Given any $\bar{x} \in X/N$ and any $\epsilon > 0$ we can find $z \in N$ such that

$$\| x - z \| \leq \| \bar{x} \| + \epsilon,$$

for if no such z existed, we would get a greater lower bound than given.

(a) We verify the three conditions for a norm. Given x_1 and x_2 and any $\epsilon > 0$, choose z_1 and z_2 in N so that

$$\| x_i - z_i \| \leq \| \bar{x}_i \| + \frac{\epsilon}{2}, \quad i = 1, 2.$$

Then,

$$\| \bar{x}_1 + \bar{x}_2 \| = \| \overline{x_1 + x_2} \| \leq \| x_1 + x_2 - z_1 - z_2 \| \leq \| x_1 - z_1 \| + \| x_2 - z_2 \| \\ \leq \| \bar{x}_1 \| + \| \bar{x}_2 \| + \epsilon,$$

thereby proving the triangle inequality since ϵ is arbitrary.
Next, fix $x \in X$ and $\alpha \in \mathcal{F}$. For any $z \in N$,

$$\| \alpha x - z \| = |\alpha| \| \frac{x - z}{\alpha} \| \geq |\alpha| \| \bar{x} \|,$$

which shows that

$$\| \alpha \bar{x} \| = \| \overline{\alpha x} \| \geq |\alpha| \| \bar{x} \| .$$

Since equality is obvious when $\alpha = 0$, assume $\alpha \neq 0$. For any $r > 1$, choose $z_0 \in N$ such that $\| x - z_0 \| \leq r \| \bar{x} \|$. Then,

$$\| \alpha \bar{x} \| = \| \overline{\alpha x} \| \leq \| \alpha x - \alpha z_0 \| = |\alpha| \| x - z_0 \| \leq r|\alpha| \| \bar{x} \| .$$

Since r is arbitrary, we know that $\| \alpha x \| \leq |\alpha| \| x \|$. Therefore, we have proved that $\| \alpha x \| = |\alpha| \| x \|$.
Finally, note that clearly $\| 0 \| = 0$. Conversely, $\| \bar{x} \| = 0$ means that $d(x, N) = 0$, so that $x \in \bar{N} = N$. Therefore, $\bar{x} = N$, the 0 element of X/N.

(b) We use the series criterion from Theorem 2.12. Suppose that $\sum_{i=1}^{\infty} \bar{x}_i$ is an absolutely convergent series in X/N. For each i, choose an element $z_i \in N$ such that $\| x_i - z_i \| \leq \| \bar{x}_i \| + 1/2^i$. Then,

$$\sum_{i=1}^{\infty} \| x_i - z_i \| \leq \sum_{i=1}^{\infty} \left(\| \bar{x}_i \| + \frac{1}{2^i} \right) < \infty,$$

which from the completeness of X shows that $\sum_{i=1}^{\infty} (x_i - z_i)$ converges. Now, $\bar{x}_i = \pi_N(x_i - z_i)$, so from the linearity and continuity of π_N, $\sum_{i=1}^{\infty} \bar{x}_i$ converges.

(c) It is clear from the definition that $\| \pi_N \| \leq 1$. Take any unit vector $\bar{x}$ in the quotient space X/N. Given any $r > 1$, choose $z \in N$ such that $\| x - z \| \leq r$. Then,

$$1 = \| \pi_N(x - z) \| \leq \| \pi_N \| r,$$

and since r is arbitrary, we have $\| \pi_N \| \geq 1$.

(d) For any unit vector $x \in X$, we have $\| \bar{x} \| \leq 1$, and

$$\| Tx \| = \| \tilde{T}\bar{x} \| \leq \| \tilde{T} \| \| \bar{x} \| \leq \| \tilde{T} \|,$$

thereby showing that $\| T \| \leq \| \tilde{T} \|$.

Conversely, given any unit vector in the quotient space X/N, it will be of the form $\bar{x}$ for some $x \in X$. Then, given any $r > 1$, choose $z \in N$ such that $\| x - z \| \leq r$. It follows that

$$\| \tilde{T}\bar{x} \| = \| T(x - z) \| \leq \| T \| r,$$

and since r is arbitrary, it follows that $\| \tilde{T} \| \leq \| T \|$. □

Remark. The final part of the proof of part (a) of the preceding theorem shows why we want N closed. If it is not, then the construction leads to only a seminorm on the quotient space. But the proof also shows that for closed N we get a norm on the quotient space even if we begin with a seminorm on the original space. This provides a convenient way to produce a normed space from a seminormed space X. Let $N = \{x \in X : \| x \| = 0\}$. We can easily show that N is a closed linear subspace, so that X/N is a normed linear space. In this case, the norm on the quotient space is particularly easy to compute since all elements in an equivalence class have the same norm; see Exercise 2.6.

2.8 FINITE-DIMENSIONAL NORMED LINEAR SPACES

2.8.1 Introduction

In this section, we demonstrate the remark made at the end of Chapter 1, namely, that the linear structure dominates topological considerations in finite-dimensional spaces. For motivation we look at the plane $\mathcal{R}^2$, on which we may put several norms. Consider the following three: For $x = (x(1), x(2))$, we have:

(i) the sup norm: $\| x \|_{\text{sup}} = \max\{|x(1)|, |x(2)|\}$;

(ii) the usual Euclidean norm: $\| x \|_2 = \sqrt{x(1)^2 + x(2)^2}$;

(iii) the absolute value deviation norm: $\| x \|_1 = |x(1)| + |x(2)|$.

See Figure 2.3, which shows unit balls in each case. We get squares, circles, and diamonds for (i), (ii), and (iii), respectively. Note that each of these shapes contains a reduced version of each of the others, which suggests, by Proposition 2.24, that all three are equivalent.

We will, in fact, show that on any finite-dimensional space all norms are equivalent and complete. It will follow that all linear subspaces are closed and all linear operators are bounded. In addition, we will give an interesting characterization of finite dimensionality in terms of compactness.

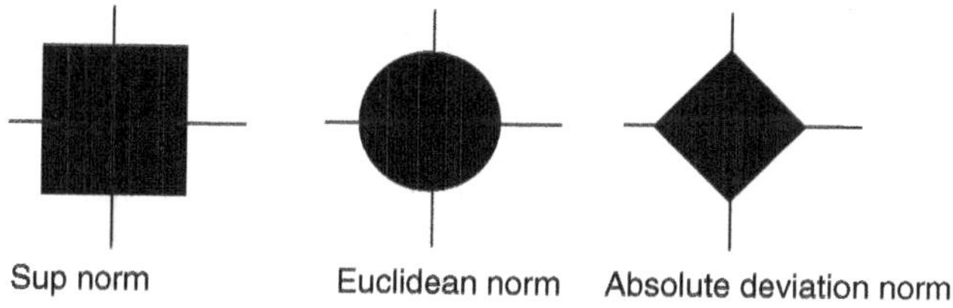

Figure 2.3. Unit balls under various norms.

2.8.2 Main Results

Throughout this section, we consider the space $B(F)$ equipped with the sup norm, in which $F = \{1, 2, \ldots, n\}$. We take the standard basis $\{\delta_i : i = 1, 2, \ldots, n\}$, as given in Section 1.4.

Proposition 2.28. *For any normed linear space Z, all elements of $\mathcal{L}(B(F), Z)$ are bounded.*

Proof. For any $T \in \mathcal{L}(B(F), Z)$ let $K = \max\{\| T\delta_i \|\}$. *Then, for any unit vector $f \in B(F), f = \sum_{i=1}^{n} f(i)\delta_i$, so that*

$$\| Tf \| \leq \sum_{i=1}^{n} |f(i)| \, \| T\delta_i \| \leq Kn. \qquad \square$$

The following theorem is well known and reminiscent of the classical Heine–Borel theorem, which is frequently stated and proved for the Euclidean norm. We need to ensure first that the result holds for the sup norm. We will derive this by assuming the conclusion in the one-dimensional case, where the Euclidean norm is exactly the same as the sup norm. For a proof of this case, see Rudin (1976, Theorem 2.4.1).

Theorem 2.29. *Closed and bounded subsets of B(F) are compact.*

Proof. Let (f_n) be any sequence in the closed and bounded set $A \subseteq B(F)$. We will show that some subsequence converges. Necessarily, the limit will be in $\bar{A} = A$. For each $i = 1, 2, \ldots, n$, we have $|f_n(i)| \leq \| f_n \|$, so that $(f_n(i))$ is a bounded sequence of scalars. Therefore, for some subsequence (f_{n_1}) of (f_n), the sequence of scalars $(f_{n_1}(1))$ must converge to a scalar, which we denote by $f(1)$. Similarly, for some subsequence (f_{n_2}) of (f_{n_1}), the sequence of scalars $(f_{n_2}(2))$ converges to a scalar $f(2)$. By continuing, we get a subsequence of the original sequence that converges pointwise to a function f. For a finite set, pointwise convergence is the same as uniform convergence, which, as we have seen, is convergence with respect to the sup norm. $\square$

Proposition 2.30. *If T is a bijective, linear operator from the normed linear space X to $B(F)$, then T is bounded.*

Proof. Suppose not; then, we can find a sequence (z_n) of unit vectors in X such that $\| Tz_n \| > n$. If $x_n = \| Tz_n \|^{-1} z_n$, then (x_n) converges to 0, and each Tx_n is of norm 1. Since the set of unit vectors in $B(F)$ is closed and bounded, we can invoke Theorem 2.29 and assume, by passing to a subsequence if necessary, that (Tx_n) converges to some unit vector y in $B(F)$. From the surjectivity of T and Proposition 2.28, T^{-1} exists as a bounded linear operator on $B(F)$, so that

$$T^{-1}y = \lim(T^{-1}Tx_n) = \lim x_n = 0,$$

a contradiction because $y \neq 0$ and T^{-1} is injective. □

We can now state our main conclusions.

Theorem 2.31

(a) *Any linear operator $T : X \to Z$, with X finite-dimensional, is bounded.*
(b) *All norms on a finite-dimensional space are equivalent and complete.*
(c) *Any finite-dimensional subspace of a normed linear space is closed.*

Proof

(a) Suppose X has a basis of n elements; then, we have a bijection J from X to $B(F)$, namely, the unique linear map that takes the nth basis element into δ_n. The composition TJ^{-1} gives a mapping from $B(F)$ to Z, and we factor T as a composition,

$$T : X \xrightarrow{J} B(F) \xrightarrow{TJ^{-1}} Z.$$

Now apply Propositions 2.28 and 2.30 together with Proposition 2.8.

(b) Given any two norms, apply (a) to the identity mappings from each space to the other to deduce the equivalence. The completeness follows from Theorems 2.14 and 2.26.

(c) Any finite-dimensional subspace can be identified as a linear space with $B(F)$. By part (b), the norm on the space is equivalent to the sup norm, which means that it is complete. From Theorem 2.16, we know the subspace must be closed. □

The one item we have not yet dealt with is the general linear operator *to* a finite-dimensional space, where the domain space is infinite-dimensional. There is a simple criterion to decide boundedness.

Theorem 2.32. *A linear operator $T : X \to Y$, with Y finite-dimensional, is bounded if and only if $N(T)$ is closed.*

Proof. If T is bounded, then $N(T)$ is the inverse image of the closed set $\{0\}$ and must be closed.

Conversely, suppose $N(T)$ is closed. Then, $X/N(T)$ must be finite-dimensional since there is an injective map from it to Y, namely, $\tilde{T}$ from Section 2.7. By applying Theorem 2.31(a) to $\tilde{T}$, we know that the composition $T = \tilde{T}\pi_{N(T)}$ is bounded. □

Where does this latter proof break down if $N(T)$ is not closed? In that case, $X/N(T)$ is only a seminormed space, in which case limits are not unique, and the argument used at the end of Proposition 2.30 to obtain the contradiction is no longer valid.

2.8.3 Topological Criterion for Finite Dimensionality

To motivate the results here, we begin by discussing *perpendiculars.* If M is a *proper subspace* (that is, not the whole space) in two- or three-dimensional real Euclidean space, we can easily visualize that there is a nonzero vector perpendicular to M. Is this true for all normed linear spaces? This question presents us a good opportunity to look at the role of questions in mathematical study. Many people think that mathematicians spend their time coming up with answers. Often, however, the trick is to come up with the right questions, which can make the answers easy.

Our query above, typical of the way a questioning procedure often starts, does not make sense. The usual notion of perpendicularity refers to angles, and in a normed linear space we do not have a definition of angles. We must first make sense of the question by defining perpendicularity in terms of length only. Visualizing this in the plane, we see that a vector is perpendicular to a subspace if and only if its distance to the subspace is as large as it could possibly be. This leads to the following formal definition.

Definition. Given a proper subspace M of the normed linear space X, a nonzero vector $x \in X$ is said to be *perpendicular* to M if $d(x, M) = \| x \|$.

The question now makes sense, but the answer is trivially no. If M is in fact dense, then $d(x, M) = 0$ for all x, so no perpendiculars exist. We want to rephrase the question to refer only to *closed* subspaces. The answer is still no, but it is not obvious; see Exercise 2.25 for an example.

Since analysis is the art of the "almost," we next ask if *almost perpendiculars exist,* and we must make sense of this by defining what we mean. This is accomplished by the following theorem, which can be interpreted intuitively to say that "almost perpendiculars" always exist. The procedure described in the proof can be easily visualized in the plane. The idea is that to obtain perpendiculars, we pick any vector not in the subspace and subtract the vector in the subspace that is closest to it; see Figure 2.4. However, there might not exist a closest element, so we get only an almost perpendicular. (The theorem is traditionally named after Frederick Riesz, a Hungarian mathematician who was one of the major early contributors to the subject of functional analysis. Like Banach, his name is attached to several results.)

Theorem 2.33 (Riesz's Lemma). *Given a closed, proper subspace M of a normed linear space X and $\epsilon > 0$, we can find a unit vector x such that $d(x, M) \geq 1 - \epsilon$.*

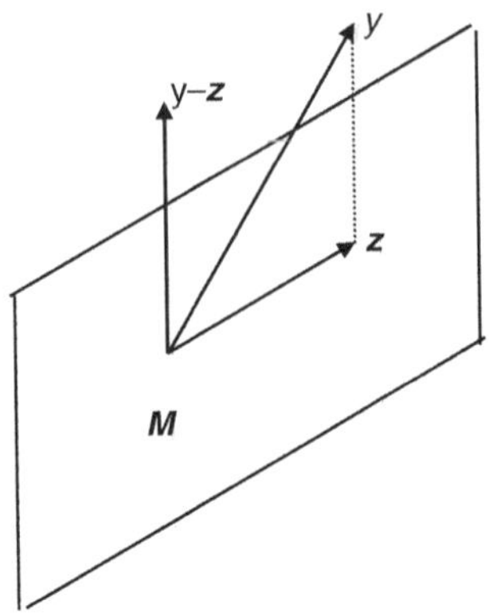

Figure 2.4. Finding a perpendicular to M.

Proof. Without loss of generality, we can assume that $0 < \epsilon < 1$. Choose any vector $y \notin M$, and let $d(y, M) = r$. Since M is closed, $r \neq 0$. Choose $z \in M$ such that $\| y - z \| \leq r/(1-\epsilon)$. For any $v \in M$, $z + v$ is also in M, so $\| y - (z + v) \| \geq r$, showing that $d(y - z, M) \geq r \geq (1-\epsilon) \| y - z \|$. By normalizing, $x = (y - z)/\| y - z \|$ is the desired unit vector. [We use the easily derived fact that $d(\alpha w, M) = \alpha d(w, M)$ for all $w \in X$ and scalars α.] □

Our final theorem of this section follows immediately.

Theorem 2.34. *A normed linear space X is finite-dimensional if and only if* $\bar{B}(1)$ *is compact.*

Proof. For finite-dimensional X, the compactness of $\bar{B}(1)$ was shown in Theorems 2.29 and 2.31(b). Conversely, suppose that X is infinite-dimensional. We inductively choose a sequence (x_n) of unit vectors as follows, starting with x_1 as arbitrary. Having chosen $\{x_1, x_2, \ldots, x_n\}$, the subspace $M_n = \operatorname{span}\{x_1, x_2, \ldots, x_n\}$ is finite-dimensional; therefore, M_n is closed and not equal to X. By Riesz's lemma, we can choose a unit vector x_{n+1} such that $d(x_{n+1}, M_n) \geq \frac{1}{2}$. The resulting sequence (x_n) cannot possibly have a convergent subsequence because any two elements are at least a distance of $\frac{1}{2}$ apart. □

This is quite a remarkable result since it establishes a connection between finite dimensionality, a purely algebraic property, and compactness, a purely topological property.

2.9 L^P SPACES

2.9.1 Introduction

It is time to add to our rather meager stock of norms. Up to this point, our only infinite-dimensional examples on spaces of functions are subspaces of $B(S)$ with the sup norm. It is, however, desirable to have more examples. One is frequently trying to define a solution to some equation or trying to find a function with certain properties by identifying a type of limit of other functions. The more modes of convergence one has, the more chance there is of success.

The uniform convergence of the sup norm is a rather stringent criterion. In order for an element to have a "small" norm, it must have small values at all points. Intuitively, however, we may want to consider a function to be small if it has small values, *except* on a small set of its domain. For example, consider a function defined on an interval of the reals that takes the value 0, except for a single point where it takes the value 1. This has sup norm 1, although one may well consider such a function as small. Norms defined by integrals achieve this flexibility. Consider the space $C[a,b]$ of continuous functions on the interval $[a,b]$. As an alternative to the sup norm, we can norm this space by defining $\| f \|_1 = \int_a^b |f(t)|dt$. This has a definite appeal since it means that a function has a small norm if the area between its graph and the horizontal axis is small. For another example, we can imitate the Euclidean norm on n-dimensional space by defining $\| f \|_2 = (\int_a^b |f(t)|^2 dt)^{1/2}$. Indeed, for any $p \geq 1$, we define the p-norm by

$$\| f \|_p = \left(\int_a^b |f(t)|^p \, dt \right)^{1/p}.$$

It is not at all obvious that these are norms, but we will prove this in due course.

One problem is that these norms are not complete on $C[a,b]$; see Exercise 2.28. As we noted above, a completion exists in an abstract sense, but it is desirable to have a more concrete representation of this space. We can, indeed, represent it by enlarging the set of functions. To fully understand these norms, we need some knowledge of measure and integration theory, which we will summarize in the following section. Detailed proofs of the material can be found in Appendix D.

2.9.2 Measure and Integration: Summary

In the early part of the twentieth century, a major problem in the theory of integration arose from the fact that continuous functions are not closed under pointwise limits. Since Riemann's definition of the integral depended to a large part on sufficient continuity, it was not clear that the integral was defined for a function obtained as such a limit. Henri Lebesgue, a French mathematician, remedied the situation.He first extended the notion of the length of an interval, so that one could talk about the "length" (or *measure* as it is called) of a large group of subsets of $\mathcal{R}$, known as *Lebesgue measurable* sets, which we will denote by $\mathcal{M}$. These sets formed what is called a *σ-algebra* (also known as a σ-field), which we define below.

For any set $A \in \mathcal{M}$, Lebesgue showed that it was possible to assign a value $\mu(A) \in [0, \infty]$, known as the *Lebesgue measure* of A, which agreed with the usual length if A is an interval, and such that the following holds: Given a finite or countably infinite sequence $A_1, A_2, \ldots, A_n, \ldots$ of pairwise disjoint sets in $\mathcal{M}$ (that is, $A_i \cap A_j = \emptyset$ for $i \neq j$), then

$$\mu\left[\bigcup_i A_i\right] = \sum_i \mu(A_i), \qquad (*)$$

a property known as *countable additivity.*

This class of sets then enabled one to choose a larger class of functions than the continuous ones which could be handled when it came to integration. We define a real-valued function on $\mathcal{R}$ to be *Lebesgue measurable* if for all $r \in R$ the set $f^{-1}(-\infty, r]$ is in $\mathcal{M}$. The measurable functions have the desirable feature that they contain all the continuous functions and are closed under pointwise limits.

The formulation of Lebesgue was subsequently generalized so as to apply to many situations. We give the relevant formal definitions.

Definition. Given any set Ω, a collection $\mathcal{S}$ of subsets of Ω is called a σ-algebra if it satisfies the following:

(a) Given any finite or countably infinite sequence $A_1, A_2, \ldots, A_n, \ldots$ of sets in $\mathcal{S}$,
   (i) their union is in $\mathcal{S}$;
   (ii) their intersection is in $\mathcal{S}$.

(b) For any $A \in \mathcal{S}$, the complement $S^c \in \mathcal{M}$.

(c) The entire set $\mathcal{S} \in \mathcal{M}$.

[Since a(ii) follows from a(i) and (b) and a(i) follows from a(ii) and (b), many authors omit one of a(i) or a(ii) in the definition of a σ-algebra.]

Definitions. A *measure* on a σ-algebra $\mathcal{S}$ is a function from $\mathcal{S}$ to the interval $[0, \infty]$ which is countably additive, as defined in $(*)$ above, and satisfies $\mu(\emptyset) = 0$. The last condition is necessary to rule out the case where all sets have measure ∞. A *measure space* consists of $(\Omega, \mathcal{S}, \mu)$, where $\mathcal{S}$ is a σ-algebra of subsets of X and μ is a measure on $\mathcal{S}$. The definition of a *measurable function* on Ω is exactly as given above for functions on the real line, with $\mathcal{S}$ replacing $\mathcal{M}$

Note that point (a) in the definition of a σ-algebra shows that the left-hand side of $(*)$ makes sense.

Any Lebesgue measurable subset A of the reals gives rise to a measure space $(A, \mathcal{M}_A, \mu)$, in which $\mathcal{M}_A$ consists of the intersection of all Lebesgue measurable sets with A and μ is the restriction of Lebesgue measure to $\mathcal{M}_A$. Whenever we speak of a subset A of the reals as a measure space, we will, unless otherwise indicated, be referring to this formulation.

Measurable functions on any measure space form an attractive class, since they are closed under almost any operation one can think up, such as addition, scalar multiplication, and multiplication. If f is measurable, so is $|f|$, an important fact in what follows. As in the case of Lebesgue measure, a pointwise limit of measurable functions is always measurable.

It is often desirable to distinguish measure spaces according to whether or not the whole space has infinite measure.

Definition. A measure space is said be *finite* if $\mu(\Omega) < \infty$.

It is important to note that a finite measure space does not mean that Ω is finite, as one may think from the terminology. An example of such is any bounded subset of the reals. Another important class of examples is given in the following.

Application: Probability Theory. We will from time to time insert notes concerning the application of the concepts we introduce to other areas of mathematics. These will be brief, as we do not expect all readers to be familiar with these fields, and the material can be omitted without loss of continuity. The purpose is to convey some flavor of the scope of the subject and in some cases to provide a reference to those who wish to pursue the ideas further.

A particularly important application of measure spaces was to axiomatize probability theory. In this case, the set Ω represents the outcomes of an experiment, the sets in the σ-algebra are combinations of outcomes known as events, and the measure μ, which in this case takes values in $[0, 1]$, denotes the probability of an event. A measure space in which the whole space has measure 1 is often called a *probability space.* Measurable functions are used to model *random variables*, which are numerical quantities dependent on the outcome of the random experiment. The defining property for a measurable function ensures that we can calculate the probability that a random variable takes a value less than or equal to r, which is an important quantity, known as the *cumulative distribution function* of the random variable.

Integral of a Nonnegative Function. We now discuss integration in this general setting. To be able to deal with infinite values conveniently, we restrict ourselves at first to *nonnegative* measurable functions. Such a function is said to be *simple* if it takes only finitely many values $c_1, c_2, \ldots, c_n$. We define the integral of such a simple function s with respect to the measure μ as

$$\int s d\mu = \sum_{i=1}^{n} c_i \mu[s^{-1}(c_i)].$$

The definitions of a σ-algebra and measurability ensure that the sets $s^{-1}(c_i)$ are in $\mathcal{S}$, so that the definition of the integral makes sense. As an aside, it is standard in measure theory to define $0 \cdot \infty = 0$ and a sum in which a summand of ∞ appears to be ∞.

As an example, on the measure space $[0, 1]$, let h be the function that takes the value of 1 on the irrationals and 0 on the rationals. This is an example of a function for which the integral does not exist under Riemann's definition. However, it is a simple function, taking only two values. The Lebesgue measure of any point (a degenerate interval) is 0, so by countable additivity, the measure of the rational points in $[0, 1]$ is 0, which by additivity again means that the measure of the irrational points in $[0, 1]$ is 1. By using the definition of the integral above, we easily see that for the Lebesgue measure $\int h\, d\mu = 1$.

Consider now the set of all measurable functions from Ω to $[0, \infty]$. (We are modifying our original definition in which measurable definitions were real valued. It is convenient to include ∞ as a possible value, since such a value often appears when we take a pointwise limit of measurable functions. The defining condition remains unchanged.) Any such function f can be approximated by a sequence of simple functions, and this leads to a definition of the integral $\int f\, d\mu$ for nonnegative measurable functions. In the case of Lebesgue measure and continuous functions, we recover the usual value of the integral, which can be interpreted as the area under the graph of the function.

Properties of Integrals. We now list six properties of the integral of nonnegative functions that we need for the material that follows. The reader unfamiliar with general integration theory should nonetheless be able to follow this material by assuming these properties as axioms for the integral. As a matter of notation, if there is only one measure μ under consideration, we often omit the $d\mu$. We first need a preliminary definition.

Definition. A set $B \in \mathcal{S}$ such that $\mu(B) = 0$ is called a *null set.* As an example, on the real line with Lebesgue measure, any countable set is null, as we pointed out above. We say that a property holds a.e. (an abbreviation for "almost everywhere") if the points where it does *not* hold form a null set. (We will sometimes write a.e. $[\mu]$ if it necessary to distinguish between different measures.)

The six properties for nonnegative measurable functions f and g are as follows:

Property 1 (additivity): $\int (f + g) = \int f + \int g$.
Property 2 (scalar property): For any $\alpha \geq 0$, $\int \alpha f = \alpha \int f$.
Property 3 (monotonicity): If $f \leq g$, then $\int f \leq \int g$.
Property 4 $\int f = 0$ if and only if $f = 0$ a.e..
Property 5 If $\int f < \infty$, then $f < \infty$ a.e..
Property 6 (Fatou's lemma): If (f_n) is a sequence of nonnegative measurable functions converging pointwise to a function f, then

$$\int f \leq \liminf \int f_n.$$

(See Appendix C for the definition of lim inf.)

Property 6 is a key one, and we now provide more discussion of it. A natural question is whether, under the conditions of Property 6, it is true that $\left(\int f_n\right)$ converges to $\int f$. This is false in general as we will demonstrate with examples from the real line with Lebesgue measure. We give two examples, the "short fat" example and the "tall skinny" example.

For the short fat example, let f_n be the function that takes the value $1/n$ on the interval $[0, n]$ and 0 elsewhere. This function is easily seen to be measurable. Then, (f_n) converges to the 0 function, which has integral 0, but $\int f_n = 1$ for all n. In the tall skinny example, $f_n = n$ on the interval $(0, 1/n]$ and 0 elsewhere. The same conclusion holds.

What we have seen is that functions that are big on small sets or small on big sets can produce large values of the integral, which diminish when we take the integral of the limiting function. However, the integral of the limit cannot become larger; that is, we must have that $\int f \le \lim \int f_n$ if this limit exists. The latter limit need not exist, but the result is true for the lim inf of the sequence.

We can add to our list many other properties that are consequence of these six. A special case of Fatou's lemma that is used repeatedly is as follows.

Property 7 (monotone convergence theorem): If (f_n) is a sequence of nonnegative measurable functions converging pointwise to a function f, and if $f_n \le f_{n+1}$ for all n, then $\int f = \lim \int f_n$.

This result follows from Fatou's lemma by noting that from Property 3 we have $\int f \ge \int f_n$ for all n.

Integral of a General Function. This completes our list of properties for the integral of a nonnegative measurable function. For a general real-valued measurable function f, we define

$$\int f = \int f^+ - \int f^-,$$

in which

$$f^+(t) = \max\{f(t), 0\} \qquad \text{and} \qquad f^-(t) = -\min\{f(t), 0\}.$$

(These are known as the *positive* and *negative parts of the function*, respectively.) The integral will be defined (possibly having values of ∞ or $-\infty$), except in the case where the integrals of both f^+ and f^- equal ∞.

Finally, we define a complex-valued function f to be measurable if both its real and imaginary parts are measurable. Recall that for a complex-valued function f we define the functions $\mathrm{Re}\, f$ and $\mathrm{Im}\, f$ in the obvious way as

$$\mathrm{Re} f(x) = \mathrm{Re}\,(f(x)), \qquad \mathrm{Im} f(x) = \mathrm{Im}(f(x)).$$

We then define

$$\int f = \int \mathrm{Re} f + i \int \mathrm{Im} f,$$

which exists provided that both integrals exist and are finite.

We will have occasion to refer to the first three properties of the integral for general functions. We note that when the functions have finite integrals, Property 1 remains exactly as given, Property 2 is true for arbitrary scalars, and Property 3 holds for all real-valued functions.

2.9.3 The p-Norms

Suppose we have a measure space $(\Omega, \mathcal{S}, \mu)$. Fix any p in the interval $[1, \infty)$. For any $\mathcal{F}$-valued measurable function f on Ω, we define

$$\| f \|_p = \left(\int |f|^p \, d\mu \right)^{1/p}.$$

We let

$$\mathcal{L}^p(\Omega, \mathcal{S}, \mu) = \{\text{measurable } f : \| f \|_p < \infty\}.$$

This notation is sometimes abbreviated to $\mathcal{L}^p$ if the measure space is fixed.

Remark. The introduction of the pth root in the definition of $\| \cdot \|_p$ is necessary in order to achieve the norm property that $\| \alpha f \|_p = |\alpha| \; \| f \|_p$. It does, however, complicate proving the triangle inequality, as we will see.

Proposition 2.35. *$\mathcal{L}^p$ is a linear space.*

Proof. Closure under scalar multiplication is indicated by the remark above. To show closure under addition, we note first that for any two nonnegative numbers $a \leq b$

$$(a + b)^p \leq (2b)^p = 2^p b^p \leq 2^p (a^p + b^p).$$

Therefore, given any $f, g \in \mathcal{L}^p$ and any $\omega \in \Omega$,

$$|f(\omega) + g(\omega)|^p \leq (|f(\omega)| + |g(\omega)|)^p \leq 2^p (|f(\omega)|^p + |g(\omega)|^p).$$

By integrating and using monotonicity (Property 3 of the integral), we get

$$\| f + g \|_p \leq 2^p (\| f \|_p{}^p + \| g \|_p{}^p),$$

thereby showing that $f + g \in \mathcal{L}^p$. □

This is not enough, however, to prove the triangle inequality for $\| \cdot \|_p$. That requires more delicate arguments, which we carry out in the next section.

2.9.4 Hölder's and Minkowski's Inequalities

In this section, we prove two classical inequalities. As with many inequality proofs, the key step is a convexity argument.

Definition. A real-valued function f defined on a interval of the reals is *convex* if for all $0 \le \alpha \le 1$ and for all a, b in the domain we have

$$f(\alpha a + (1-\alpha)b) \le \alpha f(a) + (1-\alpha)f(b).$$

In other words, the graph of the function lies beneath the line segment joining any two points on the graph. Positivity of the second derivative is a sufficient condition for convexity, showing that, in particular, the function $f(t) = e^t$ is convex.

Throughout the section, we fix a measure space $(\Omega, \mathcal{S}, \mu)$. Fix $p \in (1, \infty)$, and let q be such that $1/p + 1/q = 1$ (equivalently, $p + q = pq$). We begin with a preliminary lemma.

Lemma 2.36. *For all nonnegative numbers s and t,*

$$st \le \frac{s^p}{p} + \frac{t^q}{q}.$$

Proof. Choose a and b so that $s = e^{a/p}$ and $t = e^{b/q}$. By convexity of $f(t) = e^t$,

$$st = e^{(a/p + b/q)} \le \frac{1}{p}e^a + \frac{1}{q}e^b = \frac{1}{p}s^p + \frac{1}{q}t^q. \qquad \square$$

Theorem 2.37 (Hölder's Inequality). *For all measurable functions f and g,*

$$\| fg \|_1 \le \| f \|_p \| g \|_q .$$

Proof. If either term on the right is ∞, the result is obvious. If either term is 0, it means that the corresponding function equals 0 a.e. by applying Property 4 of the integral to the absolute value. In this case, the left side will also equal 0 by the other direction of Property 4. We can therefore assume that both terms on the right are positive and finite. It is sufficient to prove the result when both "norms" on the right side equal 1, since we can then apply that result to $f / \| f \|_p$ and $g / \| g \|_q$ for general functions by using the properties of scalar multiplication in the space and $\| \cdot \|_p$.

By applying Lemma 2.36 to $s = |f(\omega)|$ and $t = |g(\omega)|$ for each $\omega \in \Omega$, we have

$$|fg| = |f||g| \le \frac{|f|^p}{p} + \frac{|g|^q}{q},$$

and by monotonicity of the integral,

$$\| fg \|_1 \le \frac{1}{p} + \frac{1}{q} = 1,$$

completing the proof. $\square$

Theorem 2.38 (Minkowski's Inequality). *For all measurable functions f and g and for all $p \in [1, \infty)$,*

$$\| f + g \|_p \leq \| f \|_p + \| g \|_p \, .$$

Proof. For $p = 1$, the proof is direct, so assume $p > 1$. Let $h = |f + g|$, so that $h^p \leq |f| h^{p-1} + |g| \, h^{p-1}$, and from the addititivity and monotonicity of the integral, we have

$$\| h \|_p^p = \int h^p \leq \int |f| \, h^{p-1} + \int |g| h^{p-1}.$$

By applying Hölder's inequality to each term on the right, we obtain

$$\| h \|_p^p \leq (\| f \|_p + \| g \|_p) \, \| h^{p-1} \|_q \, . \qquad (*)$$

Now, note that

$$\| h^{p-1} \|_q = \left(\int h^{(p-1)q} \right)^{1/q} = \left(\int h^p \right)^{(p-1)/p} = \| h \|_p^{p-1},$$

so that dividing equation $(*)$ by $\| h \|_p^{p-1}$ gives $\| h \|_p \leq \| f \|_p + \| g \|_p$. □

We have now shown that $\| \cdot \|_p$ is a seminorm for $p \in [1, \infty)$. It is not, however, a norm because, from Property 4 of the integral, $\| f \|_p = 0$ if $f = 0$ a.e. Applying the remark at the end of Section 2.7.2, we obtain an important class of spaces.

Definition. For $1 \leq p < \infty$, let $N = \{f \in \mathcal{L}^p : \| f \|_p = 0\}$, and let $L^p(\Omega, \mathcal{S}, \mu)$ denote the quotient space $\mathcal{L}^p / N$. [This notation is sometimes shortened to $L^p(\Omega)$ or to $L^p(\mu)$ or just to L^p if the measure space is fixed.]

It has become traditional in the mathematical literature to think and talk of the L^p spaces as though they consisted of functions. In fact, however, an element of L^p is a whole equivalence class of functions. Here, f and g are equivalent if and only if $f - g$ is in the subspace N given above, that is, $f - g = 0$ a.e. In other words, f and g are equivalent if and only if the set of points at which $f(\omega) \neq g(\omega)$ has measure 0. This practice of speaking of the elements as functions usually causes no difficulty as long as one remembers the true nature of this space and can adjust one's thinking when necessary. We will point out an example of this shortly.

2.9.5 The Space L^∞

We now want to consider the space of equivalence classes of measurable functions that are bounded and to use the sup norm on that space. Here is another example of a statement that, on the face of it, does not make sense. Mathematicians come up with such statements frequently in their thought processes and their conversations. Usually such statements do not appear in books, but we like to insert these from time to time to illustrate how these thought processes work. Of course, the first task is to alter the statement so that it does make sense. What is the problem? It is that boundedness is not a property that is invariant under null set modification. For example, on the measure space $(0, 2)$, consider two functions: h is identically equal to 1, while j has a value of 1 on all irrational numbers and a value of $1/t$ if t is rational. These functions are equal a.e., but h is bounded and j is not.

The key to a proper formulation of the idea is to observe that a sup is the inf of all upper bounds, so that we can rewrite our definition of a sup norm of a function on a set S as

$$\| f \|_{\text{sup}} = \inf\{r : |f(x)| \leq r \text{ for all } s \in S\}.$$

This suggests the following definition.

Definition. The *essential supremum* of a function on a measure space is given by

$$\| f \|_{\text{ess sup}} = \inf\{r : |f(x)| \leq r \text{ for } x \text{ a.e.}\}.$$

It is clear that if $f = g$ a.e., then $\| f \|_{\text{ess sup}} = \| g \|_{\text{ess sup}}$.

To illustrate, reconsider the example of the two functions given above. Take any number greater than or equal to 1, say 1.5. Then, certainly $j \leq 1.5$ a.e. since the points where it takes a value bigger than 1.5 is a subset of the rationals and therefore of measure 0. This is not true for a number less than 1 since j takes a value of 1 on a set of positive measure. Therefore, $\| j \|_{\text{ess sup}} = 1$.

Definitions. A measurable function is said to be *essentially bounded* if $\| f \|_{\text{ess sup}} < \infty$. We let $\mathcal{L}^\infty$ denote the set of essentially bounded measurable functions. A minor modification of the argument used in Section 2.3 shows that $\mathcal{L}^\infty$ is a linear space and $\| \cdot \|_{\text{ess sup}}$ is a seminorm on $\mathcal{L}^\infty$. From this we get the normed space $L^\infty = \mathcal{L}^\infty / N$, in which N is the subspace of those functions f with $\| f \|_{\text{ess sup}} = 0$.

The exponent of ∞ was adopted since in familiar cases the resulting space can be viewed as a limit of the spaces L^p as p approaches ∞. For example, on any finite measure space, if $\| f \|_{\text{ess sup}} \leq K$, it follows that $\int |f| \leq K \mu(\Omega)$, showing that an essentially bounded measurable function is in L^p for all p. In keeping with this notation, we will, in the sequel, use $\| \cdot \|_\infty$ to denote $\| \cdot \|_{\text{ess sup}}$.

2.9.6 Completeness of L^p Spaces

We now prove that L^p spaces are complete. The key to this fact is the convergence properties of the integral, which we have expressed in the form of Fatou's lemma.

We follow the same strategy as in Section 2.5.3. In particular, we will invoke Lemma 2.13. However, to reflect the nature of our spaces as consisting of equivalence classes, we must replace pointwise convergence by a.e. convergence, which we define as follows:

Definition. A sequence (f_n) of measurable functions *converges a.e.* to a function f if the set of all points $\omega \in \Omega$ such that $(f_n(\omega))$ does not converge to $f(\omega)$ is a set of measure 0.

One must verify that a.e. convergence is well defined when applied to the equivalence classes. In other words, suppose (f_n) *converges* a.e. to f, (g_n) *converges* a.e. to g, and $g_n = f_n$ a.e. for $n = 1, 2, \ldots$. We need to know that $f = g$ a.e. This follows from the fact that a countable union of sets of measure 0 is of measure 0, a consequence of the countable additivity of the measure. Moreover, it is easy to verify that Lemma 2.13 remains valid with a.e. convergence replacing convergence. We also need the fact that Fatou's lemma holds with a.e. convergence replacing convergence.

We first show that the condition of Lemma 2.13 holds.

Lemma 2.39. *Let (f_n) be a sequence of nonnegative measurable functions in the closed unit ball of L^p for some $1 \leq p < \infty$. If (f_n) converges a.e. to a function f, then $\| f \|_p \leq 1$.*

Proof. The proof is a direct application of Fatou's lemma, Property 6. Since $(|f_n|^p)$ *converges a.e.* to $|f|^p$, we know that

$$\| f \|_p^p = \int |f|^p \leq \liminf \int |f_n|^p = \liminf \| f_n \|_p^p \leq 1. \qquad \square$$

We also want the series version of the above lemma.

Corollary 2.40. *If (h_n) is a sequence of nonnegative measurable functions such that $h = \sum_{n=1}^{\infty} h_n$ a.e., and if $\sum_{n=1}^{\infty} \| h_n \|_p \leq 1$, then $\| h \|_p \leq 1$.*

Proof. Apply Lemma 2.39 to the sequence $s_n = \sum_{i=1}^{n} h_i$, which by the general triangle inequality has norm less than 1. $\square$

Theorem 2.41. *For $1 \leq p \leq \infty$, L^p is a Banach space.*

Proof. The proof for $p = \infty$ follows by modifying the proof of completeness of $B(S)$ given in Section 2.5.3. We leave the details to the reader. We will give the proof for $p < \infty$.

From Lemmas 2.13 and 2.39, we would be done if we could show that Cauchy sequences converge a.e., but this is not true; see Exercise 2.36. However, all we need to show is that any fast Cauchy sequence (f_n) converges a.e. The definition of a fast Cauchy sequence suggests that we transform this problem from one of sequence convergence to series convergence. (This idea is often useful since the pointwise sum of a series of nonnegative functions is always defined as long as we allow a value of ∞.) We define $h_n = f_{n+1} - f_n$. Because the partial sum of the corresponding series $\sum_{i=1}^{n} h_i = f_{n+1} - f_1$, we are done if we can show that $\sum_{n=1}^{\infty} h_n$ converges a.e.

To this end, we first consider the function

$$k = \sum_{n=1}^{\infty} |h_n|.$$

To reach our conclusion, we have to rule out infinite values, except on a set of measure 0. By definition of a fast Cauchy sequence,

$$\sum_{n=1}^{\infty} \| h_n \|_p \leq \sum_{n=1}^{\infty} \frac{1}{2^n} = 1,$$

so by Corollary 2.40, $\| k \|_p \leq 1$. Now by Property 5 of the integral, the function k is finite a.e., which means that, for ω a.e., $\sum_{n=1}^{\infty} h_n(\omega)$ is an absolutely convergent sequence of scalars. Therefore, it converges to a finite sum, completing the proof. □

Remark. Refer to the beginning of this section where we looked at the space $C[a,b]$ and noticed that it was incomplete under the p-norms. We regard $[a,b]$ as a measure space, as indicated before, and one can show that, for $1 \leq p < \infty$, $C[a,b]$ is dense in L^p (Royden, 1988, Proposition 6.8). This statement makes sense even though one is a space of functions and the other a space of equivalence classes. The point is that in any class there can be only one continuous function. The result shows that L^p is the completion of $C[a,b]$ with respect to $\| \cdot \|_p$.

2.9.7 ℓ^p Spaces

A particularly easy example of a measure space is to take Ω to be the positive integers, $\mathcal{S}$ to be all subsets of Ω, and $\mu(A)$ to be the number of elements in A, including a possible value of ∞. This is known as *counting measure.* Integration for this measure just reduces to summation. Elements of the space are sequences $x = (x(1), x(2), \ldots, x(n), \ldots)$, and

$$\int x = \sum_{n=1}^{\infty} x(n).$$

The resulting L^p space in this case is denoted by ℓ^p. For $1 \leq p < \infty$, this space consists of all sequences x such that $\sum_{n=1}^{\infty} |x(n)|^p < \infty$, and the norm is given by

$$\| x \|_p = \left(\sum_{n=1}^{\infty} |x(n)|^p \right)^{1/p}.$$

The situation simplifies considerably. Since all sets are measurable, all possible sequences are measurable. We no longer have to worry about a.e. considerations since there are no nonempty sets of measure 0. So, for example, ℓ^∞ is just $B(\mathcal{N})$ with the sup norm, in which $\mathcal{N}$ is the set of positive integers. It can be seen that c_{00} is dense in ℓ^p (see Exercise 2.38), so that ℓ^p is the completion of c_{00} with respect to the p-norm, for $1 \leq p < \infty$.

Note the strict inequality in the statement above, suggesting that it is not true for $p = \infty$. Why not? In fact, the completion of c_{00} with respect to the sup norm is c_0, the space of sequences converging to 0 (see Exercise 2.7), but that by itself does not rule out our statement. For all we know, it may be that ℓ^∞ is isometric to c_0. Remember, when we say that one normed linear space is the same as another, we really mean that there exists a bijective isometry between them. We will show, however, that there is no bijective isometry between ℓ^∞ and c_0.

In general, showing that one mathematical object is not isomorphic to another can be extremely difficult. Even though the objects might appear completely different from each other, there could be some strange, hard-to-detect isomorphism lurking in the shadows. The usual procedure is to find some *invariant* that differs between the two spaces. By an invariant, we mean some feature that is preserved by any isomorphism. In our present case, it is not difficult to find one.

Definition. A normed linear space is said to be *separable* if it contains a countable dense subset.

To show a space is separable, we need only exhibit a countable set B such that span(B) is dense. Then,all the linear combinations of elements in B with rational coefficients are easily seen to constitute the required countable dense set. (We consider a complex number rational if both its real and imaginary parts are rational.) In our sequence spaces, if $B = \{\delta_n : n = 1, 2, \ldots\}$, then span($B$) is just c_{00}. This shows that c_0 and ℓ^p for are separable $1 \leq p < \infty$. The property of being separable is clearly an invariant for a bijective isometry. Our claim above, then, follows from the following proposition.

Proposition 2.42. *ℓ^∞ is not separable.*

Proof. Given any two distinct subsets A and C of $\mathcal{N}$, it is clear that $\| 1_A - 1_C \|_\infty = 1$, since for at least one point one function takes a value of 0 and the other a value of 1. This means that the open balls $B(1_A; \frac{1}{2})$ and $B(1_C; \frac{1}{2})$ in ℓ^∞ are disjoint, for if a point x were in both of them, we would have

$$\| 1_A - 1_B \|_\infty \leq \| 1_A - x \|_\infty + \| x - 1_B \|_\infty < \tfrac{1}{2} + \tfrac{1}{2} = 1,$$

a contradiction. Now, there are uncountably many subsets A of the countable set $\mathcal{N}$, and this gives rise to an uncountable family of pairwise disjoint open sets in ℓ^∞, namely $\{B(1_A;\frac{1}{2}) : A \subseteq \mathcal{N}\}$. Necessarily, any countable set in ℓ^∞ will not intersect some of these open sets; therefore, it cannot be dense. □

For the benefit of the reader with a limited background in measure theory, we give a self-contained proof that ℓ^p spaces are Banach spaces without using the general case. We will directly verify that the condition of Lemma 2.39 holds in this situation. We need only do this for $p = 1$; then, one can apply that result to $(|f_k|^p)$ and $|f|^p$ to obtain the result for general p. We show the series version as given in Corollary 2.40.

Proposition 2.43. *Let (f_k) be a sequence of nonnegative functions on the positive integers converging pointwise to a function f. If $\sum_{n=1}^{\infty} f_k(n) \leq 1$ for all k, then $\sum_{n=1}^{\infty} f(n) \leq 1$.*

Proof. We follow the familiar technique for proving inequalities in analysis, namely by looking at a relaxed version. Choose any $c > 1$. We will show that $\sum_{n=1}^{\infty} f(n) \leq c$, and since c is arbitrary, the desired inequality follows. To show this relaxed inequality, we need only show that, for all positive integers N, $\sum_{n=1}^{N} f(n) \leq c$. Since pointwise convergence on a finite set is uniform convergence, we can find K such that $f_K(n) \geq f(n)/c$ for $n = 1, 2, \ldots, N$. It follows that

$$\sum_{n=1}^{N} f(n) \leq c \sum_{n=1}^{N} f_K(n) \leq c \sum_{n=1}^{\infty} f_K(n) \leq c.$$

□

2.10 DIRECT PRODUCTS AND SUMS

We now deal with a second variation to which we alluded in Section 1.2. Our first was to notice that, in obtaining linear spaces of functions, the values of our function did not have to be in $\mathcal{F}$ but could be in any other linear space, as we discussed in Section 1.3. The second is to note that, since operations are defined pointwise, we can actually vary the space from point to point.

Suppose we have a set S and for each $s \in S$ we have a linear space X_s. Consider all functions f from S to the union of all these spaces such that $f(s) \in X_s$ for all $s \in S$. This space is known as the *product* of the spaces and denoted by $\prod_{s \in S} X_s$. It is clear that this becomes a linear space under pointwise addition and scalar multiplication. A familiar case is when S consists of two points, and this simply gives the usual Cartesian product of two spaces.

Now suppose that each X_s is a normed linear space. We can make the same definition of a norm as for $F(S)$; that is,

$$\| f \| = \sup\{\| f(s) \| : s \in S\}.$$

This defines a norm on the set of all functions in the product for which $\| f \| < \infty$. The resulting normed linear space is known as the *direct product* of the spaces X_s. If each X_s is a Banach space, we can adapt directly the proof for $B(S)$ to show that this direct product is a Banach space. When each space X_s equals $\mathcal{F}$, we just recover $B(S)$. Convergence in the direct product is, as before, uniform convergence.

The product space is equipped with natural projections π_s to X_s for each $s \in S$ defined by

$$\pi_s(f) = f(s).$$

With the given definition of the norm, it is easily seen that $\| \pi_s \| = 1$.

The subspace of the product consisting of all functions that take a value zero, except for finitely many values of s, is known as the *direct sum* of the spaces X_s. For example, when each X_s equals $\mathcal{F}$ and S is the set of positive integers, this direct sum is just c_{00}. When S is finite, the direct sum is the same as the direct product when considered as linear spaces. In general, however, the direct sum admits a greater variety of norms. In fact, we can define a general type of p-norm on the direct sum, for $1 < p < \infty$, by

$$\| f \|_p = \left(\sum_{s \in S} \| f(s) \|^p \right)^{1/p}.$$

We can define this p-norm for all f in the product of the spaces, and the set of those f with finite p-norm can be viewed as a generalized ℓ^p space, which will be a completion of the direct sum when each X_s is complete. As an example, suppose that S is the set of positive integers. The resulting general ℓ^p space consists of sequences, but the entries are not necessarily scalars and could take values in some other space, possibly varying from point to point.

We take a closer look at the case for which the index set consists of two elements. We have the direct product or sum of two spaces Y and Z. The resulting space X is denoted by either $Y \times Z$ or $Y \oplus Z$. We will generally use the first symbol when we think of this as the direct product with the sup norm and the second when we think of it as the direct sum with one of the p-norms.

The elements of X are pairs (y, z), where $y \in Y$ and $z \in Z$. We can identify Y as a subspace of X, namely those elements of the form $(y, 0)$ and similarly identify Z as those elements of the form $(0, z)$. With these identifications, we have

$$Y \cap Z = 0 \qquad \text{and} \qquad Y + Z = X. \tag{$*$}$$

In many instances, we want to consider the opposite situation. That is, we are given a space X, and we would like to write it as a direct sum of two of its subspaces. Providing we can find subspaces Y and Z satisfying the two conditions given in $(*)$, we can indeed define a bijection $i : Y \oplus Z \to X$ by $i(y, z) = y + z$. This map is surjective by the second condition. It is injective by the first condition because $i(y, z) = i(y_1, z_1)$ implies that $y - y_1 = z_1 - z \in Y \cap Z$. This shows that both $y - y_1$ and $z_1 - z$ are equal to zero. The function i is an isomorphism of linear spaces, but this leaves open the question as to when i is continuous or when it is an isometry. We will return to this idea in Chapter 4; see also Exercises 2.34 and 3.6.

2.11 SCHAUDER BASES

We will treat this subject only briefly. A good account can be found in Carothers (2005, Chapter 3). A sequence (x_n) in a normed linear space X is said to be a *Schauder basis* if every $x \in X$ can be written uniquely in the form $x = \sum_{n=1}^{\infty} \alpha_n x_n$, where each $\alpha_n \in \mathcal{F}$. This is to be distinguished from the usual linear algebra basis defined in Section 1.2, which required *finite* linear combinations. For this reason, the latter is often referred to as a *Hamel* basis. Note also that the Schauder basis depends on the norm, as it involves an infinite series, while a Hamel basis is strictly a linear space concept. Hamel bases are not very useful for infinite-dimensional spaces (see Exercise 3.4), so some writers simply use the word *basis* to mean a Schauder basis when talking about Banach spaces.

The standard example is (δ_n), which is a Schauder basis for c_0 and ℓ^p, $1 \leq p < \infty$, but not for ℓ^∞. In c_{00}, it is also a Hamel basis. The name comes from the work of Schauder (1927), who constructed a basis of piecewise linear functions for the space $C[0, 1]$. It was conjectured by Banach in 1932 that every separable Banach space had a Schauder basis. (This conjecture can be found in his famous treatise (Banach, 1955) or its English translation (Banach, 1988).) The problem was not resolved until over 40 years later when a counterexample was found by Enflo (1973).

2.12 FIXED POINTS AND CONTRACTION MAPPINGS

Many problems in mathematics deal with the following situation. We have a mapping T from a set X to itself, and we are interested in finding $x \in X$ such that $Tx = x$. In this case, x is called a *fixed point of T*. The basic problem is to determine the existence and uniqueness of such points. The situation is easily resolved when T is linear since 0 is always a fixed point, and it will be unique if and only if T is injective. Although this book deals almost exclusively with linear operators, this is one place where we make a brief excursion to discuss a more general type of mapping. The results have significant applications to differential and integral equations. One of these is discussed in Chapter 10.

Since we are dealing with nonlinear mappings here, we do not require linear spaces and the natural setting is a metric space. Such concepts as completeness, Cauchy sequences, and fast Cauchy sequences are defined in exactly the same way as for a normed linear space with $d(x, y)$ replacing $\| x - y \|$.

Definition. A mapping T from a metric space (X, d) to itself is called a *contraction mapping* if there exists a positive constant c strictly less than 1 such that for all $x, y \in X$

$$d(Tx, Ty) \leq cd(x, y).$$

A contraction mapping is easily seen to be uniformly continuous. We leave it to the reader to verify that a linear operator on a normed linear space is a contraction mapping if and only if $\| T \| < 1$; see Exercise 2.42. The following result was proved by Banach in the early days of the subject and is known as the *contraction mapping principle.*

Theorem 2.44. *A contraction mapping T from a complete metric space to itself has a unique fixed point.*

Proof. Choose any $x \in X$. It is easily verified by induction that

$$d(T^{n+1}x, T^n x) \leq c^n d(Tx, x),$$

in which $c < 1$ is the constant in the definition of a contraction. This means that $(T^n x)$ is a fast Cauchy sequence in the general sense given after the definition in Section 2.5, by using $c^n d(Tx, x)$ in place of $(\frac{1}{2})^n$. Therefore, by completeness, $(T^n x)$ converges to a point $y \in X$. Clearly the sequence $(T^{n+1}x)$ converges to the same point y, and by writing $T^{n+1}x$ as $T(T^n x)$, the continuity of T shows that $(T^{n+1}x)$ also converges to Ty. By the uniqueness of limits, it follows that $Ty = y$.

To show uniqueness, we note that for any other fixed point z

$$d(y, z) = d(Ty, Tz) \leq cd(y, z),$$

and since $c < 1$, we must have $d(y, z) = 0$. □

The above proof furnishes a technique for finding the fixed point. Starting with an arbitrary $x \in X$, we can just apply the mapping T successively, and we are guaranteed eventual convergence. As a practical matter, convergence may be quite slow if the initial value is not sufficiently close to the actual fixed point.

Following is a simple but useful generalization of the theorem above.

Corollary 2.45. *If T is a mapping from a complete metric space to itself such that T^n is a contraction for some positive integer n, then T has a unique fixed point.*

Proof. Let y be the unique fixed point of T^n. Then,

$$T^n(Ty) = T(T^n y) = Ty.$$

So, Ty is a fixed point of T^n, and by uniqueness of the fixed point of T^n, we know that $Ty = y$. This is the unique fixed point of T since $Tz = z$ implies that $T^n z = z$, which implies that $y = z$. □

EXERCISES

Sections 2.1–2.3

2.1 Show that the two definitions of continuity as given in Section 2.1 are equivalent.

2.2 Prove parts (a) and (c) of Theorem 2.3.

2.3 Prove Proposition 2.4 and the remark following.

2.4 Prove parts (a), (b), and (c) of Proposition 2.9.

2.5 Show that the five definitions of the closure of a set given in Section 2.2.2 are equivalent.

2.6 Suppose that $\| \cdot \|$ is a seminorm on X.

(a) Show that $\{x \in X : \| x \| = 0\}$ is a subspace.

(b) Show that $\| x - y \| = 0$ implies that $\| x \| = \| y \|$.

2.7 Show that in the space of bounded sequences with the sup norm the closure of c_{00} is c_0, the subspace of sequences converging to zero.

2.8 Show that in the space of bounded continuous functions on $\mathcal{R}$ with the sup norm the closure of $C_{00}(\mathcal{R})$ is $C_0(\mathcal{R})$; see Example 2.19 in Section 2.5.4 for the definition of these two sets of functions.

2.9 If Y is a subspace of a normed linear space X, show that Y is closed if and only if the set $\{y \in Y : \| y \| \leq 1\}$ is closed.

2.10 Show that a linear operator $T : X \to Z$, where X and Z are normed linear spaces, is completely determined by its values on $B(x_0; r)$ for any $x_0 \in X$ and $r > 0$.

Section 2.4

2.11 Show that a sequence (A_n) in $\mathcal{B}(X, Y)$ converges to 0 if and only if for all sequences (x_n) of unit vectors $(A_n x_n)$ converges to 0.

2.12 Suppose that a sequence (A_n) in $\mathcal{B}(X)$ satisfies:

(i) $(A_n x)$ converges to 0 for all $x \in X$.

(ii) There is a $K > 0$ such that $\| A_n \| \leq K$ for all n.

Show that if (x_n) is a sequence of unit vectors and (Tx_n) converges for some $T \in \mathcal{B}(X)$, then $(A_n T x_n)$ converges to 0.

2.13 Show that the mapping of Exercise 1.1(b) is not bounded when c_{00} has the sup norm.

2.14 Let $P[0, 1]$ denote the polynomial functions on the interval $[0, 1]$ equipped with the sup norm. Show that $D : P[0, 1] \to P[0, 1]$, given by $Df = df/dx$, is not bounded.

2.15 Some writers define the norm on operators by $\| T \| = \sup\{\| Tx \| : x \in \bar{B}(1)\}$. Show that this is the same as the definition given here.

2.16 Define $T : C[0, 2] \to C[0, 2]$ by $Tf(t) = \int_0^t f(s)ds$. If the norm on $C[0, 2]$ is the sup norm, what is $\| T \|$.

2.17 Suppose that (S_n) is a sequence in $\mathcal{B}(X, Y)$ converging to S and (T_n) is a sequence in $\mathcal{B}(Y, Z)$ converging to T. Show that $(T_n S_n)$ converges to TS.

2.18 Show that convergence with respect to the norm on $\mathcal{B}(X, Y)$ is uniform convergence on bounded sets.

2.19 Define $T \in \mathcal{B}(c_0)$ by

$$T(x(1), x(2), x(3), \ldots) = (x(2), x(3), \ldots).$$

Show that for all $x \in c_0$ the sequence of vectors $(T^n x)$ converges to 0 under the sup norm, but (T^n) does not converge to 0 with respect to the norm on $\mathcal{B}(c_0)$. (Here, $T^n = TT \cdots T$ is the composition of T with itself n times.)

Sections 2.5–2.8

2.20 Consider the sup norm on c_0 and c_{00}.

(a) Show that in c_0 the series $\sum_{i=1}^{\infty} \delta_i / i$ is convergent but not absolutely convergent.

(b) In c_{00}, find a series that is absolutely convergent but not convergent.

2.21 Consider the interval $[1, \infty)$ with the usual metric, namely $d(x, y) = |x - y|$, and the metric $d_1(x, y) = |1/x - 1/y|$. Show that these are equivalent in the sense that both have the same convergent sequences. Deduce from this that equivalence of metrics does not preserve boundedness or completeness.

2.22 Show that a norm $\| \cdot \|_1$ is weaker than a norm $\| \cdot \|_2$ if for all $r > 0$ there is an $s > 0$ such that $B_2(0; s) \subseteq B_1(0; r)$, in which B_i denotes a ball with respect to $\| \cdot \|_i$.

2.23 Show that if the norms on two spaces X and Y are each replaced with an equivalent norm, then the resulting norm on $\mathcal{B}(X, Y)$ is equivalent to the original.

2.24 Suppose that $\| \cdot \|_1$ and $\| \cdot \|_2$ are two norms on X.

(a) Show that $\| x \|_3 = \| x \|_1 + \| x \|_2$ also defines a norm. Describe in words what convergence means for $\| \cdot \|_3$.

(b) Show that $\| x \|_4 = \max\{\| x \|_1, \| x \|_2\}$ defines a norm equivalent to $\| \cdot \|_3$.

(c) Show by example that $\| x \|_5 = \min\{\| x \|_1, \| x \|_2\}$ does *not* define a norm.

2.25 Let X be the space of all continuous functions f on $[0, 1]$ such that $f(0) = 0$, equipped with the sup norm. Let $Y = \{f \in X : \int_0^1 f(t)dt = 0\}$. Show that for any $g \in X$ and positive integer n, there is a constant α such that the function $g - \alpha t^{1/n}$ is in Y. Conclude that perpendiculars to closed proper subspaces need not exist.

2.26 Suppose that T is a nonempty subset of a set S and Y is the subspace of $B(S)$ consisting of all functions that take the value 0 on T. If $B(S)$ is equipped with sup norm, show that the quotient space $B(S)/Y$ is isometric to $B(T)$ with the sup norm.

2.27 Show that if a normed linear space X has a complete subspace M such that X/M is complete, then X is complete.

2.28 Consider the space of continuous functions on $[0, 2]$ with the norm

$$\| f \| = \left(\int_0^2 |f(t)|^2 dt \right)^{1/2}.$$

(a) Show that the sequence (f_n), in which

$$f_n(t) = \begin{cases} 0, & 0 \leq t < 1, \\ n(t-1), & 1 \leq t < 1 + 1/n, \\ 1, & 1 + 1/n \leq t \leq 2, \end{cases}$$

is Cauchy but does not converge in this space.

(b) Show that the linear mapping T from this space to $\mathcal{F}$, given by $T(f) = f(1)$, is not bounded.

2.29 Let (x_n) be a sequence in a finite-dimensional space X, and let $x \in X$ be such that for all $f \in \mathcal{B}(X, \mathcal{F})$ the sequence $f(x_n)$ converges to $f(x)$. Show that x_n conveges to x.

2.30 Suppose that $T : X \to Y$, where X and Y are normed linear spaces, is bijective and bounded and has a bounded inverse. Show that we can replace the norm on Y by an equivalent norm so that X and Y are isometric.

Sections 2.9–2.12

2.31 Let $(X, \mathcal{S}, \mu)$ be a finite measure space. (So, the function that is identically equal to 1 is in L^p for all $p \geq 1$.)

(a) Show that $L^p \subseteq L^1$ for all $p \geq 1$.

(b) Show that for $1 \leq q \leq p \leq \infty$, $L^p \subseteq L^q$.

(c) Show that (b) need not hold if the measure space is not finite.

2.32 Show that for $1 \leq q \leq p \leq \infty$, $\ell^q \subseteq \ell^p$.

2.33 Show that for the measure space $[0, \infty)$, if we are given any p, q in the interval $[1, \infty)$, there is a function $f \in L^p$ that is not in L^q.

2.34 Show that the bijection $i : Y \oplus Z \to X$ given at the end of Section 2.10 is bounded when the direct sum is equipped with the p-norm for $p = 1$.

2.35 Let c be the space of all convergent sequences with the sup norm.

(a) Show that as a linear space c is isomorphic to $c_0 \oplus \mathcal{F}$.

(b) If i is the bijection given in Exercise 2.34 for this direct-sum decomposition, find the norm of i when the direct sum is equipped with (i) $\|\cdot\|_1$ (ii) $\|\cdot\|_{\text{sup}}$. In each case, c_0 has the sup norm and $\mathcal{F}$ the absolute value norm.

2.36 Define the following sequence of functions on $[0, 1]$:

$$f_1 = 1_{[0,1/2]}, \quad f_2 = 1_{[1/2,1]}, \quad f_3 = 1_{[0,1/4]}, \quad f_4 = 1_{[1/4,1/2]}, \quad f_5 = 1_{[1/2,3/4]},$$
$$f_6 = 1_{[3/4,1]}, \quad f_7 = 1_{[0,1/8]}, \quad f_8 = 1_{[1/8,1/4]}, \quad \ldots,$$

and we continue in this fashion by choosing at each step a function that takes the value 1 on some interval of length $1/2^k$ for some positive integer k. An interval ending in 1 is followed by one starting at 0.

Show that this sequence converges in the p-norm for $p = 1$ but does not converge pointwise. In fact, show it does not converge at any point of $[0, 1]$.

2.37 Verify that $\|\cdot\|_{\text{ess sup}}$ is a norm on L^∞ and with this norm, the space is complete.

2.38 Show that the set of all rational-valued sequences in c_{00} is dense in ℓ^p with respect to $\|\cdot\|_p$ for all $1 \leq p < \infty$.

2.39 Suppose that $0 < p < 1$ and q is such that $1/p + 1/q = 1$; note that $q < 0$. Show that in Lemma 2.36, Hölder's inequality, and Minkowski's inequality, the inequality sign is reversed. (For $-\infty < p < 1$ and $p \neq 0$, $\|\cdot\|_p$ is defined in exactly the same way as for $p \geq 1$. However, it is no longer a norm, as this exercise shows.)

2.40 (Decreasing monotone convergence theorem) Suppose that (f_n) is a sequence of nonnegative measurable functions converging to f and, in addition, $f_n \geq f_{n+1}$ for all n. Show that if $\int f_1 < \infty$, then $\int f = \lim \int f_n$. Give an example to show that the conclusion is not true if we omit the finiteness hypothesis.

2.41 (a) Show that (δ_n) is a Schauder basis for c_0 and for ℓ^p, $1 \leq p < \infty$.

(b) Exhibit an element in ℓ^∞ that cannot be written in the form $\sum_{n=1}^{\infty} \alpha_n \delta_n$.

(c) Show that a space with a Schauder basis must be separable.

2.42 Verify that $T \in \mathcal{B}(X, Y)$ is a contraction if and only if $\| T \| < 1$.

2.43 Let X be a Banach space and A an element of $\mathcal{B}(X)$ such that A^n is a contraction for some positive integer n. Fix $w \in X$ and define a mapping T from X to itself by $Tx = w + Ax$. Show that T has a unique fixed point, and find it.

2.44 Consider the metric space $[1, \infty)$ with the usual absolute value metric, and consider the mapping defined by $Tx = x + 1/x$. Show that $d(Tx, Ty) < d(x, y)$, but there is no fixed point.

3

MAJOR BANACH SPACE THEOREMS

3.1 INTRODUCTION

A problem that arises frequently in functional analysis is to show the boundedness of a given linear operator T. This is not only for theoretical interest but also is required in many applications. Suppose you want to compute the value of Tx for a certain value of x. It often happens that you can easily compute Tx_n for a sequence (x_n) converging to x, and you would like to be able to say that $Tx = \lim Tx_n$. However, you cannot assert this unless you know that T is bounded and therefore continuous. Even though continuity at only a single point is required, it can often be difficult to verify.

There are a group of results, all closely related to each other, that are useful for showing boundedness of operators, as well as affiliated issues, such as showing that certain sets are bounded or closed. They are regarded as the major theorems on linear operators between Banach spaces. These are known as the *open mapping theorem, bounded inverse theorem, closed graph theorem*, and *uniform boundedness principle*. Prior to proving these, we need a preliminary result of Baire (often referred to as the *Baire category theorem*, using a different definition of "category" than the present-day use of that word in mathematics).

3.2 BAIRE CATEGORY THEOREM

We begin with a general result on completeness which heuristically might be said to show that complete spaces have "lots of points." We will motivate this by first considering dense sets. A dense set S in a normed linear space is intuitively a set that is quite large since, in some sense, it almost fills up the entire space. It seems reasonable to ask if the intersection of two dense sets is dense. The answer is a

A First Course in Functional Analysis. By S. David Promislow

resounding no! On the real line, the rationals and irrationals are both dense, and yet their intersection is the empty set, which is as far removed from being dense as one can get.

Visualizing sets in the real line or plane, however, suggests that if one set is open, then this sliding-in-between process cannot occur. Indeed, the intuition is correct, and the intersection of two dense sets S and T is dense when S is open. Given any open set W, $W \cap S$ is open and nonempty since S is dense and therefore intersects the dense set T. In other words, the intersection of W with $S \cap T$ is not empty.

By induction, one can show that the intersection of n dense sets, $n - 1$ of which are open, is dense. The next obvious question is what happens with a countable intersection of open dense sets, and this is the content of Baire's theorem. We begin with a preliminary result which is proved exactly as it is on the real line.

Proposition 3.1 (Nested Set Theorem). *Given a sequence $F_1 \supseteq F_2 \supseteq F_3 \cdots$ of closed, nonempty sets in a Banach space such that* $\operatorname{diam}(F_n) \to 0$, *there is a unique point that is in F_n for all n.*

Proof. Choose a point x_n in each F_n. From the nested property and the diameter condition, (x_n) is a Cauchy sequence which converges to some point x. Now given any n, x is the limit of the sequence $(x_n, x_{n+1}, \ldots)$, all of which belong to F_n. So, x is in $\overline{F_n} = F_n$ for all n. The diameter condition also implies uniqueness of x. □

Note that if we were to remove a point in the space, thereby creating a hole, we could choose sets surrounding the hole, so that the conclusion would not hold. Another intuitive picture of a Banach space is that of one without such holes.

Theorem 3.2. (Baire's Theorem). *The intersection of countably many open and dense sets in a Banach space is dense.*

Proof. We start by observing a fact that holds in any normed linear space. Suppose we are given a nonempty open set U and $r > 0$. Choose any $x \in U$. Then, for some $q > 0, B(x; q) \subseteq U$. If we pick $s < \min\{r, q\}$, then $\overline{B(x; s)} \subseteq U$. We conclude that any nonempty open set contains a closed ball of arbitrarily small positive radius.

Given a sequence (V_n) of open, dense sets and any other nonempty, open set W, we seek a point x that is in V_n for all n and also in W. We are going to find this point by constructing a nested sequence of nonempty closed sets. Then, we will invoke Proposition 3.1.

Here, $W \cap V_1$ is open and nonempty since V_1 is dense; therefore, $W \cap V_1$ contains $\overline{B_1}$, where B_1 is an open ball of radius less than 1. Similarly, $B_1 \cap V_2$ is open and nonempty and therefore contains $\overline{B_2}$, where B_2 is an open ball of radius less than $\frac{1}{2}$. By continuing inductively, we produce open balls B_n of radius less than $1/n$, whose closures are contained in $B_{n-1} \cap V_n$. We then apply the nested set theorem to the sequence $(\overline{B_n})$ to obtain a point that is in $\overline{B_n} \subseteq V_n$ for all n and in $\overline{B_1} \subseteq W$. □

There is a dual form of Baire's theorem, which we use when we apply the result in the next section and which illustrates more aptly the intuitive statement we made at the beginning of this section. The complement of a dense set A is a set such that every point is a boundary point, so it has an empty interior. By using De Morgan's law (which states that the complement of a union is the intersection of the complements), we can formulate Baire's theorem as follows:

Corollary 3.3. (Dual Form of Baire's Theorem). *In any Banach space, a countable union of closed sets with empty interiors has an empty interior.*

In particular, then, a countable union of closed sets with empty interiors cannot fill up the whole space, and this statement by itself is often useful.

For a simple illustration of Corollary 3.3, imagine that you take some bounded square in the plane, and you draw a one-dimensional curve in this square, so it has no thickness. Suppose the curve does contain its endpoints, so it is closed. Then, you repeat and continue to draw curves. How many curves do you need in order to fill up the whole square? Obviously, you need infinitely many, but a legitimate question is whether you can do it with a countable number. The answer is no, but it is not completely obvious how to show this by elementary methods. However, the solution is immediate by Baire's theorem, which shows that the union of countably many curves is a set with no interior points, so it cannot fill up the square.

There are many applications of Baire's theorem in analysis that are tangential to our main objectives. We will discuss one particularly fascinating result. Suppose you draw a continuous function on the interval $[0, 1]$, and you try to make it as jagged as possible. Despite your efforts, the function will invariably be differentiable, except at a finite number of points where you insert corners or cusps. With a fair amount of work, one can nevertheless show the existence of a continuous function on $[0, 1]$ that is nondifferentiable at all points. At one time, mathematicians viewed such functions as being extremely strange and pathological. However, from one point of view, such functions are not at all unusual and are, in fact, typical. We can show that the set of all functions in $C[0, 1]$ that are nondifferentiable at all points is dense in $C[0, 1]$ with the sup norm. The steps are outlined in Exercise 3.14.

3.3 OPEN MAPPINGS

In this section, we prove one of the main theorems of functional analysis. The importance will become clear when we apply it in later sections. The proof is somewhat more sophisticated than those we have had up to this point. We make use of some elementary facts summarized in the following lemma, whose proof we leave to the reader in Exercise 3.1.

Lemma 3.4. *Given normed linear spaces X and Y, subsets A and C of X, $\alpha \in \mathcal{F}$, and $T \in \mathcal{L}(X, Y)$, we have:*

(a) $\alpha B(r) = B(|\alpha| r)$;
(b) $\alpha \bar{A} = \overline{\alpha A}$;
(c) $\bar{A} + \bar{C} \subseteq \overline{A + C}$;
(d) $T(\alpha A) = \alpha T(A)$.

***Theorem 3.5* (Open Mapping Theorem).** *Given a surjective $T \in \mathcal{B}(X, Y)$ in which X and Y are Banach spaces, $T(U)$ is an open set in Y for any open set $U \subseteq X$.*

Proof. We will break the proof up into three distinct steps, summarized as follows:

1. We use the linear structure to localize. That is, we simplify the proof by showing that it is sufficient to prove that 0 is an interior point of $TB(1)$.
2. We use the completeness of Y and the dual form of Baire's theorem to conclude an approximation to the above, namely that 0 is an interior point of $\overline{T(B(1))}$.
3. We use the completeness of X to remove the closure sign.

We now proceed with the three steps.

Step 1: We will use the symbols B to denote open balls in X and B' to denote open balls in Y. Suppose we have shown that $TB(1) \supseteq B'(\delta)$ for some $\delta > 0$. Then, given any open $U \subseteq X$ and $y \in T(U)$, we have $y = Tx$ for some $x \in U$, and we know that, for some $r > 0$, $U \supseteq B(x; r) = x + rB(1)$. Therefore,

$$T(U) \supseteq Tx + T[rB(1)] = y + r\,TB(1) \supseteq y + B'(\delta r) = B'(y; \delta r),$$

thereby showing that $T(U)$ is open.

Step 2: Let $V = TB(\frac{1}{2})$, and notice that V has the symmetry property that $-V = V$. We first show, in preparation for using Baire's theorem, that $Y = \bigcup_{n=1}^{\infty} nV$. Indeed, for any y in Y, we know from surjectivity that $y = Tx = 2k\,T(x/2k)$ for some $x \in X$, so if $\| x \| < k$, then $y \in 2kV$. We next invoke Corollary 3.3 to conclude that, for some n, $\overline{nV} = n\bar{V}$ has an interior point w. Then, $y = w/n$ is an interior point of $\bar{V}$, so that $y + B'(s) \subseteq \bar{V}$ for some $s > 0$. Now, suppose $z \in B'(s/2)$. Then, $-(y + z) \in -\bar{V} = \overline{-V} = \bar{V}$, so that

$$z = (y + 2z) - (y + z) \subseteq \bar{V} + \bar{V} \subseteq \overline{V + V} \subseteq \overline{TB(1)}$$

since $V + V \subset TB(1)$. We have therefore shown that $B'(s/2) \subseteq \overline{TB(1)}$.

Step 3: From step 2, $B'(r) \subseteq \overline{TB(1)}$ for some $r > 0$. We will show that $B'(r) \subseteq TB(2)$, which will show that $B'(r/2) \subseteq TB(1)$, and the conclusion from step 1 will complete the proof.

For any positive integer k, we have $B'(r/2^k) \subseteq \overline{TB(1/2^k)}$. This means that, given any $z \in B'(r/2^k)$ and $\epsilon > 0$, we can find $x \in B(1/2^k)$ such that $\| z - Tx \| < \epsilon$. Suppose, then, that $y \in B'(r)$. Choose $x_1 \in B(1)$ such that $\| y - Tx_1 \| < r/2$. Then, choose $x_2 \in B(1/2)$ such that $\| y - Tx_1 - Tx_2 \| < r/4$. By proceeding inductively, we choose x_n in $B(1/2^{n-1})$ such that

$$\| y - T(x_1 + x_2 + \cdots + x_n) \| < \frac{r}{2^n}. \qquad (*)$$

Let $x = \sum_{n=1}^{\infty} x_n$, which exists since the series is absolutely convergent. By the general triangle inequality, $\| x \| \leq \sum_{n=1}^{\infty} \| x_n \| \leq 2$.

Finally, we use the fact that T is bounded and therefore continuous. We have $Tx = T(\sum_{n=1}^{\infty} x_n) = \sum_{n=1}^{\infty} Tx_n = y$; the final equality comes from $(*)$ above. So, $y \in TB(2)$, thereby completing the proof. □

3.4 BOUNDED INVERSES

In this section, we consider an injective bounded linear operator T on a Banach space. Because T is injective, T^{-1} exists as a linear operator on $R(T)$. A natural problem is to determine when T^{-1} is bounded. For this we need a stronger condition than injectivity.

Definition. We say that a linear operator T from one normed linear space to another is *bounded below* if there is a $k > 0$ such that for all unit vectors x we have $\| Tx \| \geq k$.

It is clear that being bounded below implies that T is injective since $N(T) = 0$, but the converse is not true in infinite-dimensional spaces. Our main theorem in this section, to which we will refer many times in later chapters, is as follows:

Theorem 3.6. *Given an injective $T \in \mathcal{B}(X, Y)$ for which X and Y are both Banach spaces, the following are equivalent:*

(i) T^{-1} is bounded;

(ii) T is bounded below;

(iii) $R(T)$ is closed.

Proof. (i) $\Rightarrow$ (ii): If $\| T^{-1} \| = k < \infty$, then, for any unit vector $x \in X$, $1 = \| x \| = \| T^{-1}Tx \| \leq k \| Tx \|$, thereby showing that $\| Tx \| \geq 1/k$.

(ii) $\Rightarrow$ (iii): For $y \in \overline{R(T)}$, choose (x_n) in X such that (Tx_n) converges to y. Then, (Tx_n) is Cauchy, and since $\| Tx_n \| \geq k \| x_n \|$ for some $k > 0$, it follows that

(x_n) is Cauchy and must converge to some point x. But, then, $y = \lim Tx_n = Tx$, thereby showing that $y \in R(T)$.

(iii) $\Rightarrow$ (i): We use the open set criteria for continuity as given in Section 2.2.2. For any open set V in X, we have $(T^{-1})^{-1}V = T(V)$, which is open by the open mapping theorem 3.5. □

Remark. The particular statement (iii) $\Rightarrow$ (i) is often termed the *bounded inverse theorem* and phrased as saying that a bijective, bounded linear operator between two Banach spaces has a bounded inverse. By Theorem 2.16, this is equivalent to the given statement. The proof gives some insight into the importance of open mappings. Given a bijective map from one metric space to another, to say the map is open is equivalent to saying that the inverse is continuous.

3.5 CLOSED LINEAR OPERATORS

Definition. Given $T : X \to Y$, we define the *graph* of T, denoted by G_T, as the set of all pairs $(x, y) \in X \times Y$ such that $y = Tx$.

Remark. There is an interesting language issue involved with the above definition. Mathematicians usually give a formal definition of a function as a set of ordered pairs that satisfy a certain property. But, they do not always think of a function that way. Rather, they view it in more dynamic terms, as taking some object and assigning some other object to it or transforming the object in some way. Evidence of this is the fact that, when they really do want to refer to the particular set of ordered pairs, they do not call it the "function" but give it a completely different name, namely the "graph of the function." This is not confined to linear operators. Going back to our earliest mathematical experiences, when we draw this particular set of ordered pairs in the plane for a real-valued function on a subset of the reals, we invariably call it the graph of the function.

When X and Y are normed linear spaces, we view $X \times Y$ as the direct product of the spaces X and Y, equipped with the product norm of Section 2.10, namely $\| (x, y) \| = \max\{\| x \|, \| y \|\}$.

Definition. A linear operator T is said to be *closed* if G_T is a closed subset of $X \times Y$.

Remark. There is another quirk of language here to note. It is customary to call a linear operator "open" if it takes open sets to open sets, as in the open mapping theorem above. One might well think then that a closed operator is one that takes closed sets to closed sets, but that is not the conventional terminology. A closed operator under the definition above does not necessarily take closed sets to closed sets.

It is instructive to formulate the definition of a closed operator in terms of sequences. T is closed if and only if for any sequence (x_n) in X such that (x_n) converges to x and (Tx_n) converges to y we must have $y = Tx$. This shows that a continuous operator is closed but that the condition of being closed is weaker than being continuous. For a closed operator, we allow for the fact that (Tx_n) might not converge at all, a condition that is required for continuity. It is only when this sequence *does* converge that we must verify that it has the right limit.

Let us revisit inverses.

Theorem 3.7. *If $T \in \mathcal{L}(X, Y)$ is injective and closed, then T^{-1} is closed.*

Proof. Given such a T, take any (y, x) in the closure of $G_{T^{-1}}$. We must show that it is actually in the graph. There is a sequence (y_n, x_n) in $G_{T^{-1}}$ converging to (y, x). But, the sequence (x_n, y_n) is in G_T and converges to (x, y). Since T is closed, $y = Tx$, thereby showing that $x = T^{-1}y$, which implies that $(y, x) \in G_{T^{-1}}$. □

Therefore, given a bounded, injective operator that does not satisfy the criteria of Theorem 3.6, its inverse will not be bounded, but it will be closed. The following is a simple, but important, example.

Example 3.8. Let $X = C[0, 1]$, with the sup norm, and $Y = \{f \in C[0, 1] : f(0) = 0\}$. Define $T : X \to Y$ by

$$(Tf)(t) = \int_0^t f(s)\ ds.$$

T is injective, and by the fundamental theorem of calculus, its inverse is the operator D given by differentiation, or $(Dg)(t) = g'(t)$. Since T is bounded (see Exercise 2.16) and therefore closed, it follows that D is closed. It is well known that the operation of differentiation is discontinuous. (See Exercise 2.14 for a simple example.) This is clear intuitively since a small function can have arbitrarily large slopes. However, the differentiation operator is not completely chaotic because it is closed. This property of differentiation is important for many applications in analysis, particularly to differential equations.

The main theorem in this section is the following, which is of great importance, since as indicated above, it is much easier to show that an operator is closed than to show it is continuous. Many applications of this theorem are given in the exercises. It is good to keep in mind the fact that, whenever you are trying to prove an operator is continuous and having difficulty, first try to prove that it closed.

Theorem 3.9 (Closed Graph Theorem). *If X and Y are Banach spaces, any closed $T : X \to Y$ is bounded.*

Proof. Let P_X denote the projection from G_T to X given by $P_X(x, y) = x$, and similarly define P_Y from G_T to Y. Clearly, both projections are bounded, being

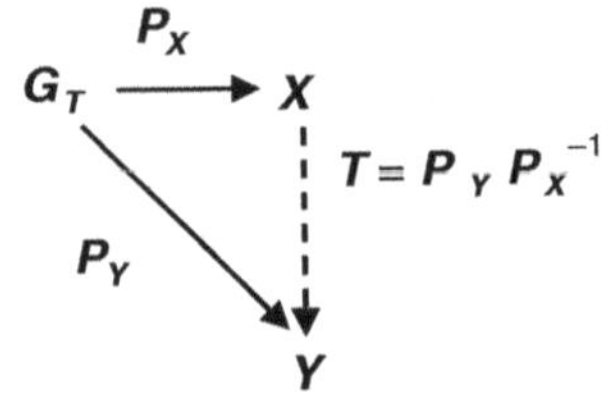

Figure 3.1. Proof of Theorem 3.9.

of norm 1. Since G_T is closed in $X \times Y$, it is a Banach space. Moreover, $R(P_X)$ is all of X. By Theorem 3.6, ${P_X}^{-1}$ is bounded, which shows that $T = P_Y {P_X}^{-1}$ is bounded. (See Fig. 3.1.) □

3.6 UNIFORM BOUNDEDNESS PRINCIPLE

There are numerous applications for which it is required to show that a given set is bounded. In this section, we present one method for doing so. We are concerned here with the particular normed linear space $\mathcal{B}(X, Y)$, but in Chapter 6, we will adapt the result to be applicable to any normed linear space.

Suppose we have a set $\mathcal{A} \subseteq \mathcal{B}(X, Y)$ that we wish to show is bounded. [The term *uniformly bounded* in often used in the literature. This terminology simply reflects the fact that each $T \in \mathcal{A}$ is automatically bounded by definition, and to say the *set* is bounded in $\mathcal{B}(X, Y)$ means that we can find $K > 0$ that is a uniform bound for all the norms of elements in $\mathcal{A}$.]

A natural approach for deducing facts about elements in $\mathcal{B}(X, Y)$ is to consider their action on the space X.

Definition. We say that the set $\mathcal{A}$ is *pointwise bounded* if for each $x \in X$ the set $\mathcal{O}x = \{Tx : T \in \mathcal{A}\}$ is bounded in Y.

Certainly a bounded set is pointwise bounded, for if $\| T \| \leq K$ for all $T \in \mathcal{A}$, then all elements in $\mathcal{O}x$ have norm less than $K \| x \|$. Our main result is the converse.

Theorem 3.10 (Uniform Boundedness Principle). *If X is complete, then a pointwise bounded subset of $\mathcal{B}(X, Y)$ is bounded.*

Proof. We can assume from the outset that Y is complete by replacing it by its completion. This will not affect the assumption of pointwise boundedness since if a set is bounded in Y, it is obviously still bounded in any space containing Y. Given a pointwise bounded set $\mathcal{A}$, form a direct-product space $\mathcal{Y}$ over the set $\mathcal{A}$ with each space equal to Y. That is, the elements of $\mathcal{Y}$ are bounded functions on the set $\mathcal{A}$ taking values in Y.

We define $\mathcal{J} : X \to \mathcal{Y}$ in a natural way. Namely, $\mathcal{J}x$ is that function on $\mathcal{A}$ given by

$$\mathcal{J}x(T) = Tx, \qquad T \in \mathcal{A}.$$

We must verify that $\mathcal{J}x$ is, in fact, a bounded function on $\mathcal{A}$, but this follows from our pointwise bounded assumption. Indeed, suppose K_x is an upper bound to $\{\|Tx\|: T \in A\}$. Then, the norm of the function $\mathcal{J}x$ in the product space, which is the supremum of all its values, is bounded above by K_x.

Suppose we show that $\mathcal{J}$ is a bounded operator with norm K. Then, we are done because it means that for any unit vector $x \in X$ and any $T \in \mathcal{A}$ we have $\| Tx \| \leq K$, and this implies that $\| T \| \leq K$ for all $T \in \mathcal{A}$, as required. To this end, we will show that $\mathcal{J}$ is closed and then invoke the closed graph theorem to conclude that $\mathcal{J}$ is bounded. Suppose that (x_n) converges to x and that $(\mathcal{J}x_n)$ converges to a function g. We must show that $g = \mathcal{J}x$, and we will do this by showing that the two functions agree for each $T \in \mathcal{A}$. Indeed,

$$g(T) = \lim(\mathcal{J}x_n)(T) = \lim\ Tx_n = Tx = \mathcal{J}x(T),$$

in which we use the boundedness of T for the third equality. □

It is instructive to note how the direct-product construction was used in the above proof. We were dealing with a collection of linear operators, which we were able to view as a single linear operator $\mathcal{J}$ by using the direct product. This is a useful idea, which we will revisit.

An alternative, more classical proof of the uniform boundedness principle is outlined in Exercise 3.11.

Many authors state the uniform boundedness principle for an indexed family (T_λ) of linear operators rather than a subset. The difference is that in the indexed family, the same operator can appear more than once, but there is obviously no essential difference in the concept or the proof.

The next theorem, which considers a family indexed on the integers, is a major application of the uniform boundedness principle. It says that the space $\mathcal{B}(X, Y)$ is closed under pointwise limits. The result is noteworthy since, in general, pointwise convergence does not preserve continuity, as we have remarked before. The conclusion does follow, however, when we combine continuity with linearity on a complete space.

Theorem 3.11. *Suppose that (T_n) is a pointwise converging sequence of bounded linear operators from a Banach space X to a normed linear space Y. That is, for each $x \in X$ the sequence (T_nx) converges to an element $Tx \in Y$. Then, T is linear and bounded.*

Proof. The linearity of T is straightforward, but the boundedness is not obvious. For each x, the sequence (T_nx) converges and, therefore, is bounded. So, the collection $\{T_n\}$ is pointwise bounded, and by the uniform boundedness principle, there is a $K > 0$ such that $\| T_n \| \leq K$ for all n. Then, for any unit vector x, we have

$$\| Tx \| = \| \lim\ T_nx \| = \lim \| T_nx \| \leq K,$$

thereby showing that $\| T \| \leq K$. □

The uniform boundedness principle and/or its corollaries (such as Theorem 3.11) are often referred to as the *Banach–Steinhaus* theorem.

Remark. The reader might wonder if the trick of replacing a space by its completion, used in Theorem 3.10, could be invoked to remove the completeness hypothesis from the other theorems of this chapter. The answer is no. In all cases, the essential hypotheses are altered. For example, we cannot replace X by its completion in the uniform boundedness principle, since we would not know that the pointwise bounded condition held for points in the completion. Consider the open mapping theorem. We could not replace Y by its completion since the map would no longer be surjective. What if we replace X by its completion? This will change some open sets. A set U that was open in X might not be so in its completion. (Going back to the discussion in Section 2.2.2, we are adding new attacking sequences.) We leave it as an exercise to the reader to show why completeness cannot be removed from either space in the statement of the closed graph theorem.

EXERCISES

3.1 Prove Lemma 3.4.

3.2 In c_{00}, find a nested sequence of closed sets with diameters approaching zero that has an empty intersection.

3.3 Find a sequence $y \in \ell^\infty$ such that the multiplication operator M_y on ℓ^∞ is injective but not bounded below.

3.4 Let X be a linear space that has a countably infinite Hamel basis. Show that there is no norm on X that will make it into a Banach space. Conclude that there is no norm on c_{00} that will make it into a Banach space.

3.5 Let Y be the subspace of ℓ^∞ consisting of all sequences x such that for some K (which may depend on x) we have $|x(n)| \leq K/n$ for all n. Show that Y is not closed by finding a closed but unbounded operator from Y onto a Banach space.

3.6 Suppose that Y and Z are closed subspaces of a Banach space X such that $Y \cap Z = 0$ and $Y + Z = X$. Define $P : X \to Y$ by $P(y + z) = y$, where $y \in Y$ and $z \in Z$. Show that P is well defined and bounded.

3.7 Let $P \in \mathcal{L}(X)$, in which X is a Banach space, be such that $P^2 = P$. (Such an operator is called a *projection*.) Show that P is bounded if and only if both $N(P)$ and $R(P)$ are closed.

3.8 Let $\|\cdot\|_1$ and $\|\cdot\|_2$ be two norms on a linear space X such that X is a Banach space under either of them. Suppose that $\|\cdot\|_1$ is weaker than $\|\cdot\|_2$. Show that the two norms are equivalent.

3.9 Given $S \in \mathcal{L}(X,Y)$ and $T \in \mathcal{L}(Y,Z)$, where X, Y, and Z are Banach spaces, such that TS is bounded and T is bounded and injective, show that S is bounded.

3.10 You are given $T \in \mathcal{B}(X,Z)$ and $U \in \mathcal{B}(Y,Z)$, where X, Y, and Z are Banach spaces, such that for all $x \in X$ there is a unique $y \in Y$ satisfying $Tx = Uy$. If V is the mapping that sends each x into this unique y, show that $V \in \mathcal{B}(X,Y)$.

3.11 Give an alternate proof to the uniform boundedness principle by applying Corollary 3.3 to the sets $F_n = \{x : \| Tx \| \leq n \text{ for all } T \in \mathcal{A}\}$. You may also want to make use of Exercise 2.10.

3.12 Equip c_{00} with the sup norm. Find a sequence $(T_n) \in \mathcal{B}(c_{00}, Y)$ for some normed linear space Y such that (T_n) converges pointwise to a linear operator T that is not bounded.

3.13 Explain why the statement of the closed graph theorem cannot be changed by allowing X or Y to be incomplete and by replacing the corresponding space with its completion in the proof.

3.14 For any positive integer n, let F_n be the set of all $f \in C[0,1]$ such that there exists a point $t \in [0,1]$ satisfying $|f(s) - f(t)| \leq n|s-t|$ for all $s \in [0,1]$. We consider the sup norm.

(a) Show that F_n is closed.

(b) Show that F_n has empty interior. *Hint*: Show that for any m you can approximate $f \in C[0,1]$ as closely as you like, with a continuous zigzag function that has a graph consisting of straight-line segments with slopes alternating between m and $-m$.

(c) Show that if f is differentiable at a point t, then $f \in F_n$ for some n.

(d) Conclude that the set of nowhere differentiable functions is dense in $C[0,1]$.

3.15 (a) If N is a closed subspace of the normed linear space X, give a direct proof that the canonical projection $\pi_N : X \to X/N$ is an open mapping.

(b) By assuming the statement of the closed graph theorem, prove the open mapping theorem.

4

HILBERT SPACES

4.1 INTRODUCTION

In this chapter, we look at a particularly well behaved type of Banach space known as a *Hilbert space*. A Hilbert space is a Banach space for which, as well as measuring length, we can measure angles. This has far-reaching consequences for the geometry of the space and the type of operators it admits. Historically, Hilbert spaces were the first type of Banach spaces to be studied.

The motivation comes from the familiar inner product in finite-dimensional spaces. We review the basic ideas. Recall that for two vectors in $\mathcal{R}^n$, $x = (x_1, x_2, \ldots, x_n)$ and $y = (y_1, y_2, \ldots, y_n)$, we define a number $\langle x, y \rangle$ by

$$\langle x, y \rangle = \sum_{i=1}^{n} x_i y_i.$$

(This expression goes by various names, such as inner product and scalar product. It is sometimes called the dot product and written as $x \cdot y$.)

We get an easy geometric interpretation by looking at two unit vectors u and v in $\mathcal{R}^2$. Suppose we project u onto v, that is, we find that multiple of the vector v that is a minimal distance from u. This multiple will be $\langle u, v \rangle$. The vector $u - \langle u, v \rangle v$ is perpendicular to v, so $\langle u, v \rangle$ gives the cosine of the angle between the two vectors; see Figure 4.1. In particular, when the two vectors are perpendicular, $\langle u, v \rangle = 0$, and when $u = v$, we get $\langle u, v \rangle = 1$. More generally, for all $x \in \mathcal{R}^n$, $\langle x, x \rangle$ gives the square of the length of x in the Euclidean norm. By abstracting the properties of the inner product, one can introduce the same type of geometry into certain infinite-dimensional spaces.

For the rest of the chapter we will take our field of scalars to be the complex numbers, unless mentioned otherwise. While it is possible to speak of real Hilbert spaces, the complex numbers are the natural choice. In the finite-dimensional case, this

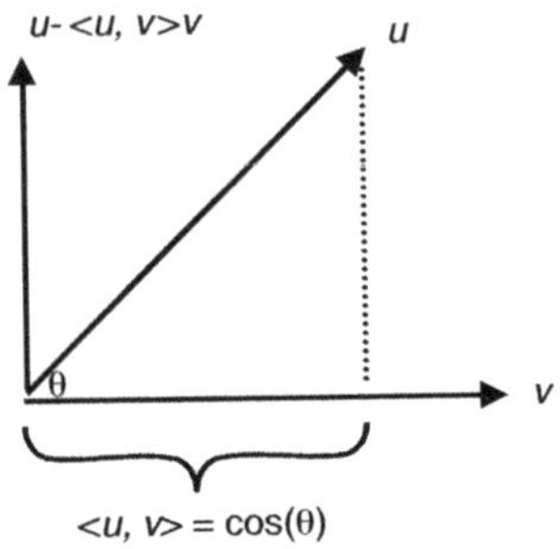

Figure 4.1. Geometric view of inner product of two unit vectors.

requires a modification of the definition of inner product given above, which, if taken as stated, would produce incongruous results. For example, if $x = (1, i)$ in $\mathcal{C}^2$, the definition above gives $\langle x, x\rangle = 0$. The remedy is that each y_i is replaced by its conjugate $\overline{y_i}$, which leaves the definition unchanged for real scalars.

4.2 SEMI-INNER PRODUCTS

4.2.1 Basic Definitions and Properties

Although we will be mainly dealing with inner products, it is sometimes useful to consider a more general concept.

Definition. A *semi-inner product* on a complex linear space X is a mapping from $X \times X$ to $\mathcal{C}$ sending the pair (x, y) to $\langle x, y\rangle$ such that for all $x, y, z \in X$ and $\alpha \in \mathcal{C}$ the following hold:

(a) Linearity in the first variable:

$$\langle x + y, z\rangle = \langle x, z\rangle + \langle y, z\rangle,$$
$$\langle \alpha x, z\rangle = \alpha\langle x, z\rangle.$$

(b) Conjugate symmetry (often known as Hermitian symmetry):

$$\langle y, x\rangle = \overline{\langle x, y\rangle}.$$

(c) Positivity:

$$\langle x, x\rangle \geq 0.$$

An immediate consequence of (a) and (b) is that $\langle z, x + y\rangle = \langle z, x\rangle + \langle z, y\rangle$ and $\langle x, \alpha z\rangle = \bar{\alpha}\langle x, z\rangle$. So, our mapping is not linear in the second variable, but rather is what is called *conjugate linear*, meaning that scalars come out of the inner product as their conjugate. Another immediate consequence is that $\langle 0, x\rangle = \langle x, 0\rangle = 0$.

Definitions. If $\langle x, x\rangle = 0$ implies that $x = 0$, then the semi-inner product is called an *inner product* (as we had in the motivating example for n-dimensional space).

A linear space equipped with a (semi-)inner product is called a *(semi-) inner product space.*

With any semi-inner product, we define the associated seminorm $\|\cdot\|$ by

$$\| x \| = \langle x, x\rangle^{1/2}.$$

Our terminology and notation suggest that this is, indeed, a seminorm. We can see immediately that the scalar rule holds since for all $\alpha \in F$

$$\| \alpha x \| = \langle \alpha x, \alpha x\rangle^{1/2} = (\alpha\bar{\alpha}\langle x, x\rangle)^{1/2} = |\alpha| \ \| x \| .$$

Moreover, we can deduce that in the case of an inner product $\| x \| = 0$ if and only if $x = 0$. It remains to prove the triangle inequality, and we will do so shortly

Example 4.1. A major example comes from the L^2 space of any measure space. For $f, g \in L^2$, Hölder's inequality with $p = q = 2$ shows that $f\bar{g} \subset L^1$, so we may define an inner product by

$$\langle f, g\rangle = \int f\bar{g}.$$

One can easily verify the inner product axioms.

This example generalizes the one that we gave at the beginning of the chapter since the space C^n can be viewed as the L^2 space of a set of n points with counting measure.

4.2.2 Identities and Inequalities for Semi-Inner Products

Throughout this section, we suppose we have a linear space X with a semi-inner product. We develop some less transparent facts. They all stem from the following.

Lemma 4.2 (Basic Identity). *For all $x, y \in X$,*

$$\| x + y \|^2 = \| x \|^2 + \| y \|^2 + 2\,\mathrm{Re}\langle x, y\rangle.$$

Proof. This expression follows directly from (a) and (b) in the definition of semi-inner product by using the linearity and expanding. □

This basic identity enables us to prove two major results. These generalize Hölder's inequality and Minkowski's inequality applied to the case for which $p = q = 2$.

Theorem 4.3 (Cauchy–Schwartz Inequality). *Given a semi-inner product on X, then for all* $x, y \in X$,

$$|\langle x, y\rangle| \leq \| x \| \, \| y \| \, .$$

In the case of an inner product, equality holds if and only if x and y are linearly dependent (i.e., one is a scalar multiple of the other).

Remark. Some geometric intuition in the case of the inner product is shown by Figure 4.1. In this case, since $|\langle u, v\rangle|$ is the length of the projected unit vector, it is clearly less than or equal to $\| u \| \, \| v \| = 1$ and only equal when the two vectors are collinear.

Proof. Note first that the inequality is not affected if we multiply x and y by non-zero scalars α and β since the effect is merely to multiply both sides by $|\alpha\beta|$.

Suppose first that at least one of $\| x \|$ or $\| y \|$, say $\| y \|$, is not zero. If $\langle x, y\rangle = 0$, the inequality is clear. If not, we can replace the vectors with scalar multiples so that both $\langle x, y\rangle$ and $\| y \|$ are equal to 1. Then, from the basic identity in Lemma 4.2, with y replaced by $-y$, we have

$$0 \leq \| x - y \|^2 = \| x \|^2 + 1 - 2,$$

showing that $\| x \| \geq 1$, which verifies the inequality.

If $\| x \| = \| y \| = 0$ but $\langle x, y\rangle \neq 0$, we can multiply by appropriate scalars so that $\langle x, y\rangle = 1$. This leads to $0 \leq \| x - y \|^2 = -2$, a contradiction. This completes the proof of the inequality.

For the second statement, we first note that if x and y are linearly independent, then equality is clear.

Conversely, suppose that equality holds in the inner product case. We can reduce to the case for which neither element is zero, so that neither norm is zero. Again, we reduce to the case for which $\langle x, y\rangle = \| y \| = 1$. We must then have $\| x \| = 1$ so that $\| x - y \|^2 = 0$, showing that $x = y$. □

Theorem 4.4 (Triangle Inequality). *For all* $x, y \in X$, $\| x + y \| \leq \| x \| + \| y \|$.

Proof. By using the basic identity above, noting that the real part of a complex number is less than its absolute value, and by using the Cauchy–Schwartz inequality, we obtain

$$\begin{aligned} \| x + y \|^2 &\leq \| x \|^2 + \| y \|^2 + 2|\langle x, y\rangle| \leq \| x \|^2 + \| y \|^2 + 2 \, \| x \| \, \| y \| \\ &= (\| x \| + \| y \|)^2. \end{aligned}$$

□

So, we now know that $\| \cdot \|$ is a seminorm, and in the case of an inner product, it is a norm.

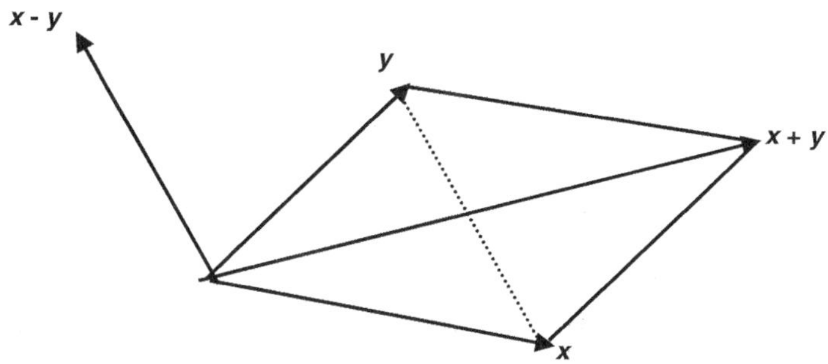

Figure 4.2. Illustration of parallelogram law.

Consider the following question. Given a norm on a linear space X, can we find an inner product that will induce that norm? The answer is no, as shown by the following rule. The name comes from the situation in $\mathcal{R}^2$ for which it captures the fact that the sum of the squared lengths of the two diagonals of a parallelogram is equal to the sum of the squared lengths of the four sides; see Figure 4.2.

Proposition 4.5 (Parallelogram Law). *For all* $x, y \in X$,

$$\| x+y \|^2 + \| x-y \|^2 = 2(\| x \|^2 + \| y \|^2).$$

Proof. Simply add to the basic identity with a vector y the same identity with y replaced by $-y$. □

Example 4.6 To see that not all norms are induced by an inner product, consider the plane with the ℓ^1-norm. Take $x = (1,0)$ and $y = (0,1)$. Then, the left-hand side of the parallelogram law is $2^2 + 2^2 = 8$, while the right-hand side is $2(1^2 + 1^2) = 4$.

The parallelogram law, in fact, characterizes norms that come from inner products. We will show this after we first provide another identity that also illustrates an important fact. Namely, suppose we know only the norm on an inner product space. We can then recapture the full inner product.

Proposition 4.7 (Polarization Identity)

$$\langle x, y\rangle = \tfrac{1}{4}(\| x+y \|^2 - \| x-y \|^2 + i \| x+iy \|^2 - i \| x-iy \|^2).$$

Proof. From the basic identity, we now *subtract* from the identity with y the same expression with y replaced by $-y$ to get

$$4\,\mathrm{Re}\langle x, y\rangle = \| x+y \|^2 - \| x-y \|^2 . \qquad (*)$$

From the fact that $\langle x,y\rangle = \text{Re}\langle x,y\rangle + i\,\text{Im}\langle x,y\rangle$, we deduce that $\text{Im}\langle x,y\rangle = \text{Re}(-i\langle x,y\rangle) = \text{Re}\langle x,iy\rangle$, so that

$$\langle x,y\rangle = \text{Re}\langle x,y\rangle + i\,\text{Re}\langle x,iy\rangle.$$

Substituting directly from $(*)$ for $\text{Re}\langle x,y\rangle$ and with y replaced by iy for $\text{Re}\langle x,iy\rangle$ yields the required identity. □

This identity can be memorized without too much trouble by noting that the coefficients of the terms (which run through the fourth roots of unity) are the same as the coefficients of the term involving y. However, its real importance is the concept that norms determine the inner product, rather than the actual details of the formula. For example, consider the question as to when we should consider two inner product spaces the same. We may well want to require that a bijection T between them should preserve the inner product. That is, $\langle Tx, Ty\rangle = \langle x,y\rangle$ for all x, y. Polarization says that this is automatic provided we have an isometry, so we do not need additional requirements.

Theorem 4.8. *If a norm $\|\cdot\|$ on a complex linear space X satisfies the parallelogram law, then it can be induced by an inner product.*

Sketch of Proof. Define the alleged inner product $\langle x,y\rangle$ by the polarization identity. We must verify that the required properties for an inner product hold. Conjugate symmetry and positivity are clear. Linearity in the first variable is not obvious but follows from some elementary, but somewhat tedious, algebra, and it is here that we must use the parallelogram law. We leave the details to the reader. □

Here is an immediate important consequence of the Cauchy–Schwartz inequality.

Theorem 4.9 (Continuity of Inner Product). *In any semi-inner product space, if the sequences $(x_n) \to x$ and $(y_n) \to y$, then $(\langle x_n, y_n\rangle) \to \langle x,y\rangle$.*

Proof. We employ, once again, the adding-and-subtracting trick to get

$$\begin{aligned}
|\langle x_n,y_n\rangle - \langle x,y\rangle| &\le |\langle x_n,y_n\rangle - \langle x_n,y\rangle| + |\langle x_n,y\rangle - \langle x,y\rangle| \\
&= |\langle x_n, y_n - y\rangle| + |\langle x_n - x, y\rangle| \\
&\le \|x_n\|\,\|y_n - y\| + \|x_n - x\|\,\|y\|,
\end{aligned}$$

where we use the Cauchy–Schwartz inequality for the last inequality. The assertion now follows by using the fact that the convergent sequence (x_n) is bounded. □

Definition. A *Hilbert space* is a complete inner product space.

As with Banach spaces, we can always complete an inner product space. Take a completion as a normed linear space. For x and y in the completion, we can define

$\langle x, y \rangle$ as $\lim \langle x_n, y_n \rangle$, where (x_n) and (y_n) are sequences in the original space converging to x and y, respectively. We must show that this isindependent of the choice of sequences, satisfies the inner product rules, and induces the norm in the larger space. We leave the details to the reader.

4.3 NEAREST POINTS AND CONVEXITY

In this section, we present a result that is a key fact in deducing the geometry of Hilbert spaces and the operators they admit. We will proceed in a somewhat more general fashion than needed for Hilbert spaces, not only to obtain the extra generality, but also to provide more geometric insight than that of some of the more conventional approaches.

A major theme that runs through all of the analysis is that of *approximation*. There are several applications in which we want to find the best approximation to some function by another that has certain specified properties. In the typical case, the set of approximating functions will form a *convex set*. We recall the definition.

Definition. A set K in a linear space X is said to be *convex* if the line segment joining any two points in K lies entirely in K. That is, for all $x, y \in K$ and any scalar $\alpha \in [0, 1]$, the point $\alpha x + (1 - \alpha)y \in K$. It follows by induction that given any n points $x_1, x_2, \ldots, x_n \in K$ and nonnegative scalars $\alpha_1, \alpha_2, \ldots, \alpha_n$ adding to 1, the point $\sum_{i=1}^{n} \alpha_i x_i \in K$.

Application: Regression Analysis. A major field in statistics is regression analysis, which seeks to establish relations between various quantities by looking at observed data. The most basic example deals with the case in which we want to model a quantity y as depending linearly on another quantity x. The problem reduces to finding that function of the form $y = ax + b$ (for some constants a and b) that is closest under some norm to an observed function $y = f(x)$. There are more advanced ideas in the subject but for the most part they can be viewed in the context of nearest point aproximation. See Brockwell and Davis (1991, Section 2.7) for more details.

We now consider the following general situation. In a Banach space X, we have a closed, convex set K and a point x; we want to find y in K that best approximates x in the sense that $\| x - y \| = d(x, K)$. We are interested in whether such a y exists and, if so, whether or not it is unique. (Of course, if x is already in K, then x itself is the unique such point.) We naturally require that K should be closed, for if x is a boundary point that is not in K, then $d(x, K) = 0$ and no such y could exist.

Consider first the uniqueness question. This can fail drastically in the absence of convexity. In $\mathcal{R}^2$, for example, if $x = 0$ and K is a circle centered at 0, then every point of K is one of best approximation. But, convexity is not enough to ensure uniqueness. In $\mathcal{R}^2$, let K be the closed, unit ball with the sup norm, and let $x = (2, 0)$. Then, $d(x, K) = 1$, and the minimum distance is assumed for all points

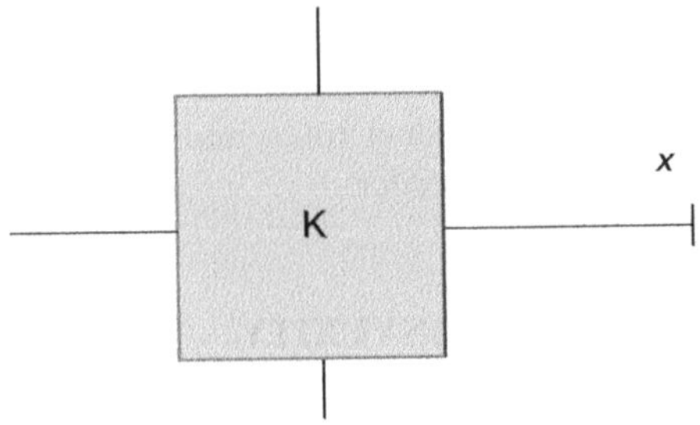

Figure 4.3. In sup norm, *d*(*x*,*K*) is minimized by all points on black line.

in the unit ball with first coordinate equal to 1; see Figure 4.3. We can find a similar example in $\mathcal{R}^2$ for the unit ball with the $\|\cdot\|_1$ norm, but not with the Euclidean norm, where we do indeed get a unique closest point from any point to the closed unit ball. The reason lies in the fact that the Euclidean unit ball is *round*, as we expect a ball to be. By capturing precisely the idea of *roundedness*, we can formulate a condition for uniqueness of a closest point.

Definition. A normed linear space X is said to be *strictly convex* if for any two distinct unit vectors x and y we have $\| (x+y)/2 \| < 1$.

Strict convexity implies that the midpoint of any two points on the boundary of the unit ball does not lie on the boundary, or to put it another way, this boundary does not contain any line segments, so we can think of it as round.

Proposition 4.10. *Suppose X is strictly convex. For any point x and convex set K, there is at most one point in K that is nearest to x.*

Proof. By translation, that is, by replacing K by $K - x$, we can assume that $x = 0$, so that we are looking for a point in K of minimal norm. This obviously cannot occur at two distinct points, since the midpoint, which is in the set by convexity, would have a smaller norm. (This is easily seen from the given definition in terms of unit vectors by multiplying by an appropriate scalar.) □

To get the existence of a best approximation, we need a stronger condition. We can phrase strict convexity as saying that if you join two points of norm 1 by a line segment and if the midpoint has norm 1, then the two points must have coincided to begin this. Looking at the closed unit ball in the plane, however, we can say more, namely, if the midpoint has norm "close" to 1, the endpoints are close to each other. The following-condition makes this precise and stipulates a uniformity to the relationship. It is convenient here to phrase it in terms of the entire ball rather than the boundary.

Definition. A normed linear space is said to be *uniformly convex*, if given any $\epsilon > 0$, there is a $\delta > 0$ such that for x and $y \in \bar{B}(1)$ we have

$$\| \tfrac{1}{2}(x+y) \| > 1 - \delta \Rightarrow \| x - y \| < \epsilon.$$

Example 4.11. Any Hilbert space is uniformly convex. Given $x, y \in \bar{B}(1)$ with $z = (x+y)/2$, we can use the parallelogram law to write

$$\|x - y\|^2 \leq 4(1 - \|z\|^2),$$

so choose δ such that $(1-\delta)^2 = 1 - \epsilon^2/4$ if $0 < \epsilon < 2$. If $\epsilon \geq 2$, then clearly any $\delta > 0$ works.

Other examples of uniformly convex spaces are L^p, $1 < p < \infty$. We will not give the proof here; see Carothers (2005, Chapter 11). Intuition as to why it might be true can be obtained by drawing pictures in the plane, where the uniform convexity is evident.

Our main result about uniform convexity is as follows.

Theorem 4.12. *Suppose X is a uniformly convex Banach space. For any point x and a nonempty, closed, convex set K, there is a unique nearest point to x in K.*

Proof. Uniqueness follows from Proposition 4.10 since uniform convexity obviously implies strict convexity. By translation and scaling, we can reduce the existence problem to showing that $d(0, K) = 1$ implies that some point in K has norm equal to 1. Given any positive integer n, $d(0, K) = 1$ implies that we can find a point y_n in K with $\| y_n \| = (1 + 1/n)$. If the sequence (y_n) converges to a point y, we necessarily have $\| y \| = 1$. Therefore, it only remains to show that the sequence is Cauchy. (The intuition behind the following derivation is that, since the midpoint of any two points of the sequence must necessarily be close to 1, it forces the points to be close.) Given $\epsilon > 0$, choose $\delta > 0$ satisfying the uniform convexity condition for $\epsilon/2$, choose $N > 1/\delta$, and suppose that $n \geq m \geq N$. Then,

$$\left\| \frac{y_m + y_n}{2} \right\| \geq 1 = \frac{N+1}{N}\left(1 - \frac{1}{N+1}\right) \geq \frac{N+1}{N}(1 - \delta).$$

If $\hat{y}_n = N/(N+1)\, y_n$ and $\hat{y}_m = N/(N+1)\, y_m$, we know that $\hat{y}_n$ and $\hat{y}_m$ lie in $\bar{B}(1)$ and that $\| (\hat{y}_n + \hat{y}_m)/2 \| > 1 - \delta$. Therefore, $\| \hat{y}_n - \hat{y}_m \| < \epsilon/2$, so that

$$\left\| y_m - y_m \right\| \leq \frac{N+1}{N}\frac{\epsilon}{2} \leq \epsilon.$$

This shows that the sequence is Cauchy, thereby completing the proof. □

A particular example of a convex set is a subspace. In the following section, we will consider closed subspaces M of a Hilbert space. For any x, we denote the unique nearest point to x in M by $P_M(x)$ and call it the *projection of x onto M*.

Application: Conditional Expectation. Refer to the material in Section 2.9.2 dealing with probability theory. Suppose we have a random variable X on a probability space $(\Omega, \mathcal{S}, P)$. (Traditional probability theory notation often uses P as the symbol for a probability measure and uses capital letters like X and Y for random variables.) We want to model the effect of being given some additional information about X that identifies events that can occur and those that cannot occur. It is logical to postulate that this collection of events will be a sub-σ-algebra $\mathcal{T}$ of $\mathcal{S}$. For example, if a number is chosen at random from the four numbers $\{1,2,3,4\}$, the information that the number is either even or odd corresponds to the σ-algebra consisting of the sets $(\{1,3\},\{2,4\},\{1,2,3,4\},\emptyset)$.

A frequently used concept is that of the *conditional expectation of X with respect to $\mathcal{T}$*. It represents our best estimate of the value taken on by X according to the given information. This conditional expectation will still be a random variable as it will depend on the particular subset(s) in $\mathcal{T}$ that occur. In the above example, the conditional expectation will take one value if the number is odd and another if the number is even. Suppose that $X \in L^2(\Omega)$. Then, this conditional expectation can be viewed as a particular application of projecting an element onto a subspace. It will be, in fact, the nearest point to X in the subspace consisting of all those elements of $L^2(\Omega)$ that are measurable with respect to $\mathcal{T}$.

In the extreme case in which we are given no information, $\mathcal{T}$ will consist of the two sets Ω and $\emptyset$. All we know is that the event Ω must occur and the event $\emptyset$ cannot occur. Therefore, the only $\mathcal{T}$ measurable functions are constant a.e., and the conditional expectation in this case is the function whose constant value a.e. is the usual *expectation* of X given by

$$E(X) = \int X\, dP.$$

See Exercise 4.18. For more details on conditional expectation see Bobrowski (2006, Section 3.2) or Brockwell and Davis (1991, Section 2.7).

4.4 ORTHOGONALITY

4.4.1 Basic Notions

Definition. Motivated by the geometry of $\mathcal{R}^n$, we consider two elements x, y of a Hilbert space to be *perpendicular*, more commonly referred to as *orthogonal*, if $\langle x, y\rangle = 0$. We sometimes write this as $x \perp y$. For any subset S of a Hilbert space H, we let $S^\perp = \{x \in H : \langle x, y\rangle = 0 \text{ for all } y \in S\}$.

We now have encountered a situation that occurs frequently in mathematics, in that we have assigned two definitions to the word "perpendicular." (Recall that in Section 2.8.3 we defined what it means for a vector to be perpendicular to a subspace in any Banach space by using only the norm.) When this happens, one must maintain consistency by showing that the definitions are indeed equivalent.

Proposition 4.13. *Suppose S is a subset of a Hilbert space that is closed under scalar multiplication, that is, for all $y \in X$ and $\alpha \in \mathcal{C}$, we have $\alpha y \in S$. Then, $S^\perp = \{x \in H : d(x,S) = \| x \|\}$.*

Proof. If $d(x,S) = \| x \|$, then for all $y \in S$ and $t \geq 0$,

$$\| x \|^2 \leq \| x \pm ty \|^2 = \| x \|^2 + t^2 \| y \|^2 \pm 2 \operatorname{Re} t\langle x, y\rangle,$$

showing that

$$\mp \operatorname{Re}\langle x, y\rangle \leq \tfrac{t}{2} \| y \|^2 .$$

Since t is arbitrary, we must have $\operatorname{Re}\langle x,y\rangle = 0$. Similarly, it follows that $\operatorname{Im}\langle x,y\rangle = \operatorname{Re}\langle x, iy\rangle$ also equals zero, so that $\langle x,y\rangle = 0$.

Conversely, if $x \in S^{\perp}$, then for all $y \in S$,

$$\| x - y \|^2 = \| x \|^2 + \| y \|^2 \geq \| x \|^2,$$

so that $d(x,S) \geq \| x \|$. Since $0 \in S$, equality must hold. □

4.4.2 Projections Onto Subspaces

An important fact about the orthogonality relationship in a Hilbert space (as opposed to a general Banach space) is that, by the linearity and continuity properties of the inner product, $S^{\perp}$ is a closed subspace for all S. This leads to a direct-sum decomposition of the space corresponding to any closed subspace. This is an indispensable tool in the study of linear operators, as shown in Section 4.6.

Theorem 4.14 (Projection Theorem). *Let M be a closed subspace of a Hilbert space H. Then:*

(*a*) $M \cap M^{\perp} = \{0\}$.

(*b*) *Any $z \in H$ can be written uniquely as the sum of an element in M and one in $M^{\perp}$. Specifically, $z = P_M z + P_{M^{\perp}} z$.*

(*c*) $M^{\perp\perp} = M$.

(*d*) *H is isometric to $M \oplus M^{\perp}$, in which the direct sum is equipped with the ℓ^2-norm.*

Proof

(a) If $x \in M \cap M^{\perp}$, then $\langle x,x\rangle = 0$, so that $x = 0$. This also shows uniqueness of the decomposition in (b), for if $y + z = y' + z'$, with $y, y' \in M$ and $z, z' \in M^{\perp}$, then $y - y' = z - z'$ is in $M \cap M^{\perp}$, so that $y = y'$ and $z = z'$.

(b) Given any $z \in H$, let $y = z - P_M z$. Then, for any $m \in M$, $\| y - m \| = \| z - (P_M z + m) \| \geq \| y \|$ since $P_M z$ is the element of M that is closest to z. We therefore have $d(y, M) = \| y \|$, and by Proposition 4.13, $y \in M^{\perp}$. We can write $z = P_M z + (z - P_M z)$ as the sum of an element in M and one in $M^{\perp}$, and it remains to show that $(z - P_M z) = P_{M^{\perp}} z$, which we will do after proving (c).

(c) It is clear that $M \subseteq M^{\perp\perp}$. Conversely, given $z \in M^{\perp\perp}$, let $z = x + y$ with $x \in M$ and $y \in M^{\perp}$. Then, $y = z - x$ is in $M^{\perp\perp}$ and in $M^{\perp}$, so by part (a) applied to $M^{\perp}$ it follows that $z = x \in M$.

We can now finish the proof of (b). By applying our conclusion in (b) with $M^{\perp}$ in place of M, we can write $z = P_{M^{\perp}} z + (z - P_{M^{\perp}} z)$ as a decomposition of z as a sum of an element in $M^{\perp}$ and one in $M^{\perp\perp} = M$. But, we also saw that $z = (z - P_M z) + P_M z$ is such a decomposition, so by uniqueness we must have $(z - P_M z) = P_{M^{\perp}} z$.

(d) The mapping that sends z to $(P_M z, P_{M^{\perp}} z)$ is a linear space isomorphism from H to $M \oplus M^{\perp}$. Since for any $x \in M$ and $y \in M^{\perp}$ we have $\langle x, y \rangle = 0$, the basic identity shows that this mapping is an isometry when $M \oplus M^{\perp}$ is given the ℓ^2-norm. □

Given any subset S of a normed linear space, we often want to consider the *closed linear span of* S, that is, $\overline{\text{span}(S)}$. This is the smallest closed linear subspace containing the set S. In a Hilbert space, we have a particularly nice way of expressing this.

Proposition 4.15. *For any $S \subseteq H$, its closed linear span is $S^{\perp\perp}$.*

Proof. Let M denote the closed linear span of S. Since $S^{\perp\perp}$ is a closed linear subspace containing S, it contains M. Conversely, since applying $\perp$ clearly reverses set inclusion,

$$S \subseteq M \text{ implies } M^{\perp} \subseteq S^{\perp} \text{ implies } S^{\perp\perp} \subseteq M^{\perp\perp} = M. \qquad \square$$

4.4.3 Orthonormal Sets

In this section, we show that any Hilbert space possesses a generating set of a particularly useful nature. A major application will be to show how one can actually find $P_M x$ for a point x and a closed subspace M.

Definition. A subset S of H is said to be *orthonormal* if for all $x, y \in S$

$$\langle x, y \rangle = 0 \text{ and } \| x \| = 1.$$

Example 4.16

(a) In ℓ^2, the set $\{\delta_n : n = 1, 2, \ldots\}$ is orthonormal.

(b) In $L^2[-\pi, \pi]$, the set

$$\left\{ \frac{1}{\sqrt{\pi}} \sin(nx), \quad \frac{1}{\sqrt{\pi}} \cos(nx), \quad \frac{1}{\sqrt{2\pi}} \right\},$$

with n ranging over the positive integers, is orthonormal. The latter functions will be familiar to some readers as those used in *Fourier analysis*, and the subsequent discussion can be looked upon as an abstract formulation of some basic principles of that subject.

In the remainder of this section, we will, for the most part, confine our attention to *separable* Hilbert spaces, those with a countable dense subset. Any orthonormal set S in such a space must be countable, since for any $x, y \in S$, we have $\| x - y \| = \sqrt{2}$, and arguing similary as in the proof of Proposition 2.4.2, we cannot have uncountably many vectors with this property in a separable space. It will therefore be convenient to represent any orthonormal set as a sequence $(x_i), i = 1, 2, \ldots$.

One mathematical fact that almost everybody is familiar with is the theorem of Pythagoras, which we may phrase in our present context as saying that, given two orthogonal vectors x and y in the plane, $\| x + y \|^2 = \| x^2 \| + \| y^2 \|$. Following is a significant generalization, our main result on orthogonality.

Theorem 4.17 (Properties of Orthonormal Sets). *Let (x_i) be orthonormal in a Hilbert space H, and let M be the closed linear span of the elements of this sequence. Then:*

(a) *Given any sequence $\alpha = (\alpha_i)$ in C, $\sum_{i=1}^{\infty} \alpha_i x_i$ converges to some element x if and only if $\alpha \in \ell^2$, and in this case*

$$\| x \| = \| \alpha \|_2 .$$

(b) *For any $z \in H$, $P_M(z) = \sum_{i=1}^{\infty} \langle z, x_i \rangle \, x_i$.*

Proof

(*a*) Let $s_n = \sum_{i=1}^{n} \alpha_i x_i$ and let $t_n = \sum_{i=1}^{n} |\alpha_i|$. Induction on the basic identity in Lemma 4.2 immediately yields that for any positive integers $m < n$

$$\| s_n - s_m \|^2 = \| \sum_{i=m+1}^{n} \alpha_i x_i \|^2 = \sum_{i=m+1}^{n} |\alpha_i|^2 = |t_n - t_m|^2.$$

Therefore, (s_n) is a Cauchy sequence in H if and only if (t_n) is a Cauchy sequence of scalars, which proves the first statement in (a). The equality of norms arises immediately from the fact that the norm is continuous.

(*b*) Let M_n denote the linear span of $x_1, x_2, \ldots, x_n$. Given any $z \in H$, let $z_n = \sum_{i=1}^{n} \langle z, x_i \rangle \, x_i$. For $i = 1, 2, \ldots, n$, we have $\langle z_n, x_i \rangle = \langle z, x_i \rangle$, thereby showing that $z - z_n \in M_n^{\perp}$. From the basic identity in Lemma 4.2,

$$\| z \|^2 = \| z_n \|^2 + \| z - z_n \|^2 \geq \| z_n \|^2 = \sum_{i=1}^{n} |\langle z, x_i \rangle|^2.$$

From this expression, we see that the sequence $(\langle z, x_i \rangle)$ lies in ℓ^2, so that by part (a), the vector

$$z' = \sum_{i=1}^{\infty} \langle z, x_i \rangle \, x_i$$

exists and is in the space M. From the continuity of the inner product, we have $\langle z, x_i \rangle = \langle z', x_i \rangle$ for all i, so that, as above, the decomposition of z relative to M and $M^\perp$ is given by $z = z' + (z - z')$, showing that $z' = P_M z$. □

It would be nice if we found an orthonormal set (x_i) whose closed linear span is all of H. Such a set is called an *orthonormal basis of* H. In that case, every element can be written as in part (b) in terms of the basis. The next theorem tells us that such a basis always exists.

Theorem 4.18. *Every Hilbert space has an orthonormal basis.*

Proof. The proof given here applies in general and does not depend on the space being separable. We are done if we find a maximal orthonormal set S. Indeed, by contradiction, if the closed linear span M of such a set is not all of H, pick any $x \notin M$ so that $P_{M^\perp} x$ is a nonzero element of $M^\perp$. Dividing this element by its norm gives a unit vector that could be added to the original set to get a larger orthonormal set.

The existence of a maximal orthonormal set follows from Zorn's lemma. (See Appendix A for the necessary background.) Let $\mathcal{P}$ be the poset whose elements are orthonormal sets ordered by by set inclusion. Clearly $\mathcal{P}$ is nonempty since any unit vector is by itself an orthonormal set. Any linearly ordered subset of $\mathcal{P}$ has an upper bound, namely just the union of all the sets. This is easily seen to be orthonormal. □

From the remarks above, it follows that any separable Hilbert space has a countable orthonormal basis.

The sets in Example 4.16 are both orthonormal bases. Indeed, the linear span of $\{\delta_n\}$ is c_{00}, which is dense in ℓ^2. The fact that the trigonometric functions, together with the constants, span all of $L^2[-\pi, \pi]$ is a much more difficult fact to prove. It is a fundamental result in Fourier analysis, and there are several proofs of it involving classical analysis. We will defer a proof of this until Chapter 10 where we outline an operator-theoretic derivation in Exercise 10.10.

In light of Theorem 4.18, there a few observations that should be made about Theorem 4.17 to show the powerful effects of orthonormality. First, suppose we have an arbitrary sequence of unit vectors (x_i), and we want to know when a series $\sum_{i=1}^{\infty} \alpha_i x_i$ converges. Without any other information, we can only affirm this in the case that (α_i) is in ℓ^1. Orthogonality allows us to assert this for all ℓ^2 sequences, a larger class. Second, given an element z, if we want to write it as $\sum_{i=1}^{\infty} \alpha_i x_i$, it might be a very difficult task to find the sequence (α_i). In the case of an orthonormal basis, we can write down the coefficients immediately as $\langle z, x_i \rangle$. The uniqueness of these coefficients shows that an orthonormal basis is, in particular, a Schauder basis as defined in Section 2.11.

It is not always easy to show that a given orthonormal sequence is a basis. Exercise 4.4 gives some criteria.

As a immediate corollary of Theorem 4.18, we have the following important result.

Theorem 4.19. *Any separable infinite-dimensional Hilbert space is isometric to ℓ^2.*

Proof. Choose an orthonormal basis (x_i). Then, the mapping that sends (α_i) to $\sum_{i=1}^{\infty} \alpha_i x_i$ is a linear isometry from ℓ^2 to H by part (a) of Theorem 4.17, and by part (b) it is surjective. □

Note that as a consequence of this theorem we see that ℓ^2 is isometric to $L^2[a, b]$, despite the fact that they seem at first glance to be very different spaces.(The latter space is separable. It has, for example, a countable orthononormal basis consisting of suitably scaled trignometric functions. See Exercise 4.15.)

Theorem 4.18 guarantees the existence of an orthonormal basis, but it is not constructive and does not tell us how to find it. There is a well-known procedure that produces orthonormal sets from any linearly independent set.

Theorem 4.20 (Gram–Schmidt Orthogonalization Process). *Given a linearly independent sequence (y_i) in H, there is an orthonormal sequence (x_i) such that, for any positive integer n,* span $\{x_1, x_2, \ldots, x_n\} = \text{span}\{y_1, y_2, \ldots, y_n\}$.

Proof. We construct the orthonormal set inductively. Let $x_1 = y_1 / \| y_1 \|$; by linear independence $y_1 \neq 0$. Then, having chosen $x_1, x_2, \ldots, x_{n-1}$, we define

$$w_n = y_n - \sum_{i=1}^{n-1} \langle y_n, x_i \rangle \, x_i.$$

It follows that w_n is orthogonal to $\{x_1, x_2, \ldots, x_{n-1}\}$. Moreover, $w_n \neq 0$, for if so, then y_n would be in span $\{x_1, x_2, \ldots x_{n-1}\}$, which by our induction hypothesis is the same as span $\{y_1, y_2, \ldots, y_{n-1}\}$, contradicting the linear independence. We define $x_n = w_n / \| w_n \|$, and it not hard to verify that the required conditions are satisfied. □

Remark. It is instructive to take a closer look at the above proof. In most applications of this theorem, we do not necessarily need the conditions on the spans of the elements. Usually, all we really want to know is that the linear span of the final sequence is the same as that of the elements we started with. This is typical of induction proofs, however. Sometimes we have to prove more than we might really want at the time, just to get a stronger induction hypothesis, which allows us to go to the next stage.

Versions of the above theorems exist in nonseparable Hilbert spaces. The only difference in Theorem 4.17 is that we sum over uncountable index sets, where we define such a sum to be α if, given any $\epsilon > 0$, we can find a finite set of indices F such that for all finite sets G containing F the sum over G is within ϵ of α. It turns out that if the sum exists, then in fact there are only countably many nonzero summands and the series converges absolutely, so there are not really any new ideas involved. In Theorem 4.19, we need a more general ℓ^2 space, which uses a counting measure on an uncountable set.

A general result, which we will not prove here, is that for any Hilbert space the cardinality of any orthonormal basis is unique and is known as the dimension of the space. Moreover, two Hilbert spaces are isometric if and only if they have the same dimension; see Dunford and Schwartz (1954, IV.4.14–16). We will prove a special case of this general result, which we make use of in later chapters.

Theorem 4.21. *If S is an orthonormal set in any separable Hilbert space H, then S is either finite or countably infinite.*

Proof. Let A be a countable dense set in H. Define a function j from S to A by taking $j(x)$ to be any element in A such that $\| j(x) - x \| < \sqrt{2}/2$. Such an element exists by the density of A. Our assertion follows if we show that j is injective. This is true since $j(x) = j(y)$ implies that

$$\| x - y \| \leq \| x - j(x) \| + \| j(y) - y \| < \sqrt{2},$$

which implies that $x = y$ since any two distinct elements from an orthonormal set have a distance between them of exactly $\sqrt{2}$. □

4.5 LINEAR FUNCTIONALS ON HILBERT SPACES

One of the major themes dominating modern mathematics is that the properties of a mathematical object cannot always be gleaned by looking at the object itself. One must also investigate how it is related to other objects of the same type by looking at appropriate mappings between the given object and others. For a linear space X, it is natural to study it by looking at linear mappings of X into the scalar field $\mathcal{F}$, the simplest nonzero linear space of them all. Such mappings are commonly called *linear functionals*. In the case of a normed linear space X, we want to consider the space $\mathcal{B}(X, \mathcal{F})$ of all bounded linear functionals on X. This is known as the *dual space* of X and will be denoted by X^*. We will carry out a more exhaustive study of dual Banach spaces in the following chapters. For a Hilbert space, the dual space turns out to be particularly nice. The following result is another one that is attributed to Riesz.

Theorem 4.22. *For any z in a Hilbert space H, the functional ψ_z defined by $\psi_z(x) = \langle x, z\rangle$ is in H^*, and $\| \psi_z \| = \| z \|$. Conversely, given any $f \in H^*$, there is a unique $z \in H$ such that $f = \psi_z$.*

Proof. The linearity of ψ_z is clear. The Cauchy–Schwartz inequality implies that $\| \psi_z \| \leq \| z \|$. For $x = z/ \| z \|$ (assuming that $z \neq 0$), we have $\psi_z(x) = \| z \|$, which implies that $\| \psi_z \| \geq \| z \|$, establishing the equality of the norms.

Obviously the 0 linear functional is equal to ψ_0, so suppose we are given a nonzero $f \in H^*$. The only possible candidates for z are those that are perpendicular to $N(f)$. Accordingly, choose a unit vector $z_0 \in N(f)^{\perp}$, and let $z = f(z_0)z_0$. Then, for any $x \in H$,

$$\phi_z(x) = \langle x, z\rangle = \left\langle x - \frac{f(x)}{f(z_0)} z_0, z \right\rangle + \frac{f(x)}{f(z_0)} \langle z_0, z\rangle.$$

But since $x - f(x)/f(z_0)z_0 \in N(f)$, the first term on the far right vanishes. The second term on the far right is equal to $f(x)$, and our claim is verified.

This z is unique since for any other vector z' we would have that $\langle z - z', x\rangle = 0$ for all $x \in H$, so that $z - z' \in H^{\perp} = 0$. □

Let J_H denote the mapping from H to H^* that sends z to ψ_z. Theorem 4.22 says that J_H is surjective and preserves length. Moreover, it is easily seen to be additive. It is not a linear map, however, but rather is *conjugate linear* because $J_H ix$ is equal to $-iJ_H x$ rather than $iJ_H x$.

Corollary 4.23. *For any Hilbert space H, the dual H^* is a Hilbert space with the same dimension as H.*

Proof. More precisely, we should say that there is an inner product on H^* that induces the norm. This inner product is defined as follows. For f and $g \in H^*$, define

$$\langle f, g\rangle = \langle y, x\rangle,$$

in which $f = \psi_x$ and $g = \psi_y$. Note the reversal of order, which is necessary to make the inner product linear, rather than conjugate linear, in the first variable. We leave it to the reader to verify that this is, indeed, an inner product inducing the norm. For the final statement, we note that for any orthonormal basis (x_n) of H, $(J_H x_n)$ is an orthonormal basis of H^*. □

Remark. We can give an explicit description of the orthonormal basis for the dual space defined in the last line of the preceding proof. For any n, $J_H x_n$ is that linear functional that maps x_n into 1 and x_m into 0 for all $m \neq n$. This is a familiar construction from linear algebra and gives us a standard way of obtaining a basis of X^* known as the *dual basis* from any given basis of X.

4.6 LINEAR OPERATORS ON HILBERT SPACES

4.6.1 Introduction

Unlike Banach spaces, the problem of classifying Hilbert spaces is completely solved, as we saw in the previous section. Mathematicians involved with Hilbert spaces invariably study operators on those spaces. There is a vast amount of research being done and many unsolved problems. In this section, we provide only a brief introduction to this topic. Throughout, we consider a Hilbert space H and $\mathcal{B}(H)$, the space of all bounded linear operators from H to itself. As well as being a linear space, $\mathcal{B}(H)$ has a multiplication through composition. Note that this is a highly noncommutative operation; that is, in general, for $S, T \in \mathcal{B}(H)$, ST is different than TS. When S and T do commute, it often means that we can say something significant about the relationship between the two operators. We let I denote the *identity* operator that maps every element x to itself. It clearly satisfies $IT = TI = T$ for all $T \in \mathcal{B}(\mathrm{H})$.

4.6.2 Sesquilinear Forms

Just as bounded linear functionals are determined by vectors in the space itself, all bounded linear operators from H to itself are determined by certain two-variable functions on the space.

Definition. A *sesquilinear form* on H is a function f from $H \times H$ to C such that for all $x, y, z \in H$ and $\alpha \in C$ we have

$$f(x + y, z) = f(x, z) + f(y, z), \qquad f(\alpha x, z) = \alpha f(x, z)$$

and

$$f(z, x + y) = f(z, x) + f(z, y), \qquad f(z, \alpha z) = \bar{\alpha} f(x, z).$$

If, in the second equation, the final $\bar{\alpha}$ were just α, then f would be linear in both variables and would be called a *bilinear form*. However, f is conjugate linear in the second variable, so it is not quite bilinear. This feature is the source of the name sesquilinear, which means one-and-one-half linear. Our major example so far is just the inner product: $f(x, y) = \langle x, y \rangle$.

Definition. For any sesquilinear form f, the function q from H to C given by $q(x) = f(x, x)$ is known as the *quadratic form* associated to f.

For any sesquilinear form, the polarization identity can be derived exactly as we did for the inner product, and we get the same result with $f(\cdot, \cdot)$ replacing $\langle \cdot, \cdot \rangle$ throughout. Recall that $\| x \|^2 = \langle x, x \rangle$, so replace $\| x \|^2$ with $f(x, x)$. From this polarization identity, it is clear that if $f(x, x)$ is real for all x, then the conjugate symmetry property holds, namely, $f(x, y) = \overline{f(y, x)}$ for all $x, y \in H$.

Given a sesquilinear form f, we define the norm of f, denoted by $\| f \|$, in a similar way to that for linear functionals. Specifically,

$$\| f \| = \sup\{|f(x,y)| : \| x \| = \| y \| = 1\}.$$

By using the usual terminology, we say that f is bounded if $\| f \|$ is finite. The reason for introducing this concept is that the sesquilinear forms of finite norm correspond precisely to the elements of $\mathcal{B}(H)$. Parallel to Theorem 4.22, we have the following.

Theorem 4.24. *Given any $T \in \mathcal{B}(H)$, the function f_T defined by*

$$f_T(x,y) = \langle Tx, y \rangle$$

is a sequilinear form with norm equal to $\| T \|$. Conversely, given any sesquilinear form f of finite norm, there is a unique $T \in \mathcal{B}(H)$ such that $f = f_T$.

Proof. It is clear that f_T is sesquilinear, and as in the linear functional proof, the Cauchy–Schwartz inequality tells us that $\| f \| \leq \| T \|$. Equality is obvious when $T = 0$. In the case that $T \neq 0$, choose a unit vector x such that $Tx \neq 0$, and let $y = Tx / \| Tx \|$. Then, $f_T(x,y) = \| Tx \|$, thereby showing that $\| f_T \| \geq \| T \|$.

Conversely, suppose we are given a sesquilinear form f of finite norm. Fix any $x \in H$, and define a linear function g on H by $g(y) = \overline{f(x,y)}$. It is clear that g is bounded with norm less than or equal to $\| x \| \| f \|$. By Theorem 4.22, there is a unique vector z such that $g = \psi_z$. Define $Tx = z$. The mapping x to Tx is easily seen to be linear. Moreover, for all $y \in H$,

$$f_T(x,y) = \langle Tx, y \rangle = \overline{\langle y, z \rangle} = \overline{\psi_z(y)} = \overline{g(y)} = f(x,y).$$

The first part of the proof shows that T has norm equal to that of f and is therefore bounded. Uniqueness follows since for any other T' we would have that, for all $x, y \in H$, $\langle Tx - T'x, y \rangle = 0$, which implies that $Tx - T'x = 0$. □

The uniqueness statement immediately yields this corollary.

Corollary 4.25. *If $\langle Tx, x \rangle = \langle Sx, x \rangle$ for all $x \in H$, then $S = T$.*

Remark. This corollary depends heavily on the fact that our scalar field is $\mathcal{C}$. It fails in a real Hilbert space. For example, in $\mathcal{R}^2$ with the usual inner product, let $T(x_1, x_2) = (-x_2, x_1)$ (rotation through 90°). Then, $\langle Tx, x \rangle = 0 = \langle 0x, x \rangle$ for all $x \in H$.

4.6.3 Hilbert Space Adjoint

In this section, we introduce the important idea of the Hilbert space adjoint operator.

Definition. Given any $T \in \mathcal{B}(H)$, its *adjoint operator*, denoted by T^*, is that unique element of $\mathcal{B}(H)$ associated by Theorem 4.24 to the sesquilinear form $f(x,y) = \langle x, Ty\rangle$.

It follows that for all $x, y \in H$

$$\langle Tx, y\rangle = \langle x, T^*y\rangle, \qquad (*)$$

which is the formula we traditionally invoke when using this concept.

Remark. It is instructive to see what this means in the finite-dimensional case. Suppose we have a Hilbert space with an orthonormal basis $\{x_1, x_2, \ldots, x_n\}$. By using this basis for both the domain and range, we can express the matrix $A = (a_{i,j})$ of T, as defined in Section 1.3.4, by

$$a_{i,j} = \langle Tx_j, x_i\rangle.$$

Consider the matrix $A^* = (a^*_{i,j})$ corresponding to T^* for the same basis;

$$a^*_{i,j} = \langle T^*x_j, x_i\rangle = \langle x_j, Tx_i\rangle = \overline{a_{j,i}}.$$

The matrix of T^* is therefore obtained by taking the transpose of the matrix of T and replacing each entry with its complex conjugate.

The familiar properties of a matrix transpose carry over to the general infinite-dimensional case as the following theorem shows. [In part (b) we must bring in the conjugation.] The significance of part (f) will be discussed in Section 8.5.

Theorem 4.26 (Properties of Hilbert Space Adjoint). *Given $S, T \in \mathcal{B}(H)$ and $\alpha \in \mathcal{C}$:*

(*a*) $(S+T)^* = S^* + T^*$;
(*b*) $(\alpha T)^* = \bar{\alpha}\, T^*$;
(*c*) $(ST)^* = T^*S^*$;
(*d*) $\| T^* \| = \| T \|$;
(*e*) $T^{**} = T$;
(*f*) $\| T^*T \| = \| T \|^2$.

Proof. Parts (a) and (b) are immediate from equation $(*)$ above. Part (d) follows from Theorem 4.24 since the norm of the sesquilinear form f in the definition of the adjoint is clearly the same as the norm of f_T. For parts (c) and (e), we have, for all $x, y \in H$,

$$\langle (ST)^* x, y\rangle = \langle x, STy\rangle = \langle S^* x, Ty\rangle = \langle T^* S^* x, y\rangle,$$

and

$$\langle T^{**} x, y\rangle = \langle x, T^* y\rangle = \langle Tx, y\rangle,$$

so the conclusions follow by the uniqueness of the sesquilinear forms; see Corollary 4.25. For part (f), note that $\| T^*T \| \leq \| T^* \| \, \| T \| = \| T \|^2$. Conversely, given any unit vector x, we have

$$\| Tx \|^2 = \langle Tx, Tx\rangle = \langle T^*Tx, x\rangle \leq \| T^*T \| .$$

Taking the supremum over unit vectors yields $\| T \|^2 \leq \| T^*T \|$. □

We now give a quick overview of the relationship between an operator and its adjoint.

Proposition 4.27. *For all* $T \in \mathcal{B}(\mathcal{H})$:

(*a*) $N(T^*) = R(T)^\perp$;
(*b*) $N(T)^\perp = \overline{R(T^*)}$.

Proof. $x \in R(T)^\perp$ if and only if for all $y \in H$ we have $\langle x, Ty\rangle = 0$, which occurs if and only if $\langle T^*x, y\rangle = 0$, which occurs if and only if $x \in N(T^*)$, which proves (a). By applying (a) with T^* in place of T and by invoking Proposition 4.15 and Theorem 4.26(e), we obtain (b). □

Definition. The *support* of $T \in \mathcal{B}(H)$, denoted by $S(T)$, is defined to be $N(T)^\perp$.

Proposition 4.27 implies that $S(T^*) = \overline{R(T)}$. By referring back to Theorem 4.14 (d), we see that for any $T \in \mathcal{B}(H)$ we get two direct-sum decompositions of H, namely, $N(T) \oplus S(T)$ and $N(T^*) \oplus S(T^*)$. The operator T maps $N(T)$ into 0 and maps $S(T)$ injectively onto a dense subspace (possibly all) of $S(T^*)$, while T^* maps $N(T^*)$ into 0 and maps $S(T^*)$ injectively onto a dense subspace of $S(T)$.

For all $T \in \mathcal{B}(H)$, we let T^r denote the restriction of T to its support. The remark above shows that T^r is an injective element in $\mathcal{B}(S(T), S(T^*))$, and T^{*r} is an injective element in $\mathcal{B}(S(T^*), S(T))$. A symbolic picture of the situation is shown in Figure 4.4.

4.6.4 Normal Operators

In this section, we introduce a particularly tractable class of operators.

Definition. An element $T \in \mathcal{B}(H)$ is said to be *normal* if $TT^* = T^*T$.

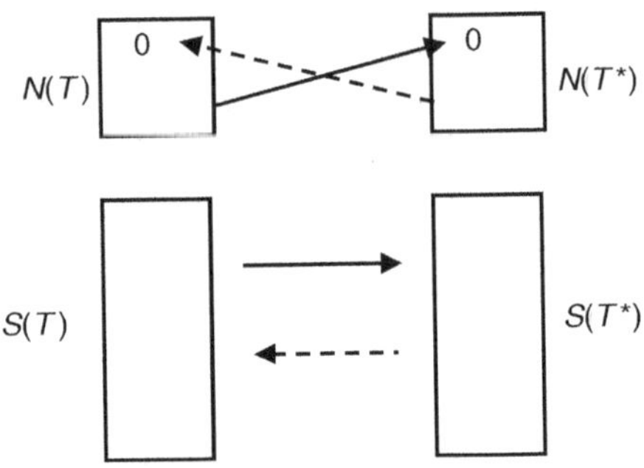

Figure 4.4. Symbolic action of $T \rightarrow$ and $T^* \leftarrow --$.

The word "normal" is used in almost every branch of mathematics to describe objects that are easy to deal with or have nice properties. It is admittedly not at all obvious from the definition why this name is warranted by operators that commute with their adjoint. We hope to make this clear in the remainder of the section. The best place to start is with an example.

Example 4.28. Consider the multiplication operator M_f, for $f \in \ell^\infty$, defined on $H = \ell^2$ by $M_f(g) = fg$; see Exercise 1.3. An easy calculation shows that $M_f^* = M_{\bar{f}}$, in which $\bar{f}$ is defined by $\bar{f}(n) = \overline{f(n)}$. Because $M_g M_h = M_h M_g$ for all g and h, it follows that M_f is normal.

This result need not be confined to sequence spaces, and the same holds where $f \in L^\infty$ and H is the L^2 space of any measure space.

Remark. This example is key because a celebrated result known as the *spectral theorem* implies that any normal operator can be viewed as a multiplication operator. We discuss this in Chapters 9 and 11. We can then identify normal operators with the functions that generate them, and it greatly simplifies their study.

We have another case here for which, after giving a definition and an example, it is desirable to give an example of something that does *not* satisfy the definition.

Example 4.29. An example of an operator that is not normal is the *right-shift operator*, which has appeared in exercises of previous chapters. Pick an orthonormal basis (x_n) of H. The right shift S is that linear operator that maps x_n to x_{n+1} for all n. It follows that

$$S\left(\sum_{i=1}^{\infty} \alpha_i x_i\right) = \sum_{i=2}^{\infty} \alpha_{i-1} x_i.$$

The *left-shift operator* T maps x_1 to 0 and x_n to x_{n-1} for $n > 1$. For any two indices m and n, we have $\langle Sx_m, x_n \rangle = \langle x_m, Tx_n \rangle = 0$, unless $m = n - 1$, in which case $\langle Sx_m, x_n \rangle = \langle x_m, Tx_n \rangle = 1$. From linearity and continuity of the inner product, it follows that for all $y, z \in H$ we have $\langle Sy, z \rangle = \langle y, Tz \rangle$, which implies that $T = S^*$. Now, S^*S is the identity operator I. On the other hand,

SS^* acts as the identity on the subspace $\{x_1\}^\perp$ but maps x_1 to 0. This shows that both S and T are not normal.

There is an alternate characterization of normality, which is often very useful.

Proposition 4.30. *T is normal if and only if $\| Tx \| = \| T^*x \|$ for all $x \in H$.*

Proof. This result follows from Corollary 4.25. We have $\langle T^*Tx, x\rangle = \langle Tx, Tx\rangle = \| Tx \|^2$, and similarly $\langle TT^*x, x\rangle = \| T^*x \|^2$. □

Remark. A consequence of the above proposition is that a normal operator T and its adjoint T^* have the same null space and therefore the same support. By referring back to the description given after Proposition 4.26, we only get one direct-sum decomposition, and each of T and T^* maps $S(T)$ onto a dense subspace of itself.

The set $\mathcal{B}(H)$ has four major operations of interest: addition, multiplication by a scalar, composition, and taking adjoints. It is of interest to determine how any given class of operators behaves with respect to these. It is obvious that the adjoint or any scalar multiple of a normal operator is normal. It is not true that the sum or composition of normal operators is normal. (One can find a counterexample already in a two-dimensional space by taking multiplication operators with respect to two different bases.)

There are several subclasses of the normal operators that are of interest.

Self-Adjoint Operators

Definition. $T \in \mathcal{B}(H)$ is said to be *self-adjoint* if $T = T^*$.

The multiplication operators that are self-adjoint are precisely those of the form M_f with f real valued; see Example 4.28.

We also have a characterization of self-adjointness in terms of the quadratic form.

Proposition 4.31. *T is self-adjoint if and only if $\langle Tx, x\rangle$ is real for all x.*

Proof. This result follows from the fact that a complex number is real if and only if it is equal to its conjugate and from

$$\langle T^*x, x\rangle = \langle x, Tx\rangle = \overline{\langle Tx, x\rangle}.$$ □

As to the combinatorial properties, the sum of two self-adjoint operators is self-adjoint, and a *real* scalar multiple of a self-adjoint operator is self-adjoint. The composition of two self-adjoint operators is self-adjoint if and only if they commute, as shown by Theorem 4.26(c).

Positive Operators

Definition. $T \in \mathcal{B}(H)$ is said to be *positive* if $\langle Tx, x\rangle \geq 0$ for all $x \in H$.

Remark. We allow $\langle Tx, x\rangle$ to be 0, so one may feel that the term *nonnegative* would be more appropriate. However, almost everyone uses the word "positive" here, and we follow this practice.

From Proposition 4.31, positive operators are necessarily self-adjoint. The positive multiplication operators are, naturally enough, those for which f takes nonnegative values. The definition shows directly that positivity is preserved by addition and by multiplying by a nonnegative scalar. There are many positive operators, as shown by the fact that for any T the operator T^*T is positive. The following is a useful result.

Theorem 4.32 *Given any positive operator T, there is a unique positive operator A such that $A^2 = T$. Moreover, A commutes with any operator that commutes with T.* □

We will give the proof of existence of such an operator A in Chapter 8, where we develop the necessary tools to do so. The proof of uniqueness is outlined in Exercise 4.11. The operator A is denoted by $T^{1/2}$. The conclusion is clear for multiplication operators, where it is obvious that $(M_f)^{1/2} = M_{f^{1/2}}$.

One immediate consequence of this theorem is the fact that the product of two commuting, positive operators S and T is positive. Indeed, since $T^{1/2}$ commutes with S, we have

$$\langle TSx, x\rangle = \langle T^{1/2}Sx, T^{1/2}x\rangle = \langle ST^{1/2}x, T^{1/2}x\rangle \geq 0.$$

Projections

Definition. $P \in \mathcal{B}(H)$ is said to be a *projection* if $P = P^*$ and $P^2 = P$.

A projection is necessarily positive since

$$\langle Px, x\rangle = \langle P^2x, x\rangle = \langle Px, Px\rangle \geq 0.$$

The multiplication operators M_f on sequence spaces that are projections are precisely those where f takes values of 0 or 1.

We have already encountered projection operators. They turn out to be precisely the operators P_M defined in Section 4.4.2, as we now show.

Proposition 4.33. *An element $P \in \mathcal{B}(H)$ is a projection if and only if there is a closed subspace M of H such that $P = P_M$.*

Proof. Given a projection P, it is easy to identify a qualifying space M. It must be precisely the support $S(P)$. For any $y = Px$ in $R(P)$, we have $Py = PPx = Px = y$. Since $R(P)$ is dense in M by the remark after Propostion 4.30, we have by continuity that $Py = y$ for all $y \in M = S(P)$. Now, P takes the value 0 on $M^\perp = N(P)$, so for any $x = y + z$ where $y \in M, z \in M^\perp$, we have $Px = y$, thereby showing that $P = P_M$.

Conversely, since P_M maps elements of M into themselves, it is clear that ${P_M}^2 = P_M$. Moreover, given $x_i = y_i + z_i$ with $y_i \in M$ and $z_i \in M^\perp$ for $i = 1, 2$, a direct calculation shows that both $\langle P_M x_1, x_2 \rangle$ and $\langle x_1, P_M x_2 \rangle$ equal $\langle y_1, y_2, \rangle$; thus, $P_M = P_M^*$. □

Mathematics is filled with examples (think of Cartesian coordinates) of describing geometric concepts algebraically. The correspondence between projections (given in algebraic terms) and closed subspaces is another example of this. We can carry this further. Suppose, for example, that $P = P_M$ and $Q = P_N$. We can verify the following facts relating a geometric statement about subspaces with an algebraic one about projection operators. We leave the verifications to the reader:

(a) $PQ = 0$ if and only if M and N are orthogonal, and in this case, $P + Q$ is a projection onto the closed subspace spanned by $M \cup N$.

(b) $PQ = P$ if and only if $M \subseteq N$, and in this case, $Q - P$ is the projection onto the subspace $N \cap M^\perp$.

These two facts are special cases of the following:

(c) PQ is a projection if and only if $PQ = QP$, and in this case, it is the projection onto the closed subspace $K = M \cap N$. This will occur if and only if the subspaces $M \cap K^\perp$ and $N \cap K^\perp$ are orthogonal.

Unitary Operators

Definition. $U \in \mathcal{B}(H)$ is said to be unitary if $U^*U = UU^* = I$.

A multiplication operator M_f is unitary if and only if f takes values with absolute value 1.

Proposition 4.34. *U is unitary if and only if it is a surjective isometry.*

Proof. Since $\langle U^*Ux, x \rangle = \langle Ux, Ux \rangle = \| Ux \|^2$, it is clear that $U^*U = I$ if and only if U is an isometry, the *if* part arising from Corollary 4.25. Moreover, if $UU^* = I$, then given any $y \in H$, we have $U(U^*y) = y$, thus showing that U is surjective.

Conversely, given that U is a surjective isometry, we know from the first part of the proof that $U^*U = I$. Additionally, for all $x \in H$, if $x = Uy$, then

$$UU^*x = UU^*Uy = Uy = x,$$

thus completing the proof that U is unitary. □

EXAMPLE **4.35.** The right shift S of Example 4.29 satisfies $S^*S = I$, but as shown, SS^* does not equal I, so S not unitary. We can also verify this from Proposition 4.34 since S is an isometry but not surjective.

On the other hand, there is a type of operator, similar to the right-shift operator, that is unitary. Suppose we have an orthonormal basis (x_n) indexed on all the integers (not just the positive ones), and we define an operator V to be the linear operator satisfying $Vx_n = x_{n+1}$ for all n. This is known as the *bilateral-shift operator*, and it is unitary.

Unitary operators play a role in defining a natural equivalence relation on $\mathcal{B}(H)$. Two operators S and T are said to be *unitarily equivalent* if $T = U^*SU$ for some unitary U. It is a similar concept to isomorphism. Two unitarily equivalent operators can be regarded as essentially the same in the sense that their actions on the elements of the space are indistinguishable. To illustrate, consider again Example 4.29, in which we defined the right-shift operator. An astute reader may want to take issue with the word *the*. There are many orthonormal bases and therefore, in one sense, many unilateral shifts. But, they are all unitarily equivalent. Any bijective mapping between the two sets of basis elements extends to a unitary operator, which will implement the equivalence.

An important example of a unitary operator arises in the case of certain transformations between measure spaces. Suppose we have a measure space $(\Omega, \mathcal{S}, \mu)$ and a mapping T from Ω to itself such that for all $B \in \mathcal{S}$ we have $T^{-1}(B) \in \mathcal{S}$. Define a new measure $\hat{\mu}$: by

$$\hat{\mu}(A) = \mu(T^{-1}(A)).$$

Integration with respect to this new measure is given in a natural way as

$$\int f\, d\hat{\mu} = \int f \circ T\, d\mu,$$

where $\circ$ denotes composition.

Example 4.36. Suppose that the transformation T given above is *measure preserving*. That is, for all $B \in \mathcal{S}$, we have $\mu(T^{-1}(B)) = \mu(B)$. In terms of our notation above, this just says that $\hat{\mu} = \mu$. Let U be the linear mapping on $L^2(\Omega)$ given by

$$Uf(x) = f(Tx) = (f \circ T)(x).$$

Then, for all $f, g \in L^2(X)$,

$$\langle Uf, Ug\rangle = \int (f \circ T)\,(\overline{g \circ T})\, d\mu = \int f\bar{g}\, d\hat{\mu} = \int f\bar{g}\, d\mu = \langle f, g\rangle,$$

thus showing that U is unitary.

Application: Ergodic Theory. Ergodic theory has its origins in statistical mechanics. In the classical model, one has k particles in three-dimensional space moving over time. The state of the system is then described by a $6k$-dimensional vector giving the position and velocity of the particles, so the collection of all these states forms a linear space. (We simplify matters by giving a discrete-time formulation.) For any state x, the postulate is that the system will be in a unique state one period later, which we denote by Tx. One is often interested in some function f of the state of the system, such as energy. The function $Uf = f \circ T$ then gives the value of the function one period later. Suppose one wants to compute the average value of the function f. We could formulate this as finding a limit in some sense of the quantity

$$\frac{1}{N}\sum_{k=0}^{N-1} U^k f. \tag{$*$}$$

Now, in many cases the state space comes equipped with a measure such that the transformation T is measure preserving. If the function is in L^2, the example above shows that U is unitary. John von Neumann, a major figure in the development of operator theory (as well as several other fields), noticed that the Hilbert space framework makes things very simple and proved the following.

Theorem 4.37 (Mean Ergodic Theorem). *If U is a unitary operator on a Hilbert space H, then for all $f \in H$ the quantity in $(*)$ converges to $P_M f$, in which $M = N(U - I)$.*

A proof is outlined in Exercise 4.17.

For more details on ergodic theory, see Reed and Simon (1972, Section II.5).

4.7 ORDER RELATION ON SELF-ADJOINT OPERATORS

The notion of positivity allows us to define a partial order (see Appendix A) on the set of all self-adjoint operators.

Given self-adjoint operators A and B, we say that $A \leq B$ if $B - A$ is positive. The fact that the sum of two positive operators is positive verifies transitivity, reflexivity is obvious, and antisymmetry follows from Corollary 4.25. In certain contexts, this order is quite natural.

For example, consider a multiplication operator M_f, in which f is real valued, acting on ℓ^2. Then, for any basis vector δ_n,

$$\langle M_f \delta_n, \delta_n \rangle = f(n),$$

which shows that

$$M_f \leq M_g \text{ if and only if } f \leq g.$$

Another example of interest concerns projections.

Proposition 4.38. *Given two closed subspaces M and N of H, we have $P_M \leq P_N$ if and only if $M \subseteq N$.*

Proof. Suppose $M \subseteq N$. Then, for any vector x,

$$\langle P_M x, x\rangle = \| P_M x \|^2 \leq \| P_N x \|^2 = \langle P_N x, x\rangle.$$

The inequality above follows since we can write $P_N x$ as an orthogonal sum

$$P_N x = P_M x + P_R x,$$

in which $R = M^{\perp} \cap N$.

Conversely, if M is not contained in N, choose a unit vector $x \in M$ but $x \notin N$. Then, $\langle P_M x, x\rangle = 1$, while $\langle P_N x, x\rangle = \| P_N x \| < 1$. □

It follows from the above proposition that the set of projection operators on a Hilbert space is a particularly nice type of poset known as a *lattice*. A lattice is a poset in which any two elements have a least upper bound and a greatest lower bound. By a least upper bound for elements x and y, we mean an element z with $x \leq z$ and $y \leq z$ such that $z \leq w$ for any other upper bound of x and y. A greatest lower bound is similarly defined with inequalities reversed. It follows by induction that any finite subset of a lattice has a greatest lower bound and a least upper bound. A lattice is said to be *complete* if *any* subset of the lattice has a least upper bound and greatest lower bound. The projection lattice is, in fact, complete. The least upper bound of any set of projections is the projection onto the closed linear span of the corresponding subspaces, and the greatest lower bound is the projection onto the intersection of the corresponding subspaces. The entire set of all self-adjoint operators is not a lattice; see Exercise 4.19.

EXERCISES

4.1 Show that in any inner product space two elements x and y are orthogonal if and only if $\| x + \alpha y \| = \| x - \alpha y \|$ for all $\alpha \in \mathcal{C}$.

4.2 Show that ℓ^p for $1 \leq p \leq \infty$ is not an inner product space, except for $p = 2$.

4.3 (Legendre polynomials) In $L^2[0, 1]$, apply the Gram–Schmidt orthogonalization procedure to the functions $\{1, t, t^2, \ldots\}$ to get an orthonormal basis $\{P_0, P_1, P_2, \ldots\}$. (Note that we start the indexing at 0.) Show that

$$P_n(t) = k_n \frac{d^n}{dt^n}(t^2 - 1)^n$$

for some constants k_n and calculate k_n.

4.4 Let (x_i) be an orthonormal set in a Hilbert space H.

(a) (Bessel's inequality) Show that for any $z \in H$

$$\| z \|^2 \geq \sum_{i=1}^{\infty} |\langle z, x_i \rangle|^2,$$

and equality holds for all z if and only if (x_n) is an orthonormal basis.

(b) (Parseval's identity) Show that for all $z, y \in H$

$$\langle z, y \rangle = \sum_{i=1}^{\infty} \langle z, x_i \rangle \langle x_i, y \rangle$$

if and only if (x_i) is an orthonormal basis.

4.5 Let (x_n) be a sequence in a Hilbert space H such that, for all $y \in H$, $\langle x_n, y \rangle$ converges to $\langle x, y \rangle$ and such that $\| x_n \|$ converges to $\| x \|$. Show that (x_n) converges to x.

4.6 (Hellinger-Toeplitz theorem) Let T be a linear operator from a Hilbert space H to itself satisfying

$$\langle Tx, y \rangle = \langle x, Ty \rangle$$

for all $x, y \in H$. Show that T is bounded.

4.7 If M is a closed subspace of the Hilbert space H, show that the space H/M with the quotient norm is a Hilbert space.

4.8 Show that U is a self-adjoint unitary operator if and only if $U = 2P - 1$ for some projection P.

4.9 An element U in $\mathcal{B}(H)$ is said to be a *partial isometry* if U^r [the restriction of U to $S(U)$] is an isometry.

(a) Show that for any partial isometry U, U^*U is the projection onto $S(U)$ and UU^* is the projection onto $R(U)$, which is closed.

(b) Show that any operator U such that U^*U is a projection is a partial isometry.

(c) Show that a partial isometry is a unitary if and only if $N(U) = 0$ and $R(U) = H$.

(d) Give an example of an injective partial isometry that is not unitary.

4.10 Let P be a positive operator in $\mathcal{B}(H)$. Define $\langle x, y \rangle_P = \langle Px, y \rangle$. Show that $\langle x, y \rangle_P$ is a semi-inner product and an inner product if P is injective.

4.11 (Uniqueness of positive square roots)

(a) Show that if C is self-adjoint, then $N(C^2) = N(C)$.

(b) Show that a self-adjoint operator C is positive if and only if C^r (the restriction to the support) is positive.

(c) Suppose that A and C are self-adjoint operators such that C commutes with A, $R(C) \subseteq R(A)$, and CA^2 is positive. Show that C is positive.

(d) Let A be a positive square root of the positive operator T such that A commutes with any operator that commutes with T. Let B be another positive square root of T. Noting that $(A - B)^2 = 2A(A - B) = 2B(B - A)$, show that $A = B$.

4.12 (Polar decomposition) For any $T \in \mathcal{B}(H)$, define the *absolute value of* T by $|T| = (T^*T)^{1/2}$.

(a) Show that $\| \, |T|x \, \| = \| Tx \|$ for all x.

(b) Show that the mapping U_0 defined on $R(|T|)$ by $U_0|T|(x) = Tx$ is well defined and extends to a partial isometry U such that $T = U|T|$.

(c) Show that if T is invertible, then U is unitary.

4.13 If T is a bijective self-adjoint operator on a Hilbert space, show that T^{-1} is self-adjoint.

4.14 Fill in the details of the proof of Theorem 4.8.

4.15 Consider the space $L^2[-\pi, \pi]$ with the orthonormal set

$$\left\{ z_n = \frac{1}{\sqrt{\pi}} \sin(nx), n = 1, 2, \ldots; \quad w_n = \frac{1}{\sqrt{\pi}} \cos(nx), n = 1, 2, \ldots, w_0 = \frac{1}{\sqrt{2\pi}} \right\}.$$

(a) Show that for any even function in the space [which means that $f(x) = f(-x)$ a.e.] $\langle f, z_n \rangle = 0$ for all n and for any odd function [which means that $f(x) = -f(-x)$ a.e.] we have $\langle f, w_n \rangle = 0$ for all n.

(b) Assuming the fact that the given set is an orthonormal basis, show that in the space $L^2[0, \pi]$ both$(\sqrt{2} z_n)$ and $(\sqrt{2} w_n)$ are orthonormal bases.

(c) Now assume that both sets given in (b) are orthonormal bases of $L^2[0, \pi]$. Use this to show that the given set is an orthonormal basis of $L^2[-\pi, \pi]$.

4.16 Let $z_n = (1/2\pi)e^{inx}, n = 0, \pm 1, \pm 2, \ldots$. Take it as given that (z_n) is an orthonormal basis for $L^2[-\pi, \pi]$. Consider the function $f(x) = x$ on $[-\pi, \pi]$.

(a) Calculate $\| f \|_2$.

(b) Calculate $\langle f, z_n \rangle$ for all n.

(c) Use your answers to (a) and (b) to derive the identity

$$\sum_{n=1}^{\infty} \frac{1}{n^2} = \frac{\pi^2}{6}.$$

4.17 (a) Show that for a unitary $U \in \mathcal{B}(H)$ and any $f \in R(U - I)$ the sum $\sum_{k=0}^{N-1} U^k(f)$ converges to 0 as N goes to ∞.

(b) Use part (a) and the remarks after Proposition 4.27 to prove Theorem 4.37.

4.18 Let $(\Omega, \mathcal{S}, \mu)$ be a measure space with $\mu(\Omega) = 1$. Let M be the subspace of $L^2(\Omega)$ consisting of those elements that are a.e. constant. Show that for all $f \in L^2(\Omega)$, the projection of f onto M is the function whose constant value a.e. is $\int f d\mu$.

4.19 Let M and N be two orthogonal one-dimensional subspaces in a two-dimensional Hilbert space. Show that, given any self-adjoint operator A such that P_M and P_N are both less than A, there is a self-adjoint $B \leq A$, but different from A, such that P_M and P_N are both less than B. Conclude that the set of all self-adjoint operators is not a lattice.

5

HAHN–BANACH THEOREM

5.1 INTRODUCTION

This chapter is devoted to the various forms of the Hahn–Banach theorem, another of the most fundamental results in functional analysis. It was proved independently by both Banach and the German mathematician Hans Hahn in the late 1920s. It differs from the theorems of Chapter 3 since it does not involve completeness and, in its most general form, does not even involve a norm.

One of the main applications of this theorem is concerned with dual spaces. In Chapter 4, we studied the dual of a Hilbert space and determined that it was a Hilbert space of the same dimension as the original. We can also determine the dual spaces for many of the basic examples of Banach spaces that we introduced in Chapter 2, and we will do so in Section 6.1 of the following chapter. However, a fact that is somewhat surprising at first is that on a general normed linear space, in contrast to a Hilbert space, there is no obvious way to see the existence of any nonzero bounded linear functional. For all we know, the dual space could reduce to the space consisting of the single element 0, and there would be no information obtained by using the dual. The Hahn–Banach theorem comes to the rescue and ensures that any normed linear space will admit plenty of bounded linear functionals, a statement that we will soon make precise. This allows a fruitful interplay between a space and its dual space, which we explore in this and later chapters.

In addition, the Hahn–Banach theorem is a key ingredient in many applications, where it is necessary to produce appropriate linear functionals. Such applications include probability theory, where expectations can be viewed as linear functionals on spaces of random variables, and economics and finance, where prices are often modeled by linear functionals.

A First Course in Functional Analysis. By S. David Promislow

5.2 BASIC VERSION OF HAHN–BANACH THEOREM

We consider now a linear space X over $\mathcal{R}$. We restrict the field in this case because we wish to use inequalities, and there is no reasonable way to order the complex numbers. We also equip the space with a certain type of real-valued function, which is more general than a norm.

Definition. A *Minkowski functional* on a linear space X is a real-valued function p such that for all $x, y \in X$:

(i) $p(x+y) \leq p(x) + p(y)$;
(ii) $p(\alpha x) = \alpha p(x)$ for all $\alpha \geq 0$.

In particular, seminorms are Minkowski functionals. A Minkowksi functional differs from a seminorm in that it can take on any real numbers as values rather than nonnegative ones, and the scalar property refers only to nonnegative values. Note that any linear functional is also a Minkowski functional.

The most basic form of the Hahn–Banach theorem is as follows.

Theorem 5.1 (Hahn–Banach Extension Theorem). *Suppose that p is a Minkowski functional on a real linear space X and f_0 is a linear functional defined on a subspace Y such that $f_0(y) \leq p(y)$ for all $y \in Y$. Then, f_0 has an extension to a linear functional f on X such that $f(x) \leq p(x)$ for all $x \in X$.*

Proof. The proof is in two stages. In the first stage, the heart of the proof, we show that we can extend the functional to a space with one more dimension. The intuitive idea is then that we just keep on extending one dimension at a time until we finally have extended the functional to the whole space. This is fine in a finite-dimensional setting, but with infinite dimensions, how do we know we will ever finish? The formal method of showing this is through Zorn's lemma.

STAGE 1: Take any $x \notin Y$. We will show how to extend f_0 to $\text{span}(Y \cup x)$. This is the space of all elements that can be written as $y + \alpha x$ for some $y \in Y$ and $\alpha \in \mathcal{R}$. Moreover, the y and α are unique since, if not, we would necessarily have $x \in Y$. There is no problem in finding general extensions in this case. The only possibilities are to choose a scalar r and to define $f(y + \alpha x) = f_0(y) + \alpha r$. The trick is to pick r so that for all $y \in Y$ and $\alpha \in \mathcal{R}$ we have

$$f_0(y) + \alpha r \leq p(y + \alpha x).$$

As a simplification, we note that from the scalar multiplication properties of both f_0 and p we only have to satisfy the above for $\alpha = \pm 1$, since if that holds, then for any $\alpha > 0$, we would have

$$f_0(y) \pm \alpha r = \alpha\left[f_0\left(\frac{y}{\alpha}\right) \pm r\right] \leq \alpha p\left(\frac{y}{\alpha} \pm x\right) = p(y \pm \alpha x).$$

Satisfying the inequality for $\alpha = \pm 1$ is accomplished by choosing r such that for all $y \in Y$

$$f_0(y) - p(y - x) \le r \le p(y + x) - f_0(y),$$

which we do by choosing any r between $\sup(A)$ and $\inf(B)$, where

$$A = \{f_0(y) - p(y - x) : y \in Y\}, \qquad B = \{p(y + x) - f_0(y) : y \in Y\}.$$

Of course, we have to ensure that $\sup(A) \le \inf(B)$. To this end, we use the triangle inequality for p. Given any y, $z \in Y$,

$$f_0(y) + f_0(z) = f_0(y + z) \le p(y + z) = p(y + x + z - x) \le p(y + x) + p(z - x),$$

which shows that any element in A is less than any element in B, thereby verifying our claim.

STAGE 2: We now must invoke Zorn's lemma. (See Appendix A.) Define an admissible extension of f_0 to consist of (Z, g), in which Z is a subspace of X containing Y and g is a linear functional on Z that extends f_0 and is dominated by p. The set of admissible extensions is clearly nonempty since it trivially contains (Y, f_0).

Define a partial order on the set of all admissible extensions by $(Z_1, g_1) \le (Z_2, g_2)$ if Z_1 is a subspace of Z_2 and g_2 extends g_1. Now, suppose we are given any linearly ordered subset $\mathcal{V}$ of the set of admissible extensions. Let $W = \bigcup\{V : (V, g) \in \mathcal{V}\}$, and define a linear functional h on W as follows. For any x in W, if $x \in V$ where $(V, g) \in \mathcal{V}$, let $h(x) = g(x)$. We must verify that this all makes sense, and this is where the linearity of the ordering comes in. We use it first to show that W is indeed a linear subspace; see Exercise 1.7. Then we use it to show that h is well defined. Indeed, if $(V_1, g_1) \le (V_2, g_2)$ and if x belongs to both V_1 and V_2, then we have defined $h(x)$ to be both $g_1(x)$ and $g_2(x)$, but these are the same since g_2 extends g_1.

Clearly, then, (W, h) is an upper bound for $\mathcal{V}$. By Zorn's lemma, there is a maximal admissible extension (Z, g). We claim that Z must be all of X, which will complete the proof. If this were not the case, we could find $x \notin Z$ and then apply the technique in stage 1 to extend g to $\text{span}(Z \cup x)$. This would yield a larger admissible extension, thereby contradicting the maximality of (Z, g).

Remark. A natural question to ask whenever we have an extension theorem is whether or not the extension is unique. Uniqueness might hold, but the proof of the theorem indicates that we should not expect it. We would need that sup $A = \inf B$ at each step, where A and B are the sets defined in stage 1. For some simple examples, see Exercise 5.9.

5.3 COMPLEX VERSION OF HAHN–BANACH THEOREM

Theorem 5.1 involves ordering and does not make sense when applied to a complex space. However, we do get a complex version by replacing the Minkowski functional with a seminorm and by replacing domination of the function with domination of its absolute value.

Let X be a linear space over $\mathcal{C}$. We must first clarify the relationship between two spaces of linear functionals. The first is X^c, the linear mappings from X to $\mathcal{C}$. But, we can also view X as a real linear space by restricting the scalar multiplication to the reals. We now get the space X^r, which are the linear mappings from X to $\mathcal{R}$, where the scalar requirement of linearity is with respect to real scalars only. How are these two spaces related? It is natural to expect that for a mapping f from X to $\mathcal{C}$ to be linear with respect to $\mathcal{C}$, we need both $\operatorname{Re} f$ and $\operatorname{Im} f$ to be linear, but that is not enough. We must add a condition as given in the following.

Proposition 5.2. *A function f from X to $\mathcal{C}$ is in X^c if and only if $\operatorname{Re} f$ and $\operatorname{Im} f$ are both in X^r and, for all $x \in X$, $\operatorname{Im} f(x) = -\operatorname{Re} f(ix)$.*

Proof. Let g and h denote $\operatorname{Re} f$ and $\operatorname{Im} f$, respectively. Suppose $f \in X^c$. It is clear that g and h are linear and, moreover, that for all $x \in X$

$$g(ix) + ih(ix) = f(ix) = if(x) = i(g(x) + ih(x)) = ig(x) - h(x).$$

Equating real parts yields the condition on g and h.

Conversely, suppose that $g, h \in X^r$ and that the condition holds. It is straightforward to show that f is additive and respects scalar multiplication for real scalars. To show that it is in X^c, it is sufficient to show that $f(ix) = if(x)$. This follows from

$$\begin{aligned} f(ix) &= g(ix) + ih(ix) = -h(x) + -ig(iix) \\ &= -h(x) + ig(x) = i[g(x) + ih(x)]. \end{aligned}$$

□

Theorem 5.3 (Complex Hahn–Banach Extension Theorem). *Suppose that $\| \cdot \|$ is a seminorm on a complex linear space X and that f_0 is a linear functional defined on a subspace Y such that $|f_0(y)| \leq \| y \|$ for all $y \in Y$. Then, f_0 has an extension to a linear functional f on X such that $|f(x)| \leq \| x \|$ for all $x \in X$.*

Proof. Let $g_0 = \operatorname{Re} f_0$. Since $g_0(y) \leq |g_0(y)| \leq |f_0(y)| \leq \| y \|$, we can, by the original theorem, extend g_0 to a real linear functional g on X satisfying $g(x) \leq \| x \|$ for all $x \in X$. Define a mapping f from X to $\mathcal{C}$ by $f(x) = g(x) + ig(-ix)$. By Proposition 5.2, f is linear and extends f_0. Given any $x \in X$, write $f(x)$ in the polar form $re^{i\Theta}$, where $r \geq 0$. Then,

$$|f(x)| = r = e^{-i\Theta} f(x) = f(e^{-i\Theta} x).$$

Because the last term is real, we can write

$$|f(x)| = g(e^{i\Theta} x) \leq \| e^{i\Theta} x \| = \| x \| .$$

□

Note that Theorem 5.3 remains true for a real space. In that case, the conclusion follows immediately from Theorem 5.1 by using the fact that

$$-f(y) = f(-y) \leq \| -y \| = \| y \| .$$

5.4 APPLICATION TO NORMED LINEAR SPACES

We now apply our theorems to the dual space X^*, where X is a normed linear space over either field, $\mathcal{R}$ or $\mathcal{C}$.

Theorem 5.4 (Normed Linear Space Version of Hahn–Banach Extension Theorem). *Suppose that f_0 is a bounded linear functional defined on a subspace Y of a normed linear space X. Then, f_0 has an extension to a bounded linear functional f on all of X such that $\| f \| = \| f_0 \|$.*

Proof. Take the seminorm defined by $x \to \| f_0 \| \| x \|$, and apply Theorem 5.3. We obtain a linear functional f such that $|f(x)| \leq \| f_0 \| \| x \|$, thereby showing that $\| f \| \leq \| f_0 \|$. Conversely, since f is an extension of f_0, we must have $\| f \| \geq \| f_0 \|$ because in the definition of $\| f \|$ we are taking the sup over a bigger set. □

The following corollaries make precise our intuitive statement that there are plenty of elements in X^* as well as indicate ways in which we can use X^* to get information about X.

Corollary 5.5. *Given any closed subspace Y of a normed linear space X and $x \notin Y$, there is $f \in X^*$ such that $f(Y) = 0$ and $f(x) = 1$.*

Proof. Define f_0 on $\text{span}(Y \cup x)$ by $f_0(y + \alpha x) = \alpha$ for all $y \in Y$. The null space of f_0 is Y, which is closed in X, so it is clearly closed in any subspace containing Y. By Theorem 2.32, f_0 is bounded. Finally, apply Theorem 5.4 to obtain f. □

A typical way of using the above corollary is as follows. Suppose we have a point x and we want to show that it is in the closure of a subspace Y. Often, the easiest method is to show that any bounded linear functional that takes the value of zero on all of Y will also take the value of zero on x.

Corollary 5.6. *Given a normed linear space X and two points $x \neq y$, there is $f \in X^*$ such that $f(x) \neq f(y)$.*

Proof. Applying Corollary 5.5 to the 0 subspace and the point $x - y$ produces f such that $f(x) - f(y) = f(x - y) \neq 0$.

The condition expressed in the last corollary is often referred to by saying that *X^* separates points in X*. This is a precise version of our intuitive comment that there are many elements in X^*.

The norm of an element in X^* is defined in terms of its value on the unit ball of X. We can now get a similar statement about the norm of an element in X, a sort of reversal of the roles of the two spaces, which is typical of duality theories in mathematics.

Corollary 5.7. Given any $x \in X$,

$$\| x \| = \sup\{|f(x)| : f \in \textit{ the closed unit ball of } X^*\}.$$

Proof. It is apparent that $\| x \|$ is greater than the right-hand side. To show it is less, define a functional f_0 on span(x) by $f_0(\alpha x) = \alpha \ \| x \|$. This functional is clearly of norm 1. By Theorem 5.4, extend f_0 to a unit vector $f \in X^*$. Therefore, the right-hand side is greater than $|f(x)| = \| x \|$.

5.5 GEOMETRIC VERSIONS OF HAHN–BANACH THEOREM

5.5.1 Introduction

Throughout this section, we will again assume that we have a real linear space so that we can use the ordering on the real line.

To motivate the results of this section, draw two disjoint sets in $\mathcal{R}^2$. Can you draw a straight line between them? A moment's reflection suggests that you indeed can if they are both convex but not necessarily if one is not convex; see Figures 5.1*a* and *b*. You are no doubt wondering what this has to do with the Hahn–Banach theorem. The basic idea is that straight lines correspond to linear functionals and certain convex sets correspond to Minkowski functionals. We develop the general theory in the following sections.

5.5.2 Hyperplanes and Linear Functionals

Recall from Section 2.7 that a *linear manifold* in a linear space X is a set of the form $x + Y$, where Y is a subspace of X.

Definition. A linear manifold for which Y is a maximal proper subspace is called a *hyperplane*. That is, Y is not the whole space, and the only subspaces containing Y are Y itself or X. For example, in $\mathcal{R}^3$, a hyperplane is a plane, while in $\mathcal{R}^2$, a hyperplane is a straight line.

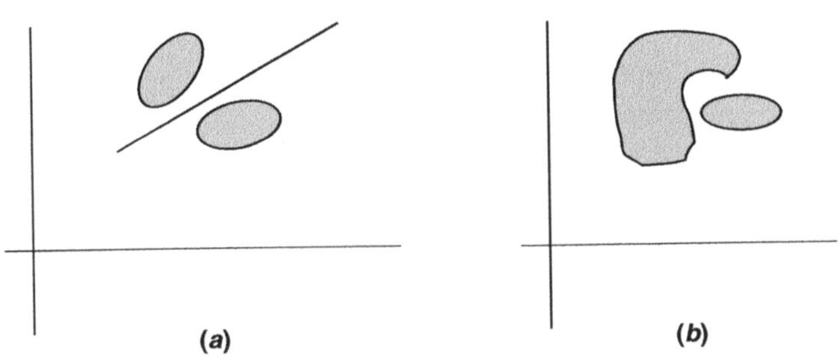

Figure 5.1 (*a*) Separation of two convex sets. (*b*) Separation is not possible.

A natural source of hyperplanes is the null space of linear functionals. Let f be a nonzero linear functional with null space N. By choosing any $x \notin N$, we can write for any $y \in X$

$$y = \left(y - \frac{f(y)}{f(x)} x \right) + \frac{f(y)}{f(x)} x.$$

Since the first term is in N, we know that $\operatorname{span}(N \cup x) = X$. Since this is true for all $x \notin N$, we have shown that N is maximal. For any $x \in X$ with $f(x) = \alpha$, the hyperplane $x + N = f^{-1}(\alpha) = \{x : f(x) = \alpha\}$.

Conversely, any hyperplane H has such a representation. If H is not a subspace, we can write $H = x + Y$, where Y is a maximal subspace and $x \notin Y$. By maximality, $\operatorname{span}(Y \cup x) = X$, and we can define a linear functional f by $f(y + \alpha x) = \alpha$ for all $y \in X$. It is clear that $H = f^{-1}(1)$. In the case for which H is a subspace, we can choose any $x \notin H$ to define f, and $H = f^{-1}(0)$.

The functional corresponding to a hyperplane is unique up to scalar multiplication. We have the two cases above to consider: (1) $H = f^{-1}(1)$ and (2) $H = f^{-1}(0)$. First, suppose that $x + Y = H = f^{-1}(1)$ also equals $g^{-1}(\beta)$ for some linear functional g. Then, for any $v, w \in Y$, $\beta = g(x + v) = g(x + w)$, which implies that g is constant on Y. Because $0 \in Y$, this constant must be 0. It follows that $g(x) = \beta$ and $g = \beta f$. Second, if H is a subspace, then for any linear functional g such that $H = g^{-1}(0)$, let $z \notin H$. Because H is maximal, we have $X = \operatorname{span}(H \cup z)$. It follows that $g = (g(z)/f(z))f$.

Definition. Given a hyperplane $H = f^{-1}(\alpha)$, consider the two subsets $\{x : f(x) \leq \alpha\}$ and $\{x : f(x) \geq \alpha\}$. These are known as the *half spaces determined by H*. If we take strict inequalities instead, the sets are called the *open* half spaces determined by H. The half spaces are well defined; see Exercise 5.1.

Definition. We say that two subsets A and B of X are *separated* by H if they lie in different half spaces and *strictly separated* if they lie in different open half spaces.

We have already seen the idea of separation. The motivating question posed at the beginning of Section 5.5.1 dealt with separation of sets by a hyperplane. Moreover, Corollary 5.5 says that in a normed linear space a closed subset and a point x not in the set can be separated by a hyperplane.

In a normed linear space, there is an interesting dichotomy of hyperplanes. Given a hyperplane $H = x + Y$, we can verify, by continuity of the operations, that $\bar{H} = x + \bar{Y}$ and that $\bar{Y}$ is a linear subspace. By maximality, $\bar{Y}$ is either Y or all of X, showing that hyperplanes are either dense (and not closed) or closed. From Theorem 2.32, we can see that in the correspondence given above with linear functionals the closed hyperplanes correspond to bounded linear functionals and the dense (not closed) hyperplanes to unbounded linear functionals.

There is an alternate proof to Theorem 2.32 that shows that linear functionals with closed null spaces are bounded, and this gives additional information about the norm of the functional; see Exercise 5.2.

5.5.3 Internal and Bounding Points

For any subset of a linear space, the linear structure alone induces a classification of points that is similar to that of the interior and boundary in normed linear spaces.

Definition. For any subset A of a linear space, a point $a \in A$ is said to be *internal* to A if given a vector x there exists $r_x > 0$ such that $a + tx \in A$ for $0 \le t \le r_x$. In other words, if you are placed at an internal point and pointed in any direction whatsoever, you can walk some positive distance in that direction and remain in the set A.

It is instructive to compare this definition with interior points in a normed linear space. If a is an interior point of A, it is clearly internal, for if $B(a; r) \subseteq A$, we can take $r_x = r/ \parallel x \parallel$ for all nonzero x. The converse is not true, however, as the following example shows.

Example 5.8. In $\mathcal{R}^2$, let B consist of all points whose polar coordinates (r, θ) satisfy $r < 1 - \cos(\theta)$. B is the inside of the familiar cardioid. Let A consist of B together with the line segment $[0, 1]$ on the $\theta = 0$ ray. Then, 0 is internal to A. However, it is not interior since r_x gets arbitrarily small as the direction gets close to the horizontal.

Internal points behave very much like interior points when the set is convex. In fact, they are the same in finite-dimensional spaces.

Proposition 5.9. *For a convex set in $\mathcal{R}^n$, all internal points are interior.*

Proof. We can equip $\mathcal{R}^n$ with any norm since we know all norms on $\mathcal{R}^n$ are equivalent by Theorem 2.31(b). In particular, we will view $\mathcal{R}^n$ as the real-valued functions on n points and assume the sup norm. Suppose that a is an internal point of a convex set A in $\mathcal{R}^n$. Let $r = \min\{r_{\delta_i}, r_{-\delta_i} : i = 1, 2, \ldots, n\}$. We claim that $B(a; r/n) \subseteq A$. If $v = \sum_{i=1}^n \alpha_i \delta_i$ has norm less than r/n, then

$$a + v = \frac{1}{n} \sum_{i=1}^{n} (a + n\alpha_i \delta_i).$$

Each summand is in A because each $n|\alpha_i| \le r$, which is less than or equal to both r_{δ_i} and $r_{-\delta_i}$. By convexity, the whole sum is in A.

It is not always true that in an infinite-dimensional normed linear space, an internal point of a convex set is interior. [See Royden (1988, Problem 10.44) for some sufficient conditions; see also Exercise 5.16.] However, in a general linear space, the internal points of convex sets satisfy a similar condition to interior points. Let A^o denote the internal points of a set A, so that A^{oo} is the set of internal points of A^o. Now, in any normed linear space the set of interior points of any set is open, so by taking the interior again, we obtain the same set. For internal points, however, A^{oo} can be strictly smaller than A^o. This is shown by the cardioid example given above, where the origin is an internal point of A but not an internal point of A^o. But, the situation is nicer with convexity.

Proposition 5.10. *If A is convex, then $A^{oo} = A^o$.*

Proof. Let $a \in A^o$, and choose r_y for each unit vector y as in the definition of internal points. We now claim that for any unit vector $y, a + ty \in A^o$ for all $0 \leq t \leq r_y/2$, and this will show that a is in A^{oo}. To show the claim, choose any unit vector z. Then, for $0 \leq s \leq r_z/2$,

$$a + ty + sz = \tfrac{1}{2}[(a + 2ty) + (a + 2sz)],$$

which is in A due to convexity. □

We close this section with two definitions that we use later.

Definitions. A point a is said to be *external* to the set A if it is internal to A^c. A point that is neither an internal point nor an external point of A is said to be a *bounding point* of A.

In a normed linear space, any bounding point is a boundary point, but the converse is not true as the cardioid example shows, since the origin is both a boundary point and internal to the given set.

5.5.4 Minkowski Functionals and Convex Sets

In this section, we establish a connection between Minkowki functionals and convex sets that have 0 as an internal point.

Proposition 5.11. *Given a Minkowski functional p, let $K_p = \{x : p(x) < 1\}$. Then, K_p is convex, and 0 is an internal point of K_p.*

Proof. Convexity follows directly from the definition of a Minkowski functional. To show the second statement, choose for $x \in X$ a number $r_x > 0$ such that $r_x p(x) < 1$. Then, for $0 \leq t \leq r_x$, if $px \geq 0$,

$$p(0 + tx) = t p(x) \leq r_x p(x) < 1,$$

so that $tx \in K_p$. This is immediate if $p(x) < 0$. □

Conversely, we can produce a Minkowski functional from convex sets. Let K be convex with 0 as internal point. For all $x \in X$, define

$$p(x) = \inf\{t > 0 : t^{-1}x \in K\}.$$

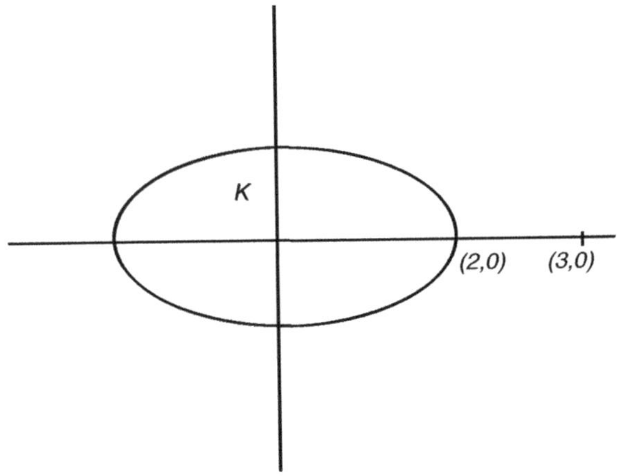

Figure 5.2 $p(3,0) = 1.5$.

In view of the fact that 0 is an internal point of K, the set of reals in question is nonempty, so that $p(x)$ is a number in $[0, \infty)$.

We can view $p(x)$ as measuring the "distance" of x from 0, but the unit of measurement varies from point to point. The unit of measurement for x is the line segment from 0 to the boundary of K in the direction of the point x. For example, suppose in $\mathcal{R}^2$ that K contains $(2,0)$ and therefore all points of the form $(t,0)$, for which $0 \leq t \leq 2$, but does not contain any other points on the positive horizontal axis. What is $p(z)$ for $z = (3,0)$? Now, $tz \in K$ for t between 0 and $\frac{2}{3}$. The set of scalars in the definition is therefore the reciprocals of these values, which is the interval $[\frac{3}{2}, \infty)$. So, $p(z) = \frac{3}{2}$. This reflects the fact that the line segment joining 0 to z is one and a half times as long as the part of the segment going from 0 to the set boundary; see Figure 5.2. Note that if K is unbounded in the direction going from 0 to x, then $p(x) = 0$.

Proposition 5.12

(a) *p is a Minkowski functional.*
(b) *$p(x) < 1$ if and only if x is an internal point of K.*
(c) *$p(x) = 1$ if and only if x is a bounding point of K.*

Proof

(a) It is clear that $p(\alpha x) = \alpha p(x)$ for all $\alpha \geq 0$. Given any $x, y \in X$, suppose we have positive t and s such that $t^{-1}x$ and $s^{-1}y \in K$. Then,

$$(t+s)^{-1}(x+y) = \frac{t}{t+s}\, t^{-1}x + \frac{s}{t+s}\, s^{-1}y,$$

which is in K by convexity. Therefore, $p(x+y) \leq t+s$. By taking the infimum over all such t and s, we obtain $p(x+y) \leq p(x)+p(y)$.

As a first step toward proving (b) and (c), we note that $x \in K$ implies, directly from the definition of p, that $p(x) \leq 1$. Conversely, if $x \notin K$, then, for all $t > 1, tx \notin K$, for if so, this would violate convexity because $0 \in K$. Hence, $s^{-1}x \in K$ implies that $s^{-1} \leq 1$, and we can conclude that $p(x) \geq 1$.

(b) If x is internal to K, choose $t > 0$ such that $x + tx \in K$. This shows that $p(x) \leq (1+t)^{-1} < 1$. Conversely, suppose that $p(x) < 1$. Given any $y \in X$, choose $r_y > 0$ so that $r_y p(y) < 1 - p(x)$. Then, for $0 \leq t \leq r_y$, we have $p(x+ty) \leq p(x) + r_y p(y) < 1$, and from the above remarks, it follows that $x + ty \in K$. Therefore, x is internal.

(c) Given (b), the statement is equivalent to proving that x is external to K if and only if $p(x) > 1$. If x is external to K, choose $0 < t < 1$ such that $x - tx \notin K$. By the remarks above, we know that $(1-t)p(x) = p((1-t)x) \geq 1$. So, $p(x) \geq 1/(1-t) > 1$. Conversely, suppose that $p(x) > 1$. For any $y \in X$, choose $r_y > 0$ so that $r_y p(-y) < p(x) - 1$. Then, for $0 \leq t \leq r_y$,

$$p(x) = p(x + ty + t(-y)) \leq p(x+ty) + tp(-y),$$

thereby showing that

$$p(x+ty) \geq p(x) - tp(-y) \geq p(x) - r_y p(-y) > 1.$$

By the remarks above, $x + ty \notin K$; therefore, x is external. □

In what follows the Minkowski functional p constructed as above from the set K will be denoted by p_K.

The following is a useful application of the results above.

Proposition 5.13. *Let K be a convex set which has some internal point, and let f be a linear functional on X. Then, $f(K^o)$ is the interior of the interval $f(K)$.*

Proof. Since K is convex, it is easy to see that $f(K)$ is a convex subset of the reals and therefore must be an interval I. Let a be the left endpoint of I and b the right endpoint of I. (These could take values of $\pm\infty$.) We will show that $f(K^o) = (a,b)$. We can assume, by translating if necessary, that 0 is internal, so that we can make use of the function p_K.

Suppose $x \in K^o$. Choose $t > 0$ such that $x + tx \in K$; then,

$$(1+t)f(x) = f((1+t)x)) \leq b,$$

thereby implying that $f(x) < b$. Similarly, we can show that $f(x) > a$, so that $f(K^o) \subseteq (a,b)$. Conversely, suppose that $c \in (a,b)$. Assume first that $c \geq 0$. Pick c_1 such that $c < c_1 < b$, and choose x in K such that $f(x) = c_1$. Let $z = (c/c_1)x$. Since $p_K(x) \leq 1$, we have

$$p_K(z) = \left(\frac{c}{c_1}\right) p_K(x) < 1.$$

If $c < 0$, pick c_1 in the interval (a, c) and argue similarly. By Proposition 5.12(b), z is internal, and $f(z) = (c/c_1)f(x) = c$. □

Note that the proposition implies, in particular, that $f(K^o)$ can never consist of a single point.

Remark. Sets K (not necessarily convex) with 0 as an internal point are sometimes called *absorbing*. The terminology comes from the fact that, given any $x \in X$, we can find $t > 0$ such that $tx \in K$, which means that $x \in t^{-1}K$. In other words, by multiplying K by a sufficient large scalar multiple, we can "absorb" any other point in the space.

5.5.5 Geometric Hahn–Banach Theorems

We can now apply our correspondences to formulate the Hahn–Banach theorem in terms of hyperplanes and convex sets. These constitute the purest form of the Hahn–Banach theorem because they do not involve norms, seminorms, or any other extraneous concerns. They make it clear that this is essentially a theorem about linear spaces only. For simplicity, we again assume real scalars in this section.

Theorem 5.14 (Geometric Hahn–Banach Extension Theorem). *Let K be a convex set with an internal point, and let P be a linear manifold such that $P \cap K^o = \emptyset$. Then, there is a hyperplane H containing P such that $H \cap K^o = \emptyset$.*

Proof. We can assume, by translating if necessary, that 0 is an internal point of K. (That is, if v is internal to K, replace K by $K - V$.) Let $P = x_0 + Y$, where Y is a subspace. We know that $x_0 \neq 0$, since if so, 0 would be in P, thereby contradicting our hypothesis.

Define a linear functional on $\text{span}(Y \cup x_0)$ by $f_0(y + \alpha x_0) = \alpha$ for all $y \in Y$ and $\alpha \in \mathcal{R}$. Then, for any $y \in Y$, we have $y + x_0 \notin K^o$, so by utilizing Proposition 5.12, $f_0(y + x_0) = 1 \leq p_K(y + x_0)$. It is obvious that $f_0(y - x_0) = -1 \leq p_K(y - x_0)$ because p_K takes only nonnegative values.

As indicated in the proof of Theorem 5.1, this is sufficient to show that for all $z \in \text{span}(Y \cup x_0)$ we have $f_0(z) \leq p_K(z)$. Extend f_0 to a linear functional f on all of X such that f is dominated by p_K. Let $H = f^{-1}(1)$. Then, H contains P, and for all $x \in K^o$, we have $f(x) \leq p_K(x) < 1$, thereby showing that $H \cap K^o$ is empty. □

This theorem is easily visualized in $\mathcal{R}^3$ by taking P to be a line or a single point; see Figure 5.3.

We can now prove some separation results, which the reader will recall motivated this section.

Theorem 5.15 (Hahn–Banach Separation Theorem). *Let K and L be convex sets such that K has some internal point and $L \cap K^o = \emptyset$. Then, there is a hyperplane separating K and L.*

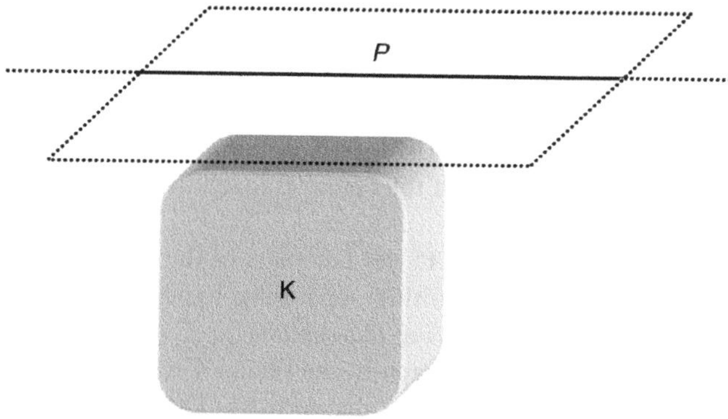

Figure 5.3 Geometric Hahn–Banach extension.

Proof. Since $K^o = p_K^{-1}[0, 1)$ for the Minkowski functional p_K constructed from K, it is easy to see that K^o is convex. It follows by a direct calculation that the set $J = K^o - L$ is convex. We want to show that $J^o = J$. To do so, take any point $y - z \in J$, where $y \in K^o$ and $z \in L$. By the double internal result (Proposition 5.10), y is internal to K^o, so for all $x \in X$, we can find r_x such that for $0 \leq \alpha \leq r_x$ we have $y + \alpha x \in K^o$. Therefore, $y - z + \alpha x \in J$, showing that $y - z \in J^o$.

By our hypothesis, $0 \notin J$, and by Theorem 5.14, we can extend the linear manifold 0 to a hyperplane H not intersecting $J = J^o$. Choose a linear functional f with $N(f) = H$. So, $N(f) \cap J = \emptyset$ means that the intervals $f(K^o)$ and $f(L)$ are disjoint. We can assume (by replacing f with $-f$ if necessary) that for some $c > 0$

$$f(K^o) < c \leq f(L),$$

implying by Proposition 5.13 that

$$f(K) \leq c \leq f(L),$$

and the hyperplane $f^{-1}(c)$ separates K and L.

Remark. Readers should note the second to the last formula, which we will refer to in later applications of this theorem. It is often important to know that the values of f on the internal points of one set are *strictly less* than the values on the other set.

Application: Fundamental Theorem of Asset Pricing. Modern mathematical finance deals with such problems as determining consistent systems of prices for financial assets given the prices of others. For example, the subject was motivated largely by the problem of determining the price of an option to buy a certain stock given the price of the stock. We will not go into detail here, but one of the most basic results of the subject, called the *fundamental theorem of asset pricing*, provides methods to arrive at such price systems, assuming that the given prices admit an internal consistency so there is no chance of making a sure profit without risk. A key ingredient in the proof of this theorem is the Hahn–Banach theorem. For a simple introduction, see Etheridge (2002, Theorem 1.5.2). Also see Scachermeyer (1992) for a proof that uses Hilbert space theory in place of the Hahn–Banach theorem.

For a special case of Theorem 5.15, let L consist of the single point $\{x\}$, where x is a bounding point of K. The resulting separating hyperplane necessarily passes through x. It is known as a *supporting hyperplane* to K.

To visualize, consider a closed n-gon in the plane. There will be infinitely many supporting hyperplanes through each of the vertices. There is exactly one supporting hyperplane through each of the edges, and for these hyperplanes, the n-gon will be precisely the intersection of all the corresponding half spaces containing it. This generalizes as follows.

Theorem 5.16. *Let K be a convex set with an internal point such that K contains all its bounding points. Then, K is the intersection of all the half spaces containing K that are determined by the supporting hyperplanes.*

Proof. We can assume that 0 is internal to K. Let K_1 be the intersection of all the half spaces as given; note that $K \subseteq K_1$ automatically. Take any point y that is not in K; we will show that y is not in K_1. Let $x = p_K(y)^{-1}y$. Then, x is a bounding point of K by Proposition 5.12(c), and we can find a hyperplane of support H to K through x.

Choose a linear functional f such that $H = f^{-1}(\alpha)$ for some $\alpha \in \mathcal{R}$. By replacing f by $-f$ if necessary, we can ensure that $f(K) \subseteq (-\infty, \alpha]$. Since $0 \in K$, we must have $\alpha > 0$. Now, $f(x) = \alpha$, and by our assumption, y cannot be a bounding point of K. Therefore, $p_K(y) > 1$, and $f(y) = p_K(y)\, f(x) > \alpha$. This shows that y is not in the half space containing K determined by H; therefore, y is not in K_1. □

In our next separation result, we deal with the case of a normed linear space.

Theorem 5.17. *If K and L are disjoint convex sets of a normed linear space X, where K is compact and L is closed, then there is a hyperplane strictly separating* K and L.

Proof. The set $K - L$ is easily seen to be convex. Moreover the compactness of K implies that $K - L$ is closed. (This is easily derived by a sequence argument.) Since $0 \notin K - L$, the open ball $B(r)$ is disjoint from $K - L$ for some $r > 0$. Apply Theorem 5.15 to find f such that $f(K - L)$ is an interval disjoint from $f(B(r))$. By replacing f by $-f$ if necessary, we can assume that $f(K - L) \subseteq (0, \infty)$.

Let $\sup f(L) = \beta$ and $\inf f(K) = \gamma$. We must have $\beta < \gamma$, for if not, we could find a $k \in K$ and $l \in L$ such that $f(k - l)$ is arbitrarily small, thus contradicting the fact (from Proposition 5.13) that $f(B(r))$ is an open interval containing 0. For any δ between β and γ, the hyperplane $f^{-1}(\delta)$ strictly separates K and L. □

Remark. The proof of our major separation result, Theorem 5.15, depended heavily on one set having an internal point, which allowed us to make use of the Minkowski functional. Exercise 5.11 shows that in general we cannot omit this hypothesis. In finite dimensions, we can, however, get a more general result. The idea is that we can always "fatten up" a set to get an internal point, at least relative to some subspace.

Theorem 5.18. *Let K and L be two disjoint convex sets in $\mathcal{R}^n$. Then, there is a hyperplane H separating K and L.*

Proof. Assume, by translating if necessary, that $0 \in K$. We will prove the result by backward induction on $\dim(K)$, which we define to be the dimension of $\mathrm{span}(K)$, or equivalently, it is equal to the maximal number of linearly independent vectors in K. If $\dim(K) = n$, then K contains an internal point (see Exercise 5.18), and our general theorem applies. Suppose $\dim(K) = m < n$. Choose any $x \notin \mathrm{span}(K)$. Let $K_+ = \{k + \alpha x : k \in K,\ 0 \leq \alpha \leq 1\}$ and $K_- = \{k - \alpha x : k \in K,\ 0 < \alpha \leq 1\}$. (We can view these as "cylinders" with one end equal to K; see Figure 5.4.) Both of these sets are easily seen to be convex and have dimension $m + 1$. The set L cannot intersect both of these, for if so, we could find points l_1, $l_2 \in L$, k_1, $k_2 \in K$, and α_1, and $\alpha_2 \in (0, 1]$ such that

$$l_1 = k_1 + \alpha_1 x, \qquad l_2 = k_2 - \alpha_2 x.$$

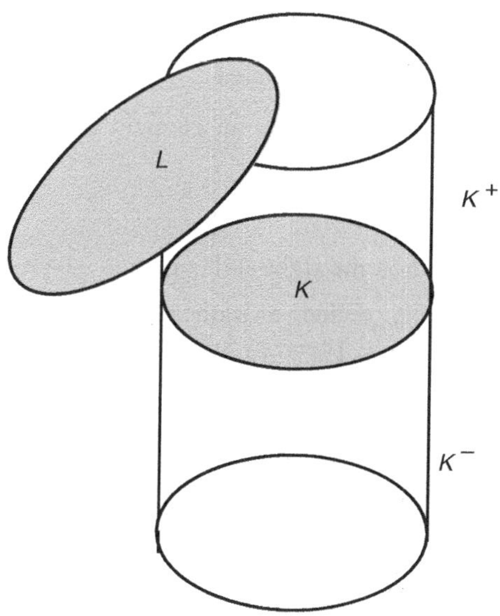

Figure 5.4 Proof of Theorem 5.18.

Now, multiply the first equation by $\beta = \alpha_2/(\alpha_1 + \alpha_2)$ and add the second equation multiplied by $1 - \beta$ to get

$$\beta l_1 + (1 - \beta)l_2 = \beta k_1 + (1 - \beta)k_2.$$

In view of the convexity, we have produced a point in $K \cap L$ contradicting our hypothesis. Supposing then that $L \cap K_-$ is empty, we can by the induction hypothesis find a hyperplane H that separates K_- and L, and it will certainly separate K and L since $K \subseteq K_-$. □

EXERCISES

5.1 Show that the half spaces determined by a hyperplane are well defined. That is, if the hyperplane H equals $f^{-1}(\alpha)$ and also equals $g^{-1}(\beta)$ with α and β both nonnegative, then $f^{-1}[\alpha, \infty) = g^{-1}[\beta, \infty)$.

5.2 Suppose that f is a linear functional on a normed linear space X such that there is a point $x \in X$ with $f(x) = 1$ and $d(x, N) = r > 0$, where N is the null space of f. Show that $\| f \| = 1/r$.

5.3 Show that if X and Y are nonzero normed linear spaces and if $y \in Y$, then there exists $T \in \mathcal{B}(X, Y)$ with $R(T) = \text{span}\{y\}$.

5.4 Suppose that X and Y are normed linear spaces with $X \neq 0$ and with both X and $\mathcal{B}(X, Y)$ complete. Show that Y is complete.

5.5 Suppose X and Y are Banach spaces. Given $T \in \mathcal{L}(X, Y)$ such that $fT \in X^*$ for all $f \in Y^*$, show that T is bounded.

5.6 (Banach limits) Show that there is an element L in $(\ell^\infty)^*$ such that for all $x \in \ell^\infty$:

(i) $\liminf x \leq Lx \leq \limsup x$;

(ii) $LSx = Lx$, where S is the right-shift operator given by $Sx(n) = x(n + 1)$.

Hint: For any $x \in \ell^\infty$, define a sequence by $y(n) = n^{-1} \sum_{i=1}^{n} x(i)$. Let $p(x) = \limsup y$, and apply Theorem 5.1 with Y equal to the space of all convergent sequences.

5.7 Show that any $T_0 \in \mathcal{B}(Y, \ell^\infty)$, where Y is a subspace of a normed linear space X, has an extension to $T \in \mathcal{B}(X, \ell^\infty)$.

5.8 In $\mathcal{R}^2$, let $K = \{(x, y) : |x| \leq 1, |y| \leq 2\}$. Find the Minkowski functional of K.

5.9 Given any z in $\mathcal{R}^2$, consider the linear functional f on $\text{span}\{z\}$ such that $f(z) = 1$. Find all norm-preserving extensions of f to the whole space for

(a) $z = (1, 0)$ and (b) $z = (1, 1)$ by assuming (i) $\|\cdot\|_{\sup}$, (ii) $\|\cdot\|_1$, and (iii) $\|\cdot\|_2$. (See Section 2.8.1 for the definition of these norms on $\mathcal{R}^2$.) Explain your answers geometrically.

5.10 In $\mathcal{R}^2$, let A be the set consisting of the whole plane with all points of the set $\{(x, y) : x^2 + y^2 = 1\}$ removed, except for $(1, 0)$. Show that $(1, 0)$ is internal to A but not interior.

5.11 In c_{00} (with real entries), let K be the set of all sequences whose last nonzero entry is positive.

(a) Show that K is convex and has no internal points.

(b) Show that $f(K)$ equals all of $\mathcal{R}$ for any nonzero linear functional f. Conclude that there is no hyperplane separating K and a point x not in K.

5.12 For each $\alpha \in \mathcal{R}$, let K_α be the set of all continuous f in $L^1[0, 1]$ (over the reals) such that $f(0) = \alpha$. Show that $\{K_\alpha : \alpha \in \mathcal{R}\}$ is a collection of pairwise disjoint, convex, dense sets. (You may use the fact that the continuous functions are dense.) Conclude, then, that for any $\alpha \neq \beta$ there is no closed hyperplane separating K_α and K_β.

5.13 (The moment problem) Let (x_n) be a sequence of elements in a normed linear space X, and let (γ_n) be a sequence in $\mathcal{F}$. Show that there is a bounded linear functional f satisfying $f(x_n) = \gamma_n$ for all n if and only if there exists $M > 0$ such that for any finite subset F of the positive integers and for any choice of scalars $\alpha_i, i \in F$, we have

$$\left| \sum_{i \in F} \alpha_i \gamma_i \right| \leq M \left\| \sum_{i \in F} \alpha_i x_i \right\|.$$

5.14 Let K be a convex set of the linear space X such that 0 is an internal point of K.

(a) What condition is needed on K so that p_K is a seminorm?

(b) What additional condition is needed on K so that p_K is a norm?

5.15 A Schauder basis (x_i) of a Banach space is said to be *monotone* if for $x = \sum_{i=1}^{\infty} \alpha_i x_i$ we have

$$\left\| \sum_{i=1}^{n} \alpha_i x_i \right\| \leq \| x \| \quad \text{for all } n.$$

(a) Give an example of a monotone basis.

(b) Show directly, without appealing to the Hahn–Banach theorem, that for a space X with a monotone basis the dual X^* separates points.

5.16 Let K be a convex set in a normed linear space such that 0 is an interior point of K.

(a) Show that p_K is continuous.
(b) Show that all internal points of K are interior.

5.17 Let m be a Minkowski functional and let p be the Minkowski functional corresponding to the convex set $K_m = \{x : m(x) < 1\}$. Under what circumstances will $p = m$?

5.18 Let K be a convex set in $\mathcal{R}^n$ that contains 0 and that contains some basis $v_1, v_2, \ldots, v_n$ for the space.

(a) Show that for any positive scalars $\alpha_1, \alpha_2, \ldots, \alpha_n$ with $\sum_{i=1}^n \alpha_i < 1$ the vector

$\sum_{i=1}^n \alpha_i v_i \in K$.

(b) Show that

$$v = \frac{1}{n+1} \sum_{i=1}^{n} v_i$$

is an internal point of K.

6

DUALITY

6.1 EXAMPLES OF DUAL SPACES

In this chapter, we explore more fully the relationship between a normed linear space and its its dual. We begin by giving some examples of dual spaces.

6.1.1 Duals of Sequence Spaces

We motivate this topic by first looking at the finite-dimensional situation. Of course, since any such space can be made into a Hilbert space, we have essentially already done this in Chapter 4, but we will slightly recast our exposition in order to fit into the general Banach space framework. Let X be the set of functions on n points. One way to get a linear functional on X is to begin with any g in X and define ϕ_g by

$$\phi_g(f) = \sum_{i=1}^{n} f(i)g(i).$$

Note that ϕ_g corresponds to $\psi_{\bar{g}}$ of Section 4.5.

Suppose we have an arbitrary linear functional ϕ. Can we find g such that $\phi = \phi_g$? There is only one possible choice since for any such g we must have

$$g(i) = \phi_g(\delta_i) = \phi(\delta_i).$$

With this definition of g, we have by linearity that for $f = \sum_{i=1}^{n} f(i)\delta_i \in X$

$$\phi(f) = \sum_{i=1}^{n} f(i)\phi(\delta_i) = \phi_g(f),$$

so that in fact *all* linear functionals are of the form ϕ_g for some g.

Now, suppose that X is a linear space of sequences. We can ask the same question: To what extent can a linear functional ϕ be obtained as ϕ_g? We now define

$$\phi_g(f) = \sum_{i=1}^{\infty} f(i)g(i).$$

The first problem is that the above sum need not converge for general infinite sequences. It will do so, however, provided we suitably restrict f and g. By basic real analysis, the sum will converge if $f \in \ell^1$ and $g \in \ell^\infty$ or similarly if $f \in \ell^\infty$ and $g \in \ell^1$. Also, by Hölder's inequality, Theorem 2.37, it will converge if $g \in \ell^p$ and $g \in \ell^q$, where $p > 1$ and $1/p + 1/q = 1$. Moreover, when we so restrict f and g, the functional ϕ_g will be linear, and the mapping that sends g to ϕ_g will be a linear mapping, as the reader can verify. In addition, when we so restrict these functions, we have from basic analysis or Hölder's inequality that

$$\| \phi_g \| \leq \| g \|_p,$$

in which the norm on g is the respective p-norm, $1 \leq p \leq \infty$.

Another problem arises when we define g by $g(i) = \phi(\delta_i)$, as we did in the finite-dimensional case. Thus defined, g is a sequence but not obviously in the desired space, and this must be verified.

With these observations, we now present the main results on sequence spaces. As usual, to say that one space "is the same" as another means that there is a surjective isometry between them.

Theorem 6.1

(a) *The dual of* c_0 *(with the sup norm) is* ℓ^1.

(b) *The dual of* ℓ^1 *is* ℓ^∞.

(c) *The dual of* ℓ^p *for* $1 < p < \infty$ *is* ℓ^q, *where* $1/p + 1/q = 1$.

Proof. In each case, we will show that the mapping that sends g to ϕ_g is a surjective isometry from Y to X^*, where X and Y are the two spaces in the order given in the statement of the theorem. In view of the observations above, we need only show the other half of the norm inequality and verify that the map is surjective. (The latter is the most difficult part.)

(a) We begin by proving surjectivity. Given ϕ in c_0^*, the only candidate for g is $g(i) = \phi(\delta_i)$, as in the finite-dimensional case, and we must show first that this is in ℓ^1. We do this by approximating g by functions in c_{00}. Fix a positive integer N, and define

$$g_N(i) = \begin{cases} g(i) & \text{if } i \leq N, \\ 0 & \text{if } i > N. \end{cases}$$

Let h be the unit vector defined by $h(i) = |g_N(i)|/g_N(i)$, where the quotient is taken to be zero if $g_N(i) = 0$. Then,

$$\| \phi \| \geq |\phi(h)| = \phi_{g_N}(h) = \sum_{i=1}^{N} |g(i)|.$$

Taking a limit as N goes to ∞ tells us that $\| g \|_1 \leq \| \phi \|$, thereby showing that $g \in \ell^1$.

We now have to show that $\phi_g = \phi$, which is not completely obvious in the case of infinite series. But, from the finite-dimensional argument, ϕ and ϕ_g are both bounded, and they agree on the dense subset c_{00}, so they agree on all of c_0.

Finally, we prove that the mapping is an isometry. Given an arbitrary $g \in \ell^1$, the above argument applied to ϕ_g in place of ϕ shows that $\| g \|_1 \leq \| \phi_g \|$, thereby completing the proof of (a).

(b) The argument is similar to part (a), but easier, since the vectors we need in order to verify the norm inequalities can be taken to be the δ_i. Given $\phi \in (\ell^1)^*$, define $g(i) = \phi(\delta_i)$. Then,

$$\| \phi \| \geq |\phi(\delta_i)| = |g(i)|,$$

thus showing immediately that $\| g \|_\infty \leq \| \phi \|$, and we then follow the same argument as in (a).

(c) The argument is basically the same, but the p's and q's make the details a bit more complex. Given $\phi \in (\ell^p)^*$, we again define $g(i) = \phi(\delta_i)$. Fix an integer N, and define g_N as in part (a). Then define a vector $h \in c_{00}$ by

$$h(i) = \frac{|g_N(i)|^q}{g_N(i)},$$

in which the quotient is taken to be zero if $g_N(i) = 0$. Then,

$$|\phi(h)| = \phi_{g_N}(h) = \sum_{i=1}^{N} |g_N(i)|^q = \| g_N \|_q^q$$

and

$$\| h \|_p^p = \sum_{i=1}^{\infty} |g_N(i)|^{(q-1)p} = \sum_{i=1}^{\infty} |g_N(i)|^q = \| g_N \|_q^q \,.$$

We then have

$$\| \phi \| \geq \frac{|\phi(h)|}{\| h \|_p} = \| g_N \|_q^{(q-q/p)} = \| g_N \|_q.$$

Taking a limit as N goes to ∞ shows that $\| g \|_q \leq \| \phi \|$, and we proceed as in part (a). □

Note that, for $q = (1 - 1/p)^{-1}$, q approaches ∞ as p approaches 1 through numbers greater than 1. One might expect, then, that the duality between ℓ^p and ℓ^q should extend to one between ℓ^1 and ℓ^∞. Indeed, part (c) of the theorem above gives us half of such a result. One might wonder why we omitted from the above theorem the claim that ℓ^1 is the dual of ℓ^∞. (The fact that ℓ^1 is already the dual of c_0 does not by itself preclude this.) The statement is false, however, as shown by the next theorem and the fact that ℓ^∞ is not separable (Proposition 2.42).

Theorem 6.2. *If X^* is separable, then X is separable.*

Proof. Let $\{f_n\}$ be a countable dense set in X^*, and for each n, choose a unit vector $x_n \in X$ such that $|f_n(x_n)| \geq \| f_n \| /2$. We claim that $M = \text{span}\{x_n\}$ is dense in X. If not, we can, by Corollary 5.5, find a unit vector $g \in X^*$ such that $g(M) = 0$. We will show, however, that no f_n can be close to such a g, which contradicts the assumed density. The idea is that since g vanishes on M, each $\| f_n \|$ has to be small, but then it cannot be close to a unit vector. Specifically, we have that for any n

$$\| f_n - g \| \geq |(f_n - g)(x_n)| = |f_n(x_n)| \geq \frac{1}{2} \| f_n \|,$$

so that

$$1 = \| g \| \leq \| g - f_n \| + \| f_n \| \leq 3 \| g - f_n \|,$$

thereby showing that $\| g - f_n \| \geq \frac{1}{3}$.

6.1.2 Radon–Nikodym Theorem

In order to extend the results of the previous section to apply to L^p for more general measure spaces, we need to develop some additional theory. To simplify, we will assume in Sections 6.1.2 through 6.1.4 that we are dealing with real scalars. The main results can be extended to the case of complex spaces by standard techniques. We will outline the ideas here and give complete proofs in Appendix D. As a first step, we consider measures that can take negative as well as positive values. We will begin by restricting our discussion to the finite case.

Definition. If $\mathcal{S}$ is a σ-algebra of subsets of Ω, a *(finite) signed measure* ν on $(\Omega, \mathcal{S})$ is a function that assigns to each A in $\mathcal{S}$ a real number $\nu(A)$ such that ν satisfies countable additivity. It is easy to see that for finite measures the measure of the empty set is automatically 0, so this need not be hypothesized separately. [In fact, all we we need is any set A with $\nu(A) < \infty$; then, $\nu(\emptyset) + \nu(A) = \nu(A)$.]

Since we can consider the sets in any order, it follows that the series of reals in the definition of countable additivity for a signed measure is absolutely convergent.

To distinguish, we will sometimes refer to a measure as we originally defined it in Chapter 2 to be a *positive* measure. An example of a signed measure is the difference of two positive measures. It is an important fact that all signed measures are obtained in this fashion. That is, given any such μ, we can find unique positive measures μ^+ and μ^- such that $\mu = \mu^+ - \mu^-$. This makes integration with respect to a signed measure quite straightforward. We simply define

$$\int f\,d\mu = \int f\,d\mu^+ - \int f\,d\mu^-$$

whenever the right-hand side exists, which occurs if at least one of the two integrals is finite.

A way of producing a whole host of signed measures from a given measure μ is to begin with any $g \in L^1(\mu)$ and define μ_g by

$$\mu_g(A) = \int 1_A g\,d\mu.$$

Property 4 of the integral in Section 2.9.2 shows that μ_g is well defined. The linearity of the integral shows that μ_g is finitely additive, and the monotone convergence theorem completes the derivation that μ_g is a signed measure. See the end of Section D.4 for more details. The decomposition given above is easy to see in this case. It is given simply by $\mu_g = \mu_{g^+} - \mu_{g^-}$.

Integration with repect to μ_g is given by

$$\int f\,d\mu_g = \int fg\,d\mu,$$

which means that knowing integrals with respect to the original measure μ allows us to compute integrals with respect to μ_g. Therefore, given a signed measure v, it would be convenient if we could find a $g \in L^1$ such that $v = \mu_g$.

This representation is not always possible since there is an obvious necessary condition.

Definition. We say that a signed measure v is *absolutely continuous* with respect to a measure μ if, for all $A \in \mathcal{S}$, $\mu(A) = 0$ implies that $v(A) = 0$.

Clearly μ_g is absolutely continuous with respect to μ.

Happily, this is a case where an obvious necessary condition turns out to be sufficient. The Radon–Nikodym theorem (see Theorem D.33) states that if v is absolutely continuous with respect to μ, then there exists $g \in L^1$ such that $v = \mu_g$. The function g is often called the *Radon–Nikodym derivative* of v with respect to μ and written as $dv/d\mu$. This notation makes things look nice since the above integral formula can be written as

$$\int f\,dv = \int f\,\frac{dv}{d\mu}\,d\mu.$$

More insight into this concept (and the notation) can be obtained when the measure space is the integers with all subets measurable. All signed measures can be identified as functions on the points. (See the beginning of Section D.7.) It is easily verified that $(d\nu/d\mu)(n) = \nu(n)/\mu(n)$, which we can define arbitrarily when $\mu(n)$ equals zero since by absolute continuity $\nu(n) = 0$ as well.

Application: Probability Density Functions. An important application occurs in probability theory. Any random variable X induces a probability measure P on the real line, where for any Lebesgue measurable set A the quantity $P(A)$ is the probability that X takes a value in A. If P is absolutely continuous with respect to the Lebesgue measure, the Radon–Nikodym derivative exists and is known as the *probability density function* of X. It is a frequently used concept in probability theory, as it allows one to describe the distribution of many random variables by a relatively simple object, namely a real-valued function defined on the reals.

6.1.3 Dual of L^p

We will outline the results here. Our treatment closely follows that of Rudin (1974, Section 6.15), to which we refer the reader for more details; also see Royden (1988, Section 6.5). To begin, take $1 \le p < \infty$, and consider a finite measure space $(\Omega, \mathcal{S}, \mu)$. For any $g \in L^q$, where $1/q + 1/p = 1$ (or $q = \infty$ for $p = 1$), we can define a linear functional ϕ_g on L^p by

$$\phi_g(f) = \int fg \, d\mu.$$

As with the sequence examples, the object is to show that the mapping $g \to \phi_g$ is a surjective isometry from L^q to $(L^p)^*$. Many of the steps are the same as with ℓ^p, but a major difficulty comes when proving surjectivity. For a general measure space, we no longer have the δ_i's available. Givena ϕ in $(L^p)^*$, how do we find the corresponding g?

The trick is to define a signed measure ν by

$$\nu(E) = \phi(1_E).$$

The additivity requirement for a finite number of summands comes from the linearity of ϕ, and the extension to full countable additivity comes from continuity of ϕ. So, ν thus defined is a signed measure. Moreover, if $\mu(E) = 0$, then the function $1_E = 0$ a.e. so that $\phi(1_E) = 0$, showing that ν is absolutely continuous with respect to μ. The desired g is given by the Radon–Nikodym derivative of ν, and this is the unique candidate for g since for any measurable set E

$$\phi_g(1_E) = \int 1_E g \, d\mu = \int 1_E \, d\nu = \nu(E) = \phi(1_E).$$

We now need an object to replace the dense subspace c_{00}, which we used in the ℓ^p case. This role will be played by L^∞, which is a subspace of L^p (see Exercise 2.31) and which can be shown to be dense. The equality $\phi_g(f) = \phi(f)$ is true for all simple functions by applying linearity to the above equation and then true for all functions in L^∞ by invoking continuity, since boundedness implies that functions in L^∞ can be uniformly approximated by simple functions. Then, it remains, as in the sequence arguments, to show that

$$\| g \|_q \leq \| \phi \| \, . \qquad (*)$$

Given the inequality in $(*)$, we know that ϕ_g is continuous on L^p, and since it agrees with ϕ on the dense set L^∞, we are done.

To show $(*)$, we first define the bounded approximating functions by

$$g_N(x) = \begin{cases} g(x) & \text{if} |g(x)| \leq N, \\ 0 & \text{if} |g(x)| > N. \end{cases}$$

We can then proceed exactly in the ℓ^p case, except for a final step we invoke the monotone convergence theorem, noting that $|g_N|$ increases to $|g|$.

This result can be extended from finite measure spaces.

Definition. We say that a measure space $(\Omega, \mathcal{S}, \mu)$ is *σ-finite* if Ω is the union of countably many sets of finite measure.

For example, the real line with Lesbesque measure is σ-finite, since $\mathcal{R}$ is the union of $\{[-n,n], n = 1, 2, \ldots\}$. For an easier example, the integers with counting measure are σ-finite.

We can extend the above proof to the σ-finite case by taking a sequence of sets of finite measure increasing to the whole space, finding a suitable function g on each of them, and then showing that the limit of these g's exists and gives the implementing function desired. We now state our conclusion, which generalizes parts (a) and (b) of Theorem 6.2.

Theorem 6.3. *Let $1 \leq p < \infty$. For a σ-finite measure space, the mapping $g \to \phi_g$ is a surjective isometry from L^q to $(L^p)^*$, where $q = (1 - 1/p)^{-1}$ for $p > 1$ or ∞ for $p = 1$.*

□

In fact, with more work the σ-finiteness assumption can be removed for $1 < p < \infty$. However, it is still required for $p = 1$. See Royden (1988, Theorem 7.30) and the remarks preceding that theorem.

6.1.4 Dual of $C[a, b]$

Here, again, we will outline the results without giving detailed proofs. We consider the space $C[a, b]$ with the sup norm. As motivation, we will first demonstrate several elements of the dual. For any $g \in L^1[a, b]$, the functional ϕ_g, as defined in the previous section, is in the dual of $C[a,b]$. In this case, the set $\{\phi_g\}$ cannot exhaust all possibilities since we have what looks to be a completely different set of linear functionals. Namely, for any $t \in [a,b]$, the evaluation functional $f \to f(t)$ is linear and bounded, with norm equal to 1. Remarkably, these two different types of linear functionals are part of the same family, which we now describe.

Those sets which are in the smallest σ-algebra containing the open sets of $\mathcal{R}$ are known as the *Borel* sets. Let M be the set of all signed measures defined on the Borel sets of $[a,b]$. It is not difficult to show that we can add two signed measures or multiply by a scalar to get other signed measures, and with these operations M is a linear space.

We also want to equip M with a norm. A clue to doing this is to note that $L^1[a,b]$ can be considered as a subspace of M via the embedding $g \longrightarrow \mu_g$, in which μ is a Lebesgue measure. We want to extend the norm $\|\cdot\|_1$ to all of M. To do so, we note that any $g \in L^1[a,b]$ is equal to $g^+ - g^-$ and $|g| = g^+ + g^-$, in which case

$$\| g \|_1 = \int g \, d\mu = \int g^+ \, d\mu + \int g^- \, d\mu = \mu_g^+[a,b] + \mu_g^-[a,b].$$

We are led, therefore, to define for any $\nu \in M$

$$\| \nu \| = \nu^+[a,b] + \nu^-[a,b].$$

One can show that this defines a norm on M and makes it into a Banach space.

There is an alternative way to define this norm on M that gives us a more general type of function and is convenient in certain proofs. This method is discussed fully in Section D.12. We will include a summary here. To motivate the discussion, we ask the following question. Given a signed measure ν, what can we say about the set function that assigns $|\nu(A)|$ to each set $A \in \mathcal{S}$? The main point to observe is that this set function is not a measure. For an example, take two disjoint sets B and C in $\mathcal{S}$ with $\nu(B) = 1$ and $\nu(C) = -1$. Suppose that we seek the smallest measure μ such that for all $A \in \mathcal{S}$ we have $|\nu(A)| \leq \mu(A)$. Such a measure exists, and it is traditionally denoted by $|\nu|$, although care must be taken not to confuse it with the first set function we gave,as the notation may suggest. It is defined as

$$|\nu|(A) = \sup \sum_{k=1}^{n} |\nu(A_i)|,$$

in which the supremum is taken over all finite collections of pairwise disjoint, measurable sets $\{A_1, A_2, \ldots, A_n\}$ with union equal to A. The measure $|\nu|$ is known as the *variation* of ν.

One can then show that in the real-valued case $\| \nu \|$ as defined above is equal to $|\nu|(\Omega)$. This is readily seen in the case that the measure space Ω is countable since both quantities are equal to

$$\sum_{\omega \in \Omega} |\nu(\omega)|,$$

which is the L^1-norm of v considered as a function on Ω.

We now note that there is a a natural mapping of M into $(C[a,b])^*$. The measure v induces the functional ψ_v given by

$$\psi_v(f) = \int f\,dv.$$

The fact that v is defined on the Borel sets implies that continuous functions are measurable so that the above integral exists. In the case for which $v = \mu_g$, it follows from the formula for integration with respect to μ_g given in Section 6.1.2 that ψ_v is just ϕ_g.

Note that the evaluation functional at t is ψ_v, in which v is the measure that assigns 1 to all Borel sets that contain t and assigns 0 to all Borel sets that do not contain t.

Let M^+ denote the set of all positive measures in M. For any $\mu \in M^+$, the linear functional ψ_μ is a so-called *positive linear functional*, that is, it assigns a nonnegative number to any nonnegative function. We can now state our main conclusion, which is one of the classical theorems of modern analysis, identifying the dual of the continuous functions as space of measures.

Theorem 6.4. (Riesz Representation Theorem). *Consider the mapping from M to $(C[a,b])^*$ that sends v to ψ_v.*

(a) *When restricted to M^+, this is a surjective mapping to the set of positive linear functionals.*

(b) *This mapping is a surjective isometry from M to $(C[a,b])^*$.*

Since this result belongs mainly in the realm of measure theory rather than functional analysis, we will give only a brief sketch of the proof. Normally part (a) is proved first, and this is used to derive (b). The passage from (a) to (b) is shown in detail in Section D.12. Consider part(a). Given a positive ϕ in the dual, how do we arrive at the inducing measure? The trick is to extend the functional from the continuous functions to a larger class of functions that includes 1_A for any Borel set A. We can then define the measure by $v(A) = \phi(1_A)$.

To get some of the flavor of the procedure, consider an open subinterval (c,d) of $[a,b]$. We can approximate $1_{(c,d)}$ with a sequence of continuous functions. For sufficiently large n, let f_n take the value 1 on $[c+1/n, d-1/n]$ and the value 0 on the invervals $[a,c]$ and $[d,b]$; then, fill in the missing values linearly; see Figure 6.1. The sequence of continuous functions (f_n) increases to $1_{(c,d)}$. Since $v((c,d)) = \int 1_{(c,d)}\,dv$, the monotone converence theorem dictates that we must define

$$v((c,d)) = \sup \phi(f_n).$$

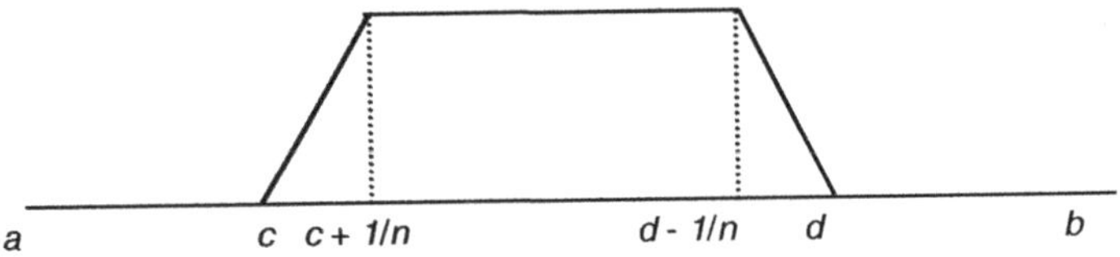

Figure 6.1. Graph of f_n.

Having thus defined the measure on open intervals, we extend it to all sets by a standard technique. The procedure is described for Lebesgue measure, where we begin with the length of an open interval, in Section D.7.

Remark. The same sequence (f_n) is used to show the density of $C[a,b]$ in $L^1[a,b]$. We will sketch the procedure briefly. Any function in $L^1[a,b]$ can be approximated by simple functions, so by using linearity, we want to show that for any measurable $A \in [a,b]$ we can approximate 1_A by a continuous function. By the definition of the measure on $[a,b]$(see Section D7), this reduces to showing the result for the case that A is open. Since any open set is the union of countably many open intervals, we are reduced to showing this for an open interval, and this is accomplished through the sequence (f_n) given above.

There is an alternative formulation of the dual of $C[a,b]$. The idea is to note that a measure v on the Borel sets is determined by its so-called *distribution function*. This is the increasing function F on $[a,b]$ given by $F(t) = v[a,t]$. For a positive measure, this is an increasing function, and for a signed measure, it will be the difference of two increasing functions, which is a function of *bounded variation*. It is possible, then, to formulate the dual as a suitable space of functions of bounded variation. For details, the reader can consult Schecter (1971, Theorem 4.4, Chapter 11). The advantage of this approach is that it can be made self-contained without a prior knowledge of measure theory. On the other hand, it is less natural, and unlike the measure formulation, it does not extend to the space of continuous functions on more general topological spaces. Some of this is described at the end of Section D.12.

For complete details on the Riesz representation theorem in a more general setting, consult Rudin (1974, Theorems 2.1 and 6.1) or Royden (1988, Theorem 13.25).

It should be noted that most writers take the words *Riesz representation theorem* to refer to Theorem 6.4 (or some generalization), as we have done. However, there are several who also use this term to refer to other theorems on the representation of dual spaces, such as Theorem 4.22 or 6.3, which are also due to Riesz.

6.2 ADJOINTS

The concept of an adjoint operator, which we saw was an important one in Hilbert space theory, can be carried over with some differences to the setting of general normed linear spaces.

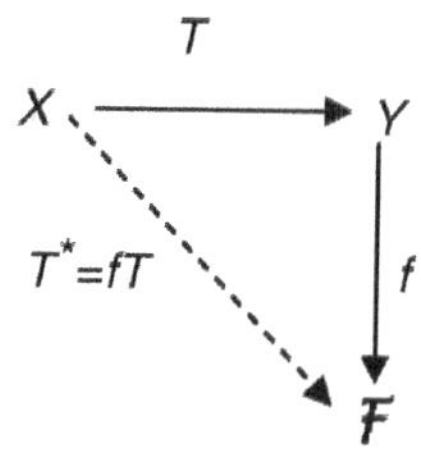

Figure 6.2. Adjoint operator.

Definition. Given $T \in \mathcal{B}(X, Y)$, in which X and Y are normed linear spaces, we define a mapping T^* from Y^* to X^* as follows. For any $f \in Y^*$,

$$(T^*f)(x) = f(Tx) \quad \text{for all } x \in X.$$

(See Figure 6.2.) We refer to T^* as the *adjoint* of T.

Of course, we must verify that T^*f is really in X^*. The linearity is clear from the linearity of T and f. The boundedness will be shown below, but first we compare this definition with the Hilbert space adjoint given in Section 4.6.3.

In the latter case, the operator T was from a space to itself, and the adjoint was defined on the same space. In this case, the original operator is between any two spaces and the adjoint is defined between dual spaces. Suppose we apply the definition of this chapter to an operator $T \in \mathcal{B}(H)$, in which H is a Hilbert space. In the case of real spaces, we would get the same result after identifying H with H^* via the map J_H of Section 4.5. For a complex Hilbert space, however, the map J_H is not linear and this accounts for the difference. We use the same terminology and notation for both types of adjoints, but this should not cause confusion, as invariably, when talking of a Hilbert space, we intend the definiton of Chapter 4.

More insight can be obtained by relating the definition of this chapter to the finite-dimensional case.

Example 6.5. Let T be a linear operator from X to Z in which both are finite-dimensional linear spaces. Choose any bases $\{b_1, b_2, \ldots, b_n\}$ of X and $\{c_1, c_2, \ldots, c_m\}$ of Z, and let A be the corresponding matrix of T as defined in Section 1.3.4. Now, take the dual bases of X^* and Y^*, as described after Corollary 4.23. (The context there was Hilbert spaces, but the definition holds for any linear space.) The matrix corresponding to T^* with respect to these bases is the transpose of A with no conjugation involved. We leave the details of verification to the reader; see Exercise 6.17.

As expected, the properties of our general adjoint parallel those given in Theorem 4.26. [Note however the difference for part(b)].

Theorem 6.6 (Properties of the Adjoint). *For all $S, T \in \mathcal{B}(X,Y)$, $A \in \mathcal{B}(Y,Z)$, and $\alpha \in \mathcal{F}$:*

(a) $(S+T)^* = S^* + T^*$;

(b) $(\alpha T)^* = \alpha T^*$;

(c) $(AT)^* = T^*A^*$;

(d) $\| T^* \| = \| T \|$.

Proof. Parts (a) and (b) are straightforward. For part (c), we must show that for any $f \in Z^*$ we have $(AT)^*f = T^*A^*f$, and to do that, we must show that these two elements of X^* agree on all elements of X. So, given any $x \in X$, we repeatedly apply the definition of adjoint to obtain

$$(AT)^*f(x) = f(ATx) = A^*f(Tx) = T^*A^*f(x).$$

For part (d), we can assume that $\| T \| = 1$. Then, for any unit vectors $f \in Y^*$ and $x \in X$, we have

$$|T^*f(x)| = |f(Tx)| \leq \| f \| \| Tx \| \leq 1,$$

thereby showing that $\| T^*f \| \leq 1$. This verifies the statement that the range of T^* is contained in X^* and also shows that $\| T^* \| \leq 1$. Conversely, given any $\epsilon > 0$, choose a unit vector $x \in X$ such that $\| Tx \| \geq 1 - \epsilon$, and then, by the proof of Corollary 5.7, choose a unit vector f in Y^* such that $|f(Tx)| = \| Tx \|$. Hence, $|T^*f(x)| = |f(Tx)| \geq 1 - \epsilon$, thus showing that $\| T^*f \| \geq 1 - \epsilon$, from which it follows that $\| T^* \| > 1 - \epsilon$. □

Just as we can sometimes use the dual space to obtain information about the original space, we can use the adjoint to obtain information about an operator. For a typical example, suppose we are given $T \in \mathcal{B}(X, Y)$, and we want to know whether or not a particular $y \in Y$ is in $R(T)$. It may be difficult to actually find some x with $Tx = y$ or to show that such x does not exist. It is often easier to use the following result, which refers to $\overline{R(T)}$, and could help answer our question if we knew that $R(T)$ was closed. We can view this as an analogous result to Proposition 4.27(b). We leave the proof to the reader; see Exercise 6.3.

Proposition 6.7. *For $T \in \mathcal{B}(X, Y)$, $y \in \overline{R(T)}$ if and only if $f(y) = 0$ for all $y \in N(T^*)\}$.*

We will make extensive use of adjoint operators throughout the remainder of the chapter.

6.3 DOUBLE DUALS AND REFLEXIVITY

6.3.1 Embedding in Double Dual

Given a normed linear space X, we let X^{**} denote the dual of the dual space, the so-called *double dual*. There is a simple but extremely useful relationship between X and its double dual. For any $x \in X$, we define an element $\hat{x}$ in X^{**} by

$$\hat{x}(f) = f(x)$$

for all $f \in X^*$.

Theorem 6.8. *The mapping $x \to \hat{x}$ is a linear isometry from X to X^{**}.*

Proof. Linearity is clear. From Corollary 5.7,

$$\| \hat{x} \| = \sup\{|\hat{x}(f)| : \| f \| = 1\} = \sup\{|f(x)| : f = 1\} = \| x \| .$$

Remark. Note that this is another case for which we obviously must use the Hahn–Banach theorem. Without it, the dual and therefore the double dual could be 0, and we would not have an isometry.

The above theorem tells us that we can regard any normed linear space X as a subspace of a space of linear operators, namely $\mathcal{B}(X^*, \mathcal{F})$. As this latter space is complete by Theorem 2.15, we have the promised proof to the existence of a completion as stated in Section 2.5.6. Moreover, we can now adapt Theorem 3.10 to characterize bounded sets in any normed linear space. The theorem below follows immediately from the uniform boundedness principle stated in Section 3.6.

Theorem 6.9. (General Uniform Boundedness Principle). *Let A be a subset of a normed linear space such that, for all $f \in X^*$, $f(A)$ is a bounded set of scalars. Then, A is bounded.*

6.3.2 Reflexive Spaces

The natural question left unanswered in the above section is whether or not the map $x \to \hat{x}$ is surjective. Obviously since X^{**} is complete, this can only be true if X is a Banach space.

Definition. Those Banach spaces for which the embedding of X into X^{**} given by $x \to \hat{x}$ is surjective are said to be *reflexive*.

Reflexive spaces are among the best behaved of all the Banach spaces, and they share some of the properties of Hilbert spaces. An example of such nice behavior is the condition outined in Exercise 6.4. (In fact, the converse to the statement of this exercise is true and a space with this property is reflexive. This is more difficult to show and will not be proved here.)

We already have an example of a nonreflexive space. From Theorem 6.1, $c_0^{**} = \ell^\infty$, which is nonseparable, so the separable space c_0 cannot be reflexive.

On the positive side, it is clear that finite-dimensional spaces are reflexive since the dual space has the same dimension (as shown by the dual basis), and so does the double dual. It is a well known fact of linear algebra that any injective map from a finite-dimensional linear space to one of the same dimension is surjective.

A major infinite-dimensional example of a reflexive space is that of a Hilbert space. It is easily verified that the natural map in that case can be obtained as the composition $J_{H^*} J_H$, where J_H is the mapping from H to H^* given in Section 4.5 and H^* is the Hilbert space described in Corollary 4.23; see Exercise 6.18. (Note that the composition of two conjugate linear mappings is, in fact, linear.) Because J_H and J_{H^*} are surjective, the composition is surjective.

It follows that L^2 is reflexive, so a natural question is to investigate the reflexivity of L^p for $1 < p < \infty$. We know from Theorem 6.3 that $(L^p)^{**} = (L^q)^* = L^p$, where $1/p + 1/q = 1$, so that L^p is isometric to its double dual. This is, however, not sufficient to show reflexivity. James (1951) provides an example of a nonreflexive Banach space that is isometric to its double dual. We must show that the natural embedding as defined at the beginning of Section 6.3.1 is an isometry, and this turns out to be true.

Theorem 6.10. *L^p is reflexive for $1 < p < \infty$.*

Proof. For any $p \in (1, \infty)$ and q such that $1/p + 1/q = 1$, let U be the isometry from L^q to L^{p*}, as given in Theorem 6.3. That is, $Ug = \phi_g$. Let V be the corresponding isometry from L^p to $(L^q)^*$. Note that U^* is a mapping from $(L^p)^{**}$ to $(L^q)^*$. We claim that the natural embedding from L^p to its double dual is $(U^*)^{-1} V$, which is certainly surjective. Indeed, for any $f \in L^p$ and $g \in L^q$, we have

$$(U^*\hat{f})(g) = \hat{f}(Ug) = (Ug)f = (Vf)(g),$$

in which the last inequality arises since both terms are equal to $\int fg$. This shows that $U^*\hat{f} = Vf$, and composing with $(U^*)^{-1}$ establishes our claim; see Figure 6.3.

The following two theorems illustrate ways of finding new reflexive spaces from given ones.

Theorem 6.11. *A closed subspace of a reflexive space is reflexive.*

Proof. Suppose that Z is a closed subspace of the reflexive space X. We are given $\phi \in Z^{**}$, and we have to find an element $x \in Z$ such that $\hat{x} = \phi$. We do this by the familiar three-pronged approach, which is used in countless mathematical proofs. First, we transfer the problem from the original setting to some other that is easier to handle. Second, we solve the problem in the easier setting; third, we go back to the orginal situation. In this situation, for the first prong we use ϕ to

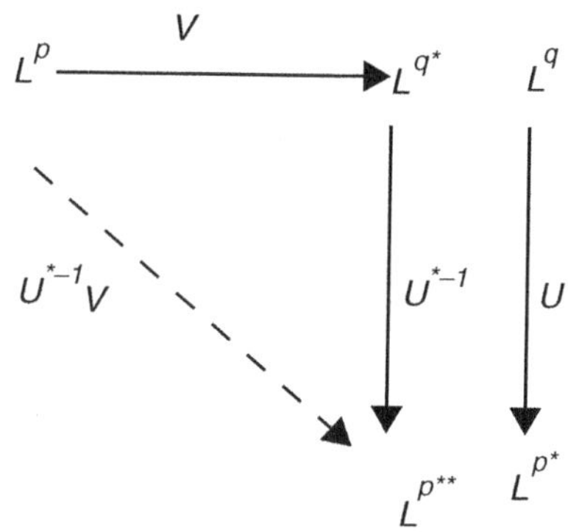

Figure 6.3. Reflexivity of L^p.

induce an element of X^{**}, second, we use the reflexivity of X to get a corresponding element $x \in X$, and third, from that we produce the element in Z that we want; see Figure 6.4. Here are the details.

Define $\eta \in X^{**}$ by $\eta(f) = \phi(f|Z)$ for all $f \in X^*$. [Here, $f|Z$ denotes the restriction of f to Z. That is, for all $z \in Z, f|Z(z) = f(z)$.] Since the norm of $f|Z$ is less than or equal to that of f, the functional η is bounded with norm less than or equal to that of ϕ. Since X is reflexive, there is an $x \in X$ such that $\hat{x} = \eta$. We now need an element in Z, and we will show that in fact x gives us such an element. Given any $f \in X^*$ with $f(Z) = 0$, we have that

$$\hat{x}(f) = \phi(f|Z) = 0,$$

so that by Corollary 5.5, $x \in \bar{Z} = Z$.

Finally, we show that $\hat{x} = \phi$, in which now the "hat" notation refers to the embedding of Z as opposed to above, where it referred to the embedding of X.

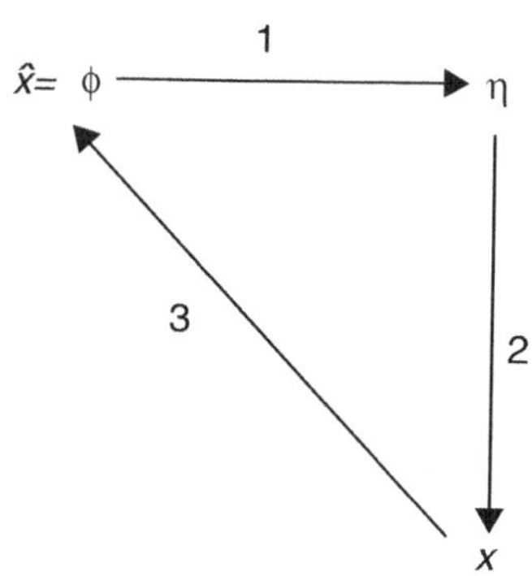

Figure 6.4. Three prongs in proof of Theorem 6.11.

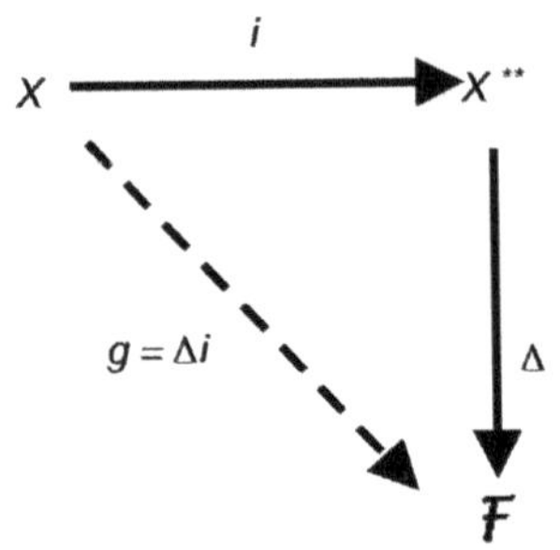

Figure 6.5. Map g of Theorem 6.12.

To this end, take any $g \in Z^*$. By the Hahn–Banach theorem, extend this to an element $f \in X^*$. Then,

$$\phi(g) = \phi(f|Z) = \eta(f) = \hat{x}(f) = f(x) = g(x) = \hat{x}(g),$$

in which the first occurrence of the hat refers to X and the second refers to Z. □

Theorem 6.12. *A Banach space X is reflexive if and only if X^* is reflexive.*

Proof. Suppose X is reflexive. Given any $\Delta \in X^{***}$, define $g \in X^*$ by $g(x) = \Delta(\hat{x})$; see Figure 6.5. We claim that $\hat{g} = \Delta$. We prove this statement by showing the two functionals agree on any $\phi \in X^{**}$. Given any such ϕ, there is $x \in X$ with $\hat{x} = \phi$. Then,

$$\Delta(\phi) = \Delta(\hat{x}) = g(x) = \hat{x}(g) = \phi(g) = \hat{g}(\phi).$$

Conversely, if X^* is reflexive, then, as just shown, X^{**} is reflexive. So X, regarded as a subspace of X^{**}, must be closed by completeness, and we apply Theorem 6.11.

6.4 WEAK AND WEAK* CONVERGENCE

6.4.1 Weak Convergence

Duality gives us some new types of convergence and more scope for finding elements with properties that we want as limits of sequences of other elements.

Definition. We say that a sequence (x_n) in the normed linear space X converges *weakly* to $x \in X$ if the sequence of scalars $(f(x_n))$ converges to $f(x)$ for all $f \in X^*$.

The terminology reflects the usage of the word "weak" as given in Section 2.6, where it meant *easier to converge*. If (x_n) converges to x in norm, it clearly converges weakly since $|f(x_n) - f(x)| \leq \| f \| \| x - x_n \|$. What about the converse? In a finite-dimensional space, weak convergence is equivalent to norm convergence (see Exercise 6.9), but this is not true in general.

Example 6.13. In ℓ^p, $1 < p < \infty$, the sequence (δ_n) is clearly not norm convergent, since the distance between any two distinct elements of the sequence is $2^{1/p}$. On the other hand, it is weakly convergent to 0. Indeed, given any $g \in \ell^q$ where $1/q + 1/p = 1$, we have $\phi_g(\delta_n) = g(n)$, which must approach 0 as shown by the fact that $\sum_{n=1}^{\infty} |g(n)|^q$ converges.

To show that weak convergence is a useful concept, we will verify the basic properties we expect for convergence in a linear space, which we exhibited for norm convergence in Chapter 2.

Proposition 6.14 (Uniqueness of Weak Limits). *If (x_n) converges weakly to both x and y, then $x = y$.*

Proof. We must clearly use the Hahn–Banach theorem here, for without it, the dual space could degenerate to 0, in which case every sequence would converge weakly to every possible point in the space. The conclusion follows immediately from Corollary 5.6. If $x \neq y$, we choose $f \in X^*$ such that $f(x) \neq f(y)$ and use the fact that limits for sequences in $\mathcal{F}$ are unique to get a contradiction.

Proposition 6.15 (Continuity of Operations). *Given sequences (x_n) converging weakly to x and (y_n) converging weakly to y and given a sequence of scalars (α_n) converging to α, it follows that:*

(a) *$(x_n + y_n)$ converges weakly to $x + y$;*

(b) *$(\alpha_n x_n)$ converges weakly to αx.*

Proof. The proof follows easily by using continuity of elements in X^* and boundedness of convergent sequences in $\mathcal{F}$.

In most infinite-dimensional spaces, weak convergence is *strictly weaker* than norm convergence. This is certainly true in ℓ^p with $p > 1$ as shown in Example 6.13. An exception is ℓ^1. We will present a proof of this by using an ingenious idea known as the "gliding hump" argument, which goes back to the early days of Banach space theory. See Carothers (2005) for other applications of this concept.

Let x be an element in ℓ^1. We will say that x has a *hump* over the interval $[c, d]$, where $c \leq d$ are positive integers, if $\sum_{i=c}^{d} |x(i)| \geq (\frac{3}{5}) \parallel x \parallel_1$. (There is nothing particularly special about the constant $\frac{3}{5}$ here. Anything over $\frac{1}{2}$ would do.) The significance of a hump is as follows: Let y be an element in ℓ^∞ with all entries of absolute value 1. Suppose that the entries of y on the hump match those of x in the sense that if $x(i) = re^{i\theta}$ for $c \leq i \leq d$, then $y(i) = e^{-i\theta}$. Now, consider the number $|\phi_y(x)|$. Over the hump, the product $y(i)x(i)$ equals $|x(i)|$. For the rest of the entries whose absolute values sum to less than or equal to $\frac{2}{5}$ of the norm, we could, in an extreme case, have "antimatching" with $y(i) = -e^{-i\theta}$ so that $y(i)x(i) = -|x(i)|$, but this still leaves at least $\frac{1}{5}$. In other words, we can say that

$$|\phi_y(x)| \geq \frac{1}{5} \parallel x \parallel_1 .$$

For example, let $x = (1, 1, 1, 2, -3, 4, -1, -1 - 1, 0, 0, \ldots)$. Then, $\parallel x \parallel = 15$, and x has a hump over the interval $[4, 6]$. Choose $y = (-1, -1, -1, 1, -1, 1, 1, 1, 1, \ldots)$, which matches on the hump and antimatches on the other entries. Then $\phi_y(x) = 3$.

Proposition 6.16. *If a sequence (x_n) in ℓ^1 converges weakly to x, then (x_n) converges to x in the norm.*

Proof. To follow this proof, it is important to keep the notation straight. Each x_n is a sequence with ith entry denoted by $x_n(i)$. By linearity, we can suppose that $x = 0$. By way of contradiction, suppose that (x_n) converges weakly to 0, but not in the norm. Passing to a subsequence if necessary, we can assume that there is an $r > 0$ such that $\parallel x_n \parallel_1 \geq 5r$ for all n. We first show that given any positive integers k and c, we can find $n \geq k$ and d such that (x_n) has a hump over $[c, d]$. To see this, choose $n > k$ large enough so that $|x_n(i)| \leq r/(c-1)$ for $i = 1, 2, \ldots, c-1$. This is possible since weak convergence to 0 implies pointwise convergence to 0; see Exercise 6.9. Then, choose d large enough so that $\sum_{i=d+1}^{\infty} |x_n(i)| < r$. It follows that

$$\sum_{i=c}^{d} |x(i)| \geq \parallel x_n \parallel_1 - 2r \geq \parallel x_n \parallel_1 - \tfrac{2}{5} \parallel x_n \parallel_1 = \tfrac{3}{5} \parallel x_n \parallel_1 .$$

By starting with $k = 0$ and repeating this procedure, we can produce a subsequence of (x_n) such that each x_{n_i} in the subsequence has a hump. Moreover, since we can push the humps out as far as we like, we can ensure that for $i \neq j$ the humps for x_{n_i} and x_{n_j} are over disjoint intervals. We can then choose a single vector y with entries of absolute value 1 that matches (as defined above) the entries on all of these humps. It follows from our preliminary remarks that for all i we have $\phi_y(x_{n_i}) \geq \frac{1}{5} \parallel x_{n_i} \parallel_1 \geq r$, thereby contradicting the fact that the original sequence converges weakly to 0. □

6.4.2 Weak* Convergence

On the dual space X^*, we have another type of convergence.

Definition. A sequence (f_n) in X^* is said to be *weak** convergent to $f \in X^*$ if $(f_n(x))$ converges to $f(x)$ for all $x \in X$.

As for weak convergence, one can verify that uniqueness of limits and continuity of operations hold for weak* convergence. Indeed, uniqueness is easier since we do not require the Hahn–Banach theorem.

On X^*, we can also speak of weak convergence as induced by X^{**}. How do the two types of convergence compare? Suppose that (f_n) converges weakly to f in X^*. Then, for all $x \in X$,

$$f_n(x) = \hat{x}(f_n) \to \hat{x}(f) = f(x),$$

which shows that f_n is weak* convergent to f. So, weak* convergence is weaker than weak convergence. Of course, on a reflexive space the argument above can be reversed to show that the two types of convergence are equivalent. Therefore, the concept of weak* convergence is only of interest when studying nonreflexive spaces.

Application: Central Limit Theorem. Suppose we take some random experiment yielding a numerical quantity (such as throwing a pair of dice), repeat it a large number, say N, times, and take the average of all the results. This average, of course, depends on chance and is another random variable. The remarkable result, however, is that no matter what distribution you begin with, as N becomes large, the distribution of the average converges in some sense to a normal distribution, one of a family of familiar bell-shaped curves. This is the simplest version of the famous central limit theorem. It is the result that justifies the extensive use of the normal curve in probability and statistics. The interesting fact from our point of view is that, to make this precise, the mode of convergence needed is precisely weak* convergence. The context is that we are looking at the space of measures on the real line, viewed as a dual of continuous functions as given in the Riesz representation theorem, Theorem 6.4 (or more precisely, by a generalization of this to the dual of the continuous functions on the whole real line). By a quirk of terminology, probabilists refer to this as *weak convergence*, but it really is weak* convergence, and since the spaces in question are not reflexive, it is strictly weaker than weak convergence in the sense of functional analysis. For more details and examples of weak* convergence, including a proof of the central limit theorem, we refer the reader to Bobrowski (2005, Section 5.5).

To obtain deeper insight, we must look not only at weak and weak* convergence but also at the corresponding topologies. We do so in the next chapter. The following section provides motivation.

6.4.3 Weak Convergence and Sequential Closure

For a set B in a normed linear space X, let $s(B)$ denote the set of all elements x for which there is a sequence (x_n) in B converging weakly to x. In particular, take $X = \ell^2$, and let B consist of the elements $x_{m,n} = \delta_m + m\delta_n$ with $m < n$. What do we know about $s(B)$? In the first place, for fixed m, the sequence $(x_{m,n})$ converges weakly to δ_m as n approaches ∞, as we can see from the fact that (δ_n) converges weakly to 0. So, all $\delta_m \in s(B)$, and consequently $0 \in s(s(B))$. We will show, however, that $0 \notin s(B)$. To this end, suppose there is a sequence in B converging weakly to some point x. Since a weakly convergent sequence is bounded (Exercise 6.12), there can be only finitely many different values of the first parameter m appearing in the sequence, and at least one such value, say m_0, must appear infinitely often. The subsequence consisting of all entries with first parameter m_0 also converges weakly to x, but the functional $\phi_{\delta_{m_0}}$ takes the value 1 on all such terms, so $x \neq 0$.

What is happening here? Examples like this caused some consternation in the early days of the subject. It shows that there is no metric that we can impose on X which corresponds to weak convergence. If there were, then $s(B)$ would be in the closure of B, so that $s(s(B))$ would be the same as $s(B)$, but it is not. This suggests that the idea behind weak convergence cannot be fully described with sequences, and we need a more general structure. This leads us directly to the idea of topological spaces, which we will discuss in the next chapter.

EXERCISES

6.1 Define $T : \ell^1 \to c_0$ by $Tx(n) = \sum_{k \geq n} x(k)$.

(a) What is $\| T \|$?

(b) Give an explicit formula for T^* considered as an element of $\mathcal{B}(\ell^1, \ell^\infty)$.

6.2 (A converse to Hölder's inequality) Let $1 < p < \infty$. Let y be a sequence such that for all $x \in \ell^p$ the sequence $(x(n)y(n)) \in \ell^1$. Show that $y \in \ell^q$, where $q = (1 - 1/p)^{-1}$. *Hint*: Consider the functionals $\{f_n : n = 1, 2, \ldots\}$ in which f_n is given by $f_n(x) = \sum_{i=1}^{n} x(i)y(i)$.

6.3 Prove Proposition 6.7.

6.4 f be a bounded linear functional on a reflexive space X. Show that $\| f \|$ achieves its maximum on the unit disk. That is, there exists a unit vector x such that $\| f \| = |f(x)|$.

6.5 Given $T \in \mathcal{B}(X, Y)$, show that $T^{**}(\hat{x}) = \widehat{T(x)}$.

6.6 For any subset M of a normed linear space X, let

$$M^\perp = \{f \in X^* : f(m) = 0 \text{ for all } m \in M\}.$$

$M^\perp$ is known as the *annihalator* of M.

(a) Show that $M^\perp$ is a closed subspace of X^*.

(b) If M is a closed subspace of X, show that $M^\perp$ is isometric to $(X/M)^*$.

6.7 Let $T \in \mathcal{B}(X, Y)$, in which X and Y are Banach spaces.

(a) Show that T injective implies that T^* is surjective, and T is surjective implies that T^* is injective.

(b) (For readers with a knowledge of matrix theory) If you apply part (a) to finite-dimensional spaces, what well-known theorem about matrices emerges?

6.8 If T is a surjective, bounded linear operator from a reflexive space X to a Banach space Y, show that Y is reflexive. (You many want to use Exercises 6.5 and 6.7.)

6.9 If X is the space of all functions on n points, show that weak convergence is the same as pointwise convergence. Conclude that on a finite-dimensional space weak convergence is the same as norm convergence.

6.10 Show that for $T \in \mathcal{B}(X, Y)$, if (x_n) converges weakly to x, then (Tx_n) converges weakly to Tx.

6.11 Show directly, without using Proposition 6.16, that (δ_n) does not converge weakly to 0 in ℓ^1.

6.12 Show that a weakly convergent sequence is bounded.

6.13 Show that in ℓ^p, for $1 < p < \infty$, a sequence converges weakly if ond only if it is bounded and converges pointwise.

6.14 Consider the sequence in $(\ell^\infty)^*$ defined by $f_n(x) = x(n)$. Show that (f_n) is bounded, but no subsequence converges weak*-ly.

6.15 Let X be any of ℓ^p for $1 < p < \infty$, or c_0. Use Propostion 6.16 to show that any injective $T \in \mathcal{B}(X, \ell_1)$ does not have a bounded inverse.

6.16 Let X be reflexive. Show that the mapping T to T^* from $\mathcal{B}(X, Y)$ to $\mathcal{B}(Y^*, X^*)$ is surjective if and only if Y is reflexive.

6.17 Verify that in the finite-dimensional case the matrix of T^* with respect to the dual basis is the transpose of the matrix of T.

6.18 Verify that for a Hilbert space H, the natural map from H to its double dual is given by $J_{H^*} J_H$ using the notation of Section 4.5.

7

TOPOLOGICAL LINEAR SPACES

7.1 REVIEW OF GENERAL TOPOLOGY

7.1.1 Basic Definitions

Examples like the one in Section 6.4.3 suggest that functional analysis requires the consideration of *topological spaces*, a more general structure than that of a metric space. The transition starts with the observation that when defining the notions of convergence, closure, and continuity, we need not make use of the actual distance measurement. All these definitions can be formulated in terms of open sets. Consequently, we can talk about these ideas in more generality by defining appropriate axioms for such sets.

Definitions. Given any set X, a *topology* on X is a collection τ of subsets of X that satisfy the following conditions:

(i) The union of any sets in τ is also in τ.
(ii) A finite intersection of any sets in τ is also in τ.
(iii) X itself and the empty set are in τ.

The sets in τ are referred to as *open sets*. A set X equipped with a topology will be called a *topological space*.

In Chapter 2, we observed that the open sets in a normed linear space (indeed, in any metric space) satisfied these three conditions, so a metric immediately induces a topology.

We can then carry over the open set formulations of Chapter 2. To review, a sequence (x_n) converges to x if the sequence is eventually in any open set containing x. A function from one topological space to another is continuous at a point x if

A First Course in Functional Analysis. By S. David Promislow

given any open set V containing $f(x)$ there is an open set W containing x such that $f(W) \subseteq V$. A set is closed if its complement is open, and the closure $\bar{A}$ of a set A is the intersection of all closed sets containing A. Equivalently, $x \in \bar{A}$ if and only if any open set containing x intersects A.

While the above definition of a topological space is convenient for proving basic facts, it is not always the easiest way to describe a specific topology. This is especially true for our ultimate goal of considering topologies on linear spaces. Rather, a more "local" approach is preferred.

Definition. Given a point x in a topological space, any open set containing x is called a *neighborhood* of x.

Some authors define a neighborhood of x to be any set containing an open set containing x. We find it more convenient to restrict neighborhoods to be open.

Definition. A *neighborhood base at* x is a collection $\mathcal{V}_x$ of neighborhoods of x such that any neighborhood of x contains a set in $\mathcal{V}_x$. Once such a base is specified, elements of $\mathcal{V}_x$ are referred to as *basic neighborhoods.*

Specifying a neighborhood base at each point completely determines the topology since we know, then, that a set G is open if and only if for all $x \in G$ we have $V \subseteq G$ for some $V \in \mathcal{V}_x$. In fact, without mentioning open sets, we can directly use the neighborhood base in the definitions of convergence, continuity, and closure given above by replacing the words "open set containing x" with " a set in $\mathcal{V}_x$." This was precisely the procedure that we followed with metric spaces, for which we essentially took the collection $\{B(x; r) : r > 0\}$ as a neighborhood base of a point x. Next, we investigate the properties of neighborhood bases.

Suppose we are given a neighborhood base at each point x of a topological space. It follows that:

1. For any x and any U, $V \in \mathcal{V}_x$, $U \cap V$ contains an element of $\mathcal{V}_x$.
2. If $y \in V \in \mathcal{V}_x$, then there is $W \in \mathcal{V}_y$ such that $W \subseteq V$.

The first statement follows from the fact that the intersection of two open sets is open, the second directly from the definitions of neighborhood and neighborhood base.

We are interested in a converse statement to the above.

Theorem 7.1 *Suppose for each point x of a set X we are given a collection $\mathcal{V}_x$ of subsets containing x such that* 1 and 2 *are satisfied. Define a set G to be open if for all $x \in X$ there is $V \in \mathcal{V}_x$, such that $V \subseteq G$. Then, the collection of open sets forms a topology, and for all $x \in X$, $\mathcal{V}_x$ is a neighborhood base at x with respect to this topology.*

Proof. The proof is straightforward. Closure of open sets under arbitrary unions follows directly from the definition of open set without regard to any properties of

$\mathcal{V}_x$. Closure under finite intersections follows from 1. Point 2 is needed to ensure that the given collection of neighborhoods at each point actually turn out to be open in the resulting topology. □

The natural mappings between two topological spaces are continuous ones, so an isomorphism is a bijective continuous map whose inverse is continuous. Such a mapping is given the special name of *homeomorphism*. Note that we must include the inverse condition here. There are continuous bijections whose inverse is not continuous. (We already encountered this in Example 3.8.)

A desirable property that does not follow from the definition of a topological space is uniqueness of limits. For most applications in analysis one wants what is known as a *Haudsorff* topological space. This means that given any two distinct points x and y we can find a neigborhood U of x and a neighborhood V of y such that $U \cap V = \emptyset$. This condition ensures that limits of sequences are unique. In a Hausdorff topological space, a sequence (x_n) cannot possibly converge to two distinct points x and y as it eventually must be inside any neighborhood of either.

We can compare different topologies on the same set, as we did with norms. A topology τ_1 is weaker than another τ_2 if the collection $\tau_1 \subseteq \tau_2$. (Our definition of "weaker" allows for the topologies to be the same. We will say *strictly weaker* if the inclusion is strict.) The main conclusions are the same as for norms. In the weaker topology, it is easier to converge, closures are bigger, there are fewer continuous functions from the space, and there are more continuous functions to the space.

Some writers use the word *coarser* in place of *weaker* and *finer* in place of *stronger*. One can remember these terms by thinking of the finest possible topology, called the *discrete topology*. This is the topology in which every set is open. The only convergent sequences are those that converge by virture of being eventually constant. The coarsest possible topology, called the *indiscrete topology*, is one in which only the whole space and $\emptyset$ are open. Every sequence converges to every point.

If $Y \subseteq X$, in which X is a topological space, then there is a natural way of considering Y as a topological space. We take the open sets to be those of the form $V \cap Y$ where V is open in X. This is known as the *subspace topology*.

7.1.2 Role of Sequences

As demonstrated by the example in Section 6.4.3, we must take care when we generalize statements about sequences from a metric space setting to that of a general topological space. The source of the difficulty can be traced to a property involving countability. In a metric space, we can choose a countable neighborhood base at each point x, namely $\{B(x; 1/n) : n = 1, 2, \ldots\}$. If a topological space has a countable neighborhood base at each point, then all the statements about sequences that we made for metric spaces still hold under the same arguments. Such a space is called *locally countable*. (It is also called *first countable*, which is not the best terminology since one has to remember to distinguish this from *second countable*. The latter means that we can choose a countable family of sets such that a neighborhood of *any point* contains a set in this family—for example, the open intervals with rational endpoints in the set of real numbers.)

For a topological space that is not locally countable, we should be aware of the following facts about sequences, which we will discuss under various headings.

Closure. If a sequence (x_n) of points in a set A converges to x, then $x \in \bar{A}$, the closure of A, as in metric spaces. However, the converse is not true. There can be points in the closure of a set but no sequences in the set converging to that point.

Continuity. If a function f from one topological space to another is continuous at a point x, then for any sequence (x_n) converging to x the sequence $(f(x_n))$ converges to $f(x)$. This condition, however, is not sufficient to guarantee continuity.

Compactness. Compactness in a general topological space is defined as in Section 2.2.3. It is customary to formulate this definition to apply first to spaces rather than sets. A topological space is said to be *compact* if every open cover has a finite subcover. A subset of a topological space is said to be compact if it is a compact space with the subspace topology. This is equivalent to the definition we gave before for a subset of a normed linear space. A space in which every sequence has a convergent subsequence is called *sequentially compact*. In a metric space (or in any locally countable space), these definitions are equivalent. However, in the general case, we do not even have a one-sided condition. There are compact spaces that are not sequentially compact and sequentially compact spaces that are not compact.

The main properties about compact sets that we stated for metric spaces remain true, although the proofs can be slightly more involved. We will present one of these to indicate the nature of the typical proof involving compactness.

Proposition 7.2. *In a Hausdorff space, a compact set is closed.*

Proof. Suppose that K is compact. We will show that if $x \notin K$, then $x \notin \bar{K}$ by producing a neighborhood U of x that is disjoint from K. For each $y \in K$, choose disjoint neighborhoods U_y and V_y of x and y, respectively. Choose $y_1, y_2, \ldots, y_n$ in K so that $K \subseteq \bigcup_{i=1}^{n} V_{y_i}$. Let $U = \bigcap_{i=1}^{n} U_{y_i}$. Then, U is the desired neighborhood of x since any point in U is not in any of the V_{y_i} and therefore not in K. □

Nets. There is generalization of a sequence, known as a *net*, that in many instances allows one to imitate sequence arguments in a general topological space. We do not use nets in this book. The interested reader can find an account in any introductory book on topology.

7.1.3 Local Compactness

There is a useful generalization of the concept of compactness.

Definition. A topological space is said to be *locally compact* if every point has a neighborhood whose closure is compact.

One natural example is $\mathcal{R}^n$ with the norm topology. (Since all norms are equivalent, there is a unique such topology.) Moreover, any discrete space is locally compact since any point consitutes a compact neighborhood of itself.

There are various normed linear spaces of continuous functions associated with compact and locally compact topological spaces. In the case of $C[a,b]$ with the sup norm, we can replace $[a,b]$ by any compact topological space. The spaces of Example 2.19 can similarly be generalized. For any locally compact space T, we have the space $C_{00}(T)$ of continuous functions on T with compact support and the space $C_0(T)$ of continuous functions vanishing at ∞, defined as in Example 2.19 with $\mathcal{R}$ replaced by T. Note that the spaces c_{00} and c_0 are particular cases of these spaces when we take T to be the nonnegative integers.

7.1.4 Product Spaces

We introduce a concept here that we will make much use of in later sections.

Suppose we have two topological spaces X and Y. We would like to make the Cartesian product $X \times Y = \{(x,y) : x \in X, y \in Y\}$ into a topological space. If we are given neighborhood bases in each space, a natural way of doing this is to take as a neighborhood base of the point (x,y) the sets $V \times W$, in which V is a basic neighborhood of x and W is a basic neighborhood of y. As an elementary example, by taking the usual neighborhood base of open intervals for each point in the real line, the resulting neighborhood base for points in the plane consists of open rectangles around each point, giving us the usual topology on $\mathcal{R}^2$.

We generalize this to the case of n topological spaces $X_1, X_2, \ldots, X_n$. The *product topology* is the one for which we take a neighborhood base of the point $(x(1), x(2), \ldots, x(n))$ in the space $X_1 \times X_2 \times \cdots \times X_n$ to be all sets of the form $V_1 \times V_2 \times \cdots \times V_n$, in which each V_i is a basic neighborhood of $x(i)$. It is easily seen that these sets satisfy the two axioms for a neighborhood system that we gave prior to stating Theorem 7.1 and also that the topology does not depend on the particular neighborhood base we choose in each of the component spaces. This definition has the desirable feature of making the projection maps p_i continuous, where p_i is the mapping from the product space to X_i that sends $(x(1), x(2), \ldots, x(n))$ to $x(i)$.

When it comes to a product of infinitely many spaces, we modify the procedure. We could define the basic neighborhoods exactly as above, but this gives a topology that is too strong for most applications. Instead, we require that in forming a neighborhood of a point in the product all but finitely many V_i's have to be the entire space X_i. It is convenient to view the product space as a space of functions, similar to the point of view we took in Section 2.10. The formal definition from that point of view is as follows.

Definition. Let S be any set, and suppose for each $s \in S$ we are given a topological space X_s. Let Y be the set of all functions f on S such that $f(s)$ is an element of X_s. For each fixed f consider set V of the following form. Choose finitely many points $s_1, s_2, \ldots, s_m \in S$, and for each such point choose a neighborhood V_i of $f(s_i)$ in X_{s_i}. Let

$$V = \{g \in Y : g(s_i) \in V_i,\ s = 1, 2, \ldots, m\}.$$

Define a neighborhood base at f to be the collection of all sets V formed in this way for all possible finite sets and neighborhoods as given. The topology defined by this collection of basic neighborhoods of each point is known as the *product topology*

We leave it to the reader to verify that the definition does satisfy the requirements for a set of basic neighborhoods, that it agrees with the definition above for a finite index set S, and that it preserves the requirement that the projection maps are continuous. The set Y is usually referred to as the *product* of the given spaces and sometimes written as $\prod_{s \in S} X_s$.

Another easily shown fact is that convergence in the product topology is precisely that of pointwise convergence of functions. If we had not restricted the neighborhood bases as we did, the converence would be strictly stronger than pointwise convergence in the case of an infinite index set.

Still another result that would not be true for infinite index sets, had we not restricted the neighborhood bases, is Tychonoff's theorem, which is no doubt the best known and most widely used fact about the product topology. We will use it in a later section of the chapter when discussing the weak* topology. A proof is given in Appendix E.

Theorem 7.3 (Tychonoff's Theorem). *A product of compact spaces is compact.*

Application: Graph Theory. Product topologies have found their way into studies of discrete mathematical objects, where one might think there was no application of topological notions. Take an infinite collection of sets, with no topological structure at all, so we can consider in effect that they each have the discrete topology. It is not hard to see that the product of all these sets, with the product topology, is not discrete, and topological methods can be brought to bear. For one of many examples, see Bollobas (1998, Exercise III.31) for an application of Tychonoff's theorem to graph theory.

7.2 TOPOLOGIES ON LINEAR SPACES

We now return to our main theme of linear spaces. We want to consider a *topological linear space* (also known as a *topological vector space*). This is a linear space X equipped with a topology such that the operations of addition and scalar multiplication are continuous. This structure will be referred to as a *linear topology.* We have to clarify what we mean by continuity here. In the particular case of a normed linear space, we phrased these conditions in terms of sequences (Theorem 2.3). For a general topological vector space, we will assume that a neighborhood base at each point has been chosen, and we give continuity conditions in terms of these bases.

Continuity of addition means that the mapping from $X \times X$ to X that sends (x, y) to $x + y$ is continuous at all points when we put the product topology on $X \times X$. In other words, given any basic neighborhood W of a point $x_0 + y_0$, we can find basic neighborhoods U and V of x_0 and y_0, respectively, such for all $x \in U$ and $y \in V$ we have $x + y \in W$.

Continuity of scalar multiplication means that the mapping from $\mathcal{F} \times X$ to X that sends (α, x) to αx is continuous when we put the usual topology on $\mathcal{F}$ and the product topology on $\mathcal{F} \times X$. In other words, given any basic neighborhood W of a point $\alpha_0 x_0$, there is an $\epsilon > 0$ and a basic neighborhood U of x_0 such that if $|\alpha - \alpha_0| < \epsilon$ and if $x \in U$, then $\alpha x \in W$.

It follows that in any topological linear space, if we fix $x_0 \in X$, then the mapping from X to itself that sends x to $x + x_0$ is continuous, and it has a continuous inverse (sending x to $x - x_0$), so it is a homeomorphism. A major conclusion of this is that the topology can be given by specifying a neighborhood base $\mathcal{V}$ at the point 0 only. We can then take the collection $\{x + V : V \in \mathcal{V}\}$ for a neighborhood base at x.

Suppose we are given a collection of subsets $\mathcal{V}$ containing 0 such that the induced neighborhood bases at each point satisfy conditions 1 and 2 for neighborhood bases as given in Section 7.1.1. Our main problem now is to determine the additional conditions needed on $\mathcal{V}$ in order that we have a linear topology. We can get some insight by looking at properties that the neighborhoods of 0 (sometimes called 0-neighborhoods) must satisfy in a topological linear space.

Proposition 7.4 (Properties of 0-Neighborhoods). *Let U be a neighborhood of 0 in a topological linear space. Then:*

(*a*) *U contains a 0-neighborhood V such that $V + V \subseteq U$;*

(*b*) *0 is an internal point of U;*

(*c*) *U contains a 0-neighborhood W with the property that $\alpha W \subseteq W$ for all $|\alpha| \le 1$. (Any set with this property is said to be balanced.)*

Remark. It is instructive to note that these properties are simple to verify in a normed linear space. Clearly any open ball is balanced, and any 0-neighborhood U contains $B(\epsilon)$ for some $\epsilon > 0$, so (c) holds. We can then take $V = B(\epsilon/2)$ to satisfy (a). In Section 5.5.3, we noted that (b) holds.

Proof

(a) From the continuity of addition at $(0, 0)$, we can choose 0-neighborhoods V_1 and V_2 such that $V_1 + V_2 \subseteq U$. Then, $V = V_1 \cap V_2$ satisfies the required condition.

(b) For any $y \in X$, the continuity of the map from $\mathcal{F}$ to X that sends α to αy is continuous at 0, so we can find $\rho > 0$ such that for $|\alpha| \leq \rho$ we have $\alpha y \in U$, thereby showing that 0 is internal.

(c) From the continuity of scalar multiplication at $(0, 0)$ (the first zero here being the one in $\mathcal{F}$), we can find $\rho > 0$ and a 0-neighborhood V such that for $|\alpha| < \rho$ we have $\alpha V \subseteq U$. Let W be the union of the sets αV for $|\alpha| < \rho$. W is clearly balanced and a 0-neighborhood contained in U. □

It turns out that these three necessary properties, together with the two properties given before for a general space, are sufficient for a neighborhood base at 0 in a topological linear space.

Theorem 7.5. *Let X be a linear space and let $\mathcal{V}$ be a collection of balanced subsets, each containing 0 as an internal point and satisfying the following:*

(i) The intersection of any two sets in $\mathcal{V}$ contains a set in $\mathcal{V}$.
(ii) If $y \in U \in \mathcal{V}$, then for some $V \in \mathcal{V}$, we have $y + V \subseteq U$.
(iii) For each $U \in \mathcal{V}$, there is a $V \in \mathcal{V}$ such that $V + V \subseteq U$.

Then, the topology in which the neighborhood base at each $x \in X$ is $\{x + V : V \in \mathcal{V}\}$ makes X into a topological linear space.

Proof. From Theorem 7.1, conditions (i) and (ii) ensure that we get a topolocial space with the stipulated neighborhood bases. We must now prove the continuity of the operations. For any $x_0, y_0 \in X$, a basic neighborhood of $x_0 + y_0$ is of the form $x_0 + y_0 + U$ for some $U \in \mathcal{V}$. By choosing V to satisfy (iii), we obtain that if $x \in x_0 + V$ and $y \in y_0 + V$, then $x + y \in x_0 + y_0 + V + V \subseteq x_0 + y_0 + U$, thereby verifying the continuity of addition.

To prove continuity of scalar multiplication, we first establish the following fact:

$$\text{Given } U \in \mathcal{V} \text{ and } k > 0, \text{ there is } W \in \mathcal{V} \text{ such that } |\alpha| \leq k \text{ implies that } \alpha W \subseteq U. \quad (*)$$

To this end, first choose n so that $2^n > k$. Then, by iteration on (iii), we can find $W \in \mathcal{V}$ such that the sum of 2^n copies of W is in U. Statement $(*)$ then follows from the fact that W and U are balanced. (For example, take $k = 3$ and $\alpha = -2.6$. Choose V such that $V + V \subseteq U$; then, choose W such that $W + W \subseteq V$. Then, for any $w \in W$, we have $2.6w = w + w + 0.6w + 0 \in W + W + W + W \subseteq V + V \subseteq U$, and since U is balanced, we have $-2.6w \in U$.)

We now complete the proof of continuity. Given any $\alpha_0 \in \mathcal{F}$ and $x_0 \in X$, a basic neighborhood of $\alpha_0 x_0$ is of the form $\alpha_0 x_0 + U$ for some $U \in \mathcal{V}$. We have to find $r > 0$ and $W \in \mathcal{V}$ such that for $|\alpha - \alpha_0| < r$ and $x \in x_0 + W$ we have $\alpha x - \alpha_0 x_0 \in U$. By using two-variable splitting, we obtain

$$\alpha x - \alpha_0 x_0 = \alpha(x - x_0) + (\alpha - \alpha_0)x_0. \quad (**)$$

After choosing $V \in \mathcal{V}$ such that $V + V \subseteq U$, we then pick r and W so that each of the two terms on the right side of $(**)$ is in V, which will complete the proof. To do so, we first choose r so that for $|\rho| < r$ we have $\rho x_0 \in V$, which is possible since 0 is internal to the balanced set V. This shows that the second term on the right side of $(**)$ is in V. We use $(*)$ above to choose W so that for $|\alpha| < |\alpha_0 + r|$ we have $\alpha W \subseteq V$; thus, the first term is in V. □

There is another property that is often imposed on the collection $\mathcal{V}$ in Theorem 7.5 to ensure that the resulting topology is Hausdorff. Namely, we want the intersection of all sets in $\mathcal{V}$ to be 0. Then, given $x \neq y$, choose $U \in \mathcal{V}$ such that $x - y \notin U$, and then choose $V \in \mathcal{V}$ with $V + V \subseteq U$. The result is that $x + V$ and $y + V$ are disjoint neighborhoods of x and y, respectively. (We use the fact that V is balanced so that $-V$ is V.) Conversely, in any Hausdorff space, the intersection of all 0-neighborhoods is 0, for if $x \neq 0$, we can find disjoint neighborhoods of x and 0, so that x is not in the intersection of all 0-neighborhoods. Compare this with the case of a seminorm. For the collection $\mathcal{V} = \{B(r) : r > 0\}$, the intersection of all sets is $\{x : \| x \| = 0\}$, so our condition is met precisely when the seminorm is a norm.

As with normed spaces, we often want to compare topologies on a linear space, and this can be done by a direct analogue of Proposition 2.24. Suppose we have two topologies given by the neighborhood bases at 0 of $\mathcal{V}$ and $\mathcal{W}$, respectively. If every $W \in \mathcal{W}$ contains some $V \in \mathcal{V}$, then the topology determined by $\mathcal{W}$ is weaker. If every $W \in \mathcal{W}$ contains some $V \in \mathcal{V}$ and also every $V \in \mathcal{V}$ contains some $W \in \mathcal{W}$, then the topologies are the same.

7.3 LINEAR FUNCTIONALS ON TOPOLOGICAL LINEAR SPACES

Do our results for linear functionals on a normed linear space carry forward to the general topological linear space? The answer is that some do and some do not. The continuity of operations tells us that, as in a normed linear space, $N(f)$ is either closed or dense. Moreover, we get the following result, as we did in the normed space context. (See Theorem 2.32.)

Theorem 7.6. *If f is a linear functional on a topological vector space such that $N(f)$ is closed, then f is continuous.*

Proof. Suppose $N(f)$ is closed. As with a normed linear space, we need only prove that f is continuous at 0. Given any $\epsilon > 0$, we must find a 0-neighborhood U such that for $y \in U$ we have $|f(y)| < \epsilon$. To do so, choose x such that $0 < |f(x)| < \epsilon$. (We can assume f is not the 0 functional.) Then, since x is not in the closed set $N(f)$, there is a balanced 0-neighborhood U such that $x + U$ does not intersect $N(f)$. If our assertion is false, there exists $y \in U$ with $|f(y)| \geq \epsilon$. Let $\alpha = -f(x)/f(y)$. So $|\alpha| < 1$, and therefore $x + \alpha y \in (x + U) \cap N(f)$, a contradiction. □

When it comes to the existence of linear functionals, however, we can run into the dreaded situation we mentioned at the beginning of Chapter 5.

Example 7.7. When discussing L^p spaces, we restricted p to be greater than 1. We now consider the case for which $p < 1$. We consider the space of real-valued, measurable functions on the interval $[0, 1]$ and define a "norm" by

$$\| f \| = \int_0^1 |f(s)|^p \, ds.$$

As before, let $L^p[0, 1]$ consist of those functions with finite norm.

We do not put in the $1/p$ power since that would destroy the triangle inequality which goes the other way (see Exercise 2.39). Therefore, we do not get the necessary scaling, and $\| \cdot \|$ is not a norm. However, we do have a translation-invariant metric defined by $d(x, y) = \| x - y \|$, and this defines a perfectly good topological linear space with a neighborhood base at 0 given as usual by the sets $\{B(r)\}$ for $r > 0$.

Take any nonzero linear functional ϕ on $L^p[0, 1]$, and choose an element $f \notin N(\phi)$ with $\| f \| = 1$. By replacing f with $-f$ if necessary, we can assume that $\phi(f) > 0$. We are going to define a sequence of functions (h_n) in the space such that $h_n \to 0$ but $\phi(h_n) \geq \phi(f)$ for all n. This means that ϕ is not continuous. So, our space does not admit *any* nonzero continuous linear functional!

We first define a nonnegative function σ on $[0, 1]$ by $\sigma(t) = \int_0^t |f(s)|^p \, ds$. We can show that σ is continuous. Since σ increases from 0 to 1, the intermediate-value theorem tells us that we can find, for any n, points $0 = t_0 < t_1 < \cdots < t_n = 1$ such that $\sigma(t_k) = k/n$. Let g_k be the function that agrees with f on $(t_{k-1}, t_k]$ and is 0 elsewhere. Thus, $f = g_1 + g_2 + \cdots + g_n$ a.e. By linearity, we must have $\phi(g_k) \geq \phi(f)/n$ for at least one value of k. Let $h_n = ng_k$; then, clearly $\phi(h_n) \geq \phi(f)$. However,

$$\| h_n \| = \int_{t_{k-1}}^{t_k} |ng_k|^p(s) ds = n^p [\sigma(t_k) - \sigma(t_{k-1})] = n^{p-1} \to 0.$$

Recalling the geometric form of the Hahn–Banach theorem gives us a clue as to where the difficulty lies with such spaces. The problem is that balls are not convex; see Figure 7.1 for a picture in $\mathcal{R}^2$. In the presence of suitable convexity requirements, as we have in the case of a norm, we recover some of the conclusions of the Hahn–Banach theorem.

Definition. A topological linear space is said to be *locally convex* if every 0-neighborhood contains a convex 0-neighborhood.

A normed linear space is clearly locally convex since $B(r)$ is convex.

Remark. Our previous separation result (Theorem 5.15) was stated for a real linear space. There are a few theorems in the remainder of this chapter in which we want to apply this theorem to a topological linear space which may be over $\mathcal{C}$. We use the

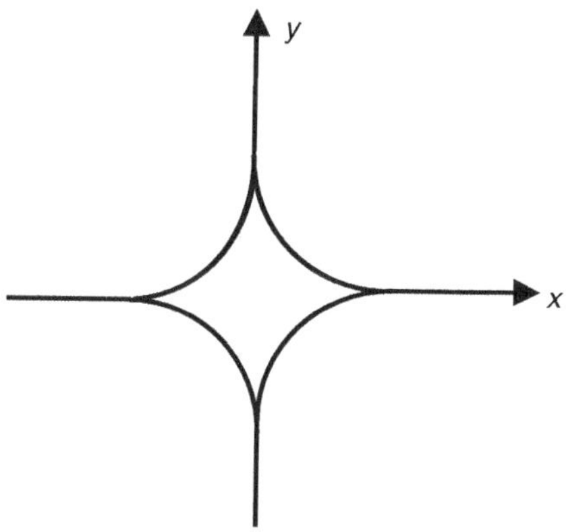

Figure 7.1. Unit ball for $\frac{1}{2}$ norm.

following argument to do so. View the space as a real linear space, and find a real linear functional g such that the associated hyperplane separates two given convex sets. From the discussion in Section 5.3, we know there is a unique complex functional f such that $\text{Re} f = g$. Moreover, from continuity of operations, we can verify that g is continuous if and only if f is continuous.

Following is a major consequence of local convexity.

Theorem 7.8. *Given any two points $x \neq y$ in a locally convex, Hausdorff topological linear space, there is a continuous linear functional f such that $f(x) \neq f(y)$.*

Proof. We note first that in a topological linear space, as in a normed linear space, it follows from Proposition 7.4(b) that any interior point of a set is internal. Pick convex 0-neighborhoods U and V such that $x + U$ and $y + V$ are disjoint. By Theorem 5.15, we can find a hyperplane separating the two convex sets $x + U$ and $y + V$. This means that, for some linear functional f and scalar α, $\text{Re} f(x + u) \leq \alpha \leq \text{Re} f(y + v)$ for all $u \in U$ and $v \in V$. Since all points of the sets are internal, we can, by Proposition 5.13, take the inequality signs as strict, so that $\text{Re} f(x) \neq \text{Re} f(y)$. Now, $\text{Re} f^{-1}(\alpha)$ is not dense since it does not intersect the open set $x + U$. By Theorem 7.6 and our remark before the theorem, f is continuous. □

7.4 WEAK TOPOLOGY

It is now time to give an example of a locally convex linear topology other than that coming from a norm. We start with a normed linear space X. We want to define a topology that is associated with the notion of weak convergence discussed in Section 6.4. For any finite subset F of unit vectors in X^* and any $\epsilon > 0$, let

$$V(F, \epsilon) = \{x \in X : |f(x)| < \epsilon \text{ for all } f \in F\}.$$

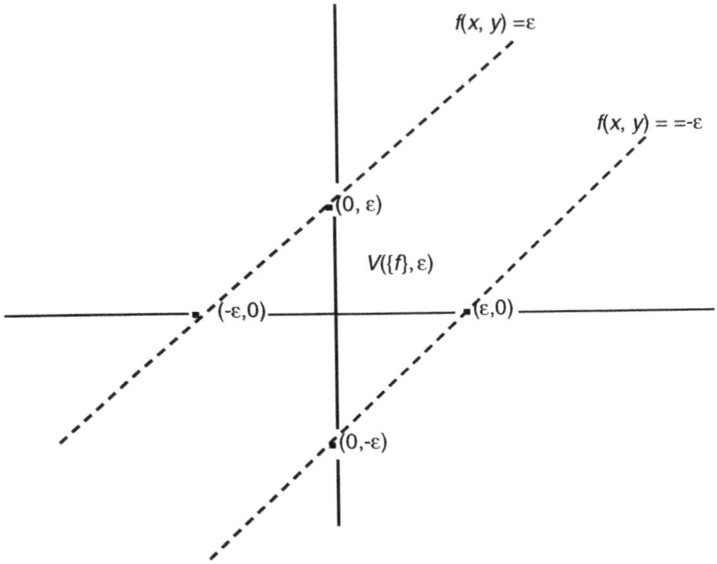

Figure 7.2. Weak neighborhood of 0 $f(x,y) = y - x$.

For a simple illustration in $\mathcal{R}^2$, see Figure 7.2, which shows such a set in which F consists of the single functional $f(x,y) = y - x$.

Theorem 7.9. *The sets $\{V(F,\epsilon)\}$, as F ranges over the finite subsets of unit vectors in X^* and ϵ over the positive numbers, satisfy the conditions of Theorem 7.5 and therefore consitute a neighborhood base at 0 for a linear topology on X, known as the weak topology. This topology is locally convex and Hausdorff. A sequence (x_n) converges in this topology if and only if it converges weakly as previously defined.*

Proof. We verify the conditions of Theorem 7.5. The sets are clearly balanced. Moreover, given $V(F,\epsilon)$, for all $x \in X$ define

$$k_x = \max\{|f(x)| : f \in F\} \qquad \text{and} \qquad r_x = \frac{\epsilon}{k_x}.$$

Then, for $0 \le \rho < r_x$, we have $|f(\rho x)| < \epsilon$ for all $f \in F$. This shows that 0 is an internal point of $V(F,\epsilon)$. We now verify conditions (i) to (iii) of Theorem 7.5.

(i) $V(F,\epsilon) \cap V(G,\delta) \supseteq V(F \cup G, \min(\epsilon,\delta))$.

(ii) For any $y \in V(F,\epsilon)$, it follows from the definitions that $k_y < \epsilon$. Now, if $w \in V(F, \epsilon - k_y)$, then for all $f \in F$

$$|f(y+w)| \le |f(y)| + |f(w)| < k_y + \epsilon - k_y = \epsilon.$$

We have therefore shown that

$$y + V(F, \epsilon - k_y) \subseteq V(F,\epsilon).$$

(iii) $V(F,\epsilon/2) + V(F,\epsilon/2) \subseteq V(F,\epsilon)$, thereby completing the derivation that we have a linear topology.

It is straightforward to see that $V(F,\epsilon)$ is convex, so this is a locally convex topology. Finally, we note that this is a Hausdorff topology since for any $x \neq 0$ we can find by the Hahn–Banach theorem a continuous linear functional with $f(x) \neq 0$, and by multiplying a suitable scalar, we can assume $f(x) = \alpha > 0$. This means that $x \notin V(\{f\},\alpha)$.

We leave the final statement of the theorem, referring to weak convergence, as an exercise. □

Another point to notice is that if $c = \max\{\| f \| : f \in F\}$, then $V(F,\epsilon) \supset B(\epsilon/c)$, as suggested by Figure 7.2. This shows that the weak topology is weaker than the norm topology, justifying the terminology. We have seen cases (Example 6.13) where it is strictly weaker, and this will in fact hold for all infinite-dimensional spaces; see Exercise 7.7. In finite-dimensional spaces, the weak topology is the same as the norm topology. Insight to this is also provided by Figure 7.2. If we added to F another function g so that f and g were linearly independent, the weak neighborhood would be the intersection of the strip shown in Figure 7.2 with another non parallel strip. For small enough ϵ, the intersection would be contained in any norm neighborhood. In general, taking F to consist of n linear independent functionals yields such a weak neighborhood in $\mathcal{R}^n$.

As indicated above, when we have a weaker topology, neighborhoods get larger, and it is harder for a function from the space to be continuous. Thus, a natural question is, as we pass from the norm topology to the weak topology on a normed linear space X, which elements of X^* remain continuous? The answer is that all of them do.

Theorem 7.10. *When we equip X with the weak topology, all elements of X^* are continuous. Moreover, this is the weakest linear topology on X that has this property.*

Proof. To show that a given $f \in X^*$ is continuous, we need only show it is continuous at 0, as we noted in the proof of Theorem 7.6. Given $\epsilon > 0$, we must show there is a weak 0-neighborhood U with $|f(x)| < \epsilon$ for all $x \in U$. This is immediate since we can take $U = V(\{f\},\epsilon)$.

Now, suppose we have a linear topology on X such that all elements of X^* are continuous. Let $\mathcal{W}$ denote the collection of 0-neighborhoods for such a topology. We want to show that any $V(F,\epsilon)$ contains an element of $\mathcal{W}$. For each $f \in F$, choose $U_f \in \mathcal{W}$ such that $|f(x)| < \epsilon$ for all $x \in U_f$. Then $\bigcap_{f\in F} U_f$ is an element of $\mathcal{W}$ that is contained in $V(F,\epsilon)$. □

An important application of the above theorem is as follows. Suppose we are faced with the problem, which arises repeatedly in applications, of showing that a point x is in the closure of another set A. Suppose also that we succeed in showing that it is in the weak closure of A. In general, this will not help at all. When you close a set under a weaker topology, you draw in more elements, so a point may well be in the weak closure but not in the norm closure. However, when A is convex, the closures are the same.

Theorem 7.11. *The weak closure of a convex set A in a normed linear space is the same as the norm closure* $\bar{A}$.

Proof. For convenience we assume a real space. The extension to complex spaces follows in the standard way. Suppose $x \notin \bar{A}$. Then, for some $\epsilon > 0$, the ball $B(x;\epsilon)$ does not intersect $\bar{A}$. By the Hahn–Banach separation theorem (Theorem 5.15), we can find $f \in X^*$ and a real number c such that $f(x) > c$ and $A \subseteq f^{-1}(-\infty, c]$. Since f is weakly continuous, $f^{-1}(-\infty, c]$ is a weakly closed set containing A but not x. Therefore x is not in the weak closure of A. □

Some geometric insight into the last theorem is furnished by Theorem 5.16. The closure of K is the intersection of the closed half spaces containing K. These are the same in both topologies since from Theorem 7.11 we can infer that the closed hyperplanes are the same.

7.5 WEAK* TOPOLOGY

We now discuss the topology which is associated with the weak* convergence of Section 6.4.

7.5.1 Basic Definitions

Given a normed linear space X, we define the weak* topology on the dual space X^* as the one with a neighborhood base at 0 consisting of the sets $V^*(F,\epsilon)$ defined as follows. For a finite subset F of unit vectors in X and $\epsilon > 0$,

$$V^*(F,\epsilon) = \{f \in X^* : |f(x)| < \epsilon \text{ for all } x \in F\}.$$

The following theorem summarizes the facts. We omit the proof because the statements and derivations parallel almost exactly those of Theorem 7.9.

Theorem 7.12. *The sets* $\{V^*(F,\epsilon)\}$ *satisfy the conditions of Theorem 7.5 and, therefore, constitute a neighborhood base at 0 for a linear topology on* X^*, *known as the weak* topology. This topology is locally convex and Hausdorff. A sequence* (f_n) *converges in this topology if and only if it converges weak*-*ly, *as previously defined.*

Of course, X^* also has a weak topology as determined by X^{**}. How do these compare? Letting V refer to a basic weak neighborhood of X^*, it is clear that

$$V^*(F,\epsilon) = V(\hat{F},\epsilon),$$

in which the "hat" denotes the natural embedding of X into its double dual. This shows that the weak* topology is *weaker* than the weak topology. In the case of a reflexive space, every finite subset of X^{**} is of the form $\hat{F}$ for some $F \subseteq X$, so the topologies coincide. The weak* topology is therefore essentially a tool designed to deal with nonflexive spaces.

7.5.2 Weak* Compactness

Here we present one of the major results concerning the weak* topology. It can be looked upon as recovering in some sense the loss of compactness of closed and bounded sets when we move from finite-dimensional to infinite-dimensional spaces.

Theorem 7.13 (Alaogulu's Theorem). *The closed unit ball of X^* is compact in the weak* topology.*

Proof. Let $\bar{B}^*(1)$ denote the closed unit ball of X^*. Let $Y = \{f : X \to \mathcal{F} : |f(x)| \leq \| x \|\}$. We can then view Y as the product of the spaces D_x as x ranges over X, in which D_x is the set of all scalars of absolute value less than or equal to $\| x \|$. Equip Y with the product topology. By Tychonoff's theorem (Theorem 7.3), Y is compact. Observe that $\bar{B}^*(1)$ is a subspace of Y consisting precisely of those functions f that are linear. Moreover, by comparing the definitions, we see that the subspace topologies on $\bar{B}^*(1)$, induced first by the product topology on Y and then by the weak* topology on X^*, are the same. Finally, it follows from the continuity of operations that $\bar{B}^*(1)$ is closed in Y and therefore compact; see Exercise 7.1(a). □

7.5.3 Embedding X into X^{**} in Nonreflexive Case

Let us return to comparing the weak and weak* topologies on X^* when X is nonreflexive. Under the weak topology, all elements of X^{**} are continuous. We can ask a question analogous to that asked above for the weak topology on X. When we weaken the topology on X^* from the norm to the weak* topology, which elements of X^{**} remain continuous? This will clearly be some subspace of X^{**}. Aside from the whole space, there is only one other reasonable choice of distinguished subspace, and this does indeed turn out to be what we want.

Theorem 7.14. *The weak* continuous linear functionals on X^* are precisely the elements in $\hat{X}$.*

As a preliminary step in the proof, we present a strictly algebraic result, which is of interest in its own right. Suppose we are given a collection of linear functionals $f_1, f_2, \ldots, f_n$ on a linear space, and we want to know whether a particular linear functional f can be written as a linear combination of the given ones. An obvious necessary condition is that given a point x for which $f_i(x) = 0$ for all i we must have $f(x) = 0$. It turns out that this is sufficient as well.

Lemma 7.15. *Given linear functionals $f_1, f_2, \ldots, f_n$ on a linear space X, any linear functional f such that $N(f) \supseteq \bigcap_{i=1}^{n} N(f_i)$ can be written as a linear combination of the given functionals.*

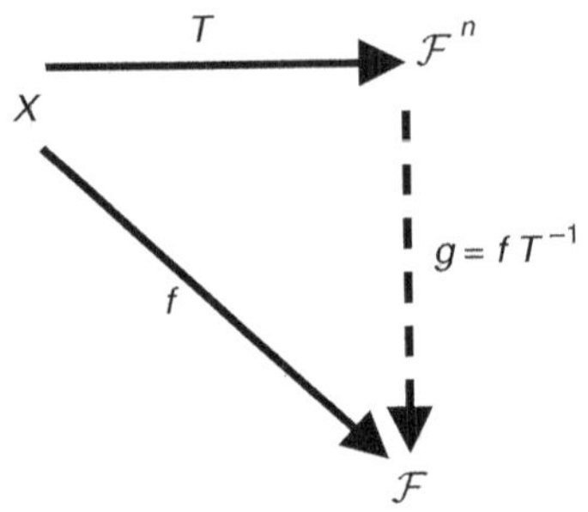

Figure 7.3. Map g of Lemma 7.15.

Proof. Define a linear transformation $T : X \to \mathcal{F}^n$ by

$$T(x) = (f_1(x), f_2(x), \ldots, f_n(x)).$$

Define a linear functional g on $R(T)$ by $g(y) = f\,T^{-1}y$; see Figure 7.3. Note that T^{-1} here could be multivalued, but g is well defined, since if $Tx_1 = Tx_2$, then $x_1 - x_2$ is in the intersection of the null spaces of all the f_i's, and by our hypothesis $f(x_1) = f(x_2)$. As we showed at the beginning of Section 6.1, there exists a vector $z = (z(1), z(2), \ldots, z(n)) \in R(T)$ such that for all $y = (y(1), y(2) \ldots y(n)) \in R(T)$ we have

$$g(y) = \sum_{i=1}^{n} z(i)\, y(i).$$

Then, for all $x \in X$,

$$f(x) = g(T(x)) = \sum_{i=1}^{n} z(i) f_i(x),$$

which shows that $f = \sum_{i=1}^{n} z(i) f_i$. □

Proof of Theorem 7.14. Let ϕ be a weak* continuous linear functional on X^*. Then, we can find a basic weak* neighborhood $V^* = V^*(F, \epsilon)$ such that $|\phi(f)| < 1$ for all $f \in V^*$. Suppose g is a point in X^* such that $\hat{x}(g) = 0$ for all $x \in F$. This implies that $kg \in V^*$ for all positive k, so that $|k\phi(g)| \leq 1$. But since k can be chosen arbitrarily large, we must have $\phi(g) = 0$. By Lemma 7.15, ϕ is a linear combination of elements in $\hat{X}$ and therefore belongs to $\hat{X}$. □

Remark. The above proof can be adapted to yield a general result that encompasses both Theorems 7.10 and 7.14. Take any linear space X and a subspace M of the linear functionals on X. Take the weakest linear topology on X that makes all the elements of M continuous. Then, the only continuous linear functionals on X with respect to this topology are those in M. In other words, by choosing the weakest topology that we can get away with, we do not let any other functionals qualify as continuous.

We now give a key application of Theorem 7.14. When X is nonreflexive, there are elements in the double dual that do not come from X. We can show, however, that all elements can be approximated by elements of X with respect to the weak* topology. This turns out to be a useful tool when dealing with nonreflexive spaces. We actually will show a somewhat stronger result, namely, that we can also control the norm of the approximating element, which is often important to do. The precise statement is as follows.

Theorem 7.16. *The closed unit ball of $\hat{X}$ is dense in the closed unit ball of X^{**} when the double dual is equipped with the weak* topology.*

Proof. By way of contradiction, assume the assertion is not true. We can then find $\phi \in \widehat{\bar{B}^{**}}(1)$ (the closed unit ball in the double dual) that is not in the weak* closure of $\widehat{B(1)}$. Therefore, we can find a weak* basic neigborhood U such that $\phi + U$ does not intersect $\widehat{B(1)}$. Viewing the double dual as a real linear space, we can find a hyperplane H separating the two sets. By the remark after Theorem 5.15, we can find a linear functional v on X^{**} such that $H = (\text{Re}\, v)^{-1}(1)$. Since $v(0) = 0$, we conclude that $\| x \| \leq 1$ implies $\text{Re}\, v(\hat{x}) \leq 1$ and $\text{Re}\, v$ takes values greater than 1 on $\phi + U$. Moreover, H is not weak* dense since it does not intersect the weak* open set $\phi + U$, so it must be closed. Then, by Theorem 7.6, $\text{Re}\, v$ is weak* continuous, implying that v is weak* continuous. From Theorem 7.14, $v = \hat{f}$ for some $f \in X^*$. For all $x \in X$, we have

$$f(x) = \hat{x}f = \hat{f}\hat{x} = v(\hat{x}).$$

It follows that, for $\| x \| \leq 1$, $\text{Re}\, f(x) \leq 1$, and since the closed unit ball is balanced, we must in fact have that $|f(x)| \leq 1$. We conclude that

$$|v(\phi)| = |\hat{f}(\phi)| = |\phi(f)| \leq 1,$$

contradicting the fact that $\text{Re}\, v(\phi) > 1$. □

As a corollary, we get an interesting companion to Theorem 2.34.

Theorem 7.17. *A normed linear space X is reflexive if and only if its closed unit ball is compact in the weak topology.*

Proof. If X is reflexive, we may identify it with X^{**}, and its weak topology is exactly the same as the weak* topology on the latter. From Theorem 7.13, its closed unit ball is weakly compact.

Conversely, suppose that $\bar{B}(1)$ is weakly compact. The mapping that sends x to $\hat{x}$ is easily seen to be continuous when X has the weak topology and X^{**} has the weak* topology. Therefore, the image of $\bar{B}(1)$ under this mapping is compact and closed; see Exercise 7.1. But, since it is also dense, it must be all of the unit ball of X^{**}. This shows that the embedding is onto. □

7.6 EXTREME POINTS AND KREIN–MILMAN THEOREM

In this section, we present another major theorem of functional analysis which, like the Hahn–Banach theorem, is of a geometric character.

7.6.1 Introduction

Given a mathematical object of some type, one often searches for a "generating set," which we want to be as small as possible. We do not want to define this term precisely. It often takes the form of a subset such that the smallest set of a given type containing that subset is the whole set with which we started. For example, in a linear space, a basis forms such a set since the smallest subspace containing the basis is the whole space. In a topological space, a dense subset has the property that the smallest closed subset containing that set is the whole space. Information about a generating set often gives us the information we desire about the whole space. For example, the values of a linear operator on a basis of a linear space determine the operator completely. Similarly, the values of a continuous function on a dense set determine the function completely.

In this section, we are concerned with convex sets. Take, for example, a triangle in the plane. The smallest convex subset containing the three vertices is the whole triangle. So, these three vertices may be thought of a generating set with regard to convexity. The question is, then, in a general convex set, how do we select a subset that will play the role of vertices?

7.6.2 Extreme Points

Definition. A point x in a convex set K is said to be an *extreme point* of K if whenever $x = ty + (1-t)z$ for some t in the interval $(0, 1)$ and $y, z \in K$, we must have $y = z = x$. In other words, in a real space, there is no line segment in K containing x, other than the degenerate segment consisting of the point x itself.

As examples, for any closed polygon in the plane, the vertices are extreme. For a disc in the plane, all boundary points are extreme.

Definition. For any subset A of a linear space we define the *convex hull* of A, denoted by $\text{co}(A)$, to be the intersection of all convex sets containing A. Here, $\text{co}(A)$ is convex and therefore the smallest convex subset containing A. It can also be characterized as the set of all points that can be written in the form $\sum_{i=1}^{n} t_i a_i$ for some positive scalars t_i adding to 1 and points a_i in A.

Our problem as stated in the previous section is as follows. Given a convex set K, for what sets A will $\mathrm{co}(A) = K$? It seems clear that we must include the extreme points in A, for there is no way that they can appear as a linear combination of other points. The question then is, as we observed for the triangle, is it enough to take the extreme points, or do we need others?

A first observation is that no internal point of a convex set can be extreme, so if we are in a topological vector space, the extreme points can only occur on the boundary. Moreover, we can see from simple examples in the plane that unbounded sets will not be the convex hull of their extreme points. For this reason, the question makes sense only for a compact convex set. For such a set K in a finite-dimensional space, it is indeed true that $K = \mathrm{co}(E)$, in which E is the set of extreme points of K. The Krein–Milman theorem extends this to infinite-dimensional spaces. We must, however, make a few modifications. There is no guarantee in the infinite-dimensional case that $\mathrm{co}(E)$ is closed, so we want to show that K equals the closure of $\mathrm{co}(E)$. We also need some restrictions on the space. The formal statement is as follows.

Theorem 7.18 (Krein–Milman Theorem). *If K is a compact, convex set in a locally convex, Hausdorff topological linear space, then K is the closure of $\mathrm{co}(E)$, in which E is the set of extreme points of K.*

We will give the proof after first developing some background.

7.6.3 Faces

A face of a convex set is intuitively a type of "extreme" set which generalizes the idea of an extreme point. Precisely, we have the following.

Definition. A *face* of a convex set K is a subset F such that given any x, y in K and $t \in (0, 1)$

$$tx + (1 - t)y \in F \text{ if and only if } x, y \in F.$$

A face, then, is a convex subset of K such that if the set intersects the interior of a line segment, it also contains the endpoints of the segment. In the case of a triangle, the three sides are faces (as, of course, is the whole triangle). Extreme points are faces that consist of a single point.

Following are three elementary facts about faces.

Proposition 7.19

(*a*) *If F is a face of G and G is a face of K, then F is a face of K.*

(*b*) *Any nonempty intersection of a collection of faces of a set is a face of that set.*

(c) *Let f be a real linear functional on the whole space. Suppose that, on the convex set K, f assumes a maximum value M. Then, $K_f = \{x \in K : f(x) = M\}$ is a face of K.*}

Proof. We will prove (c) and leave (a) and (b) to the reader. We have $tx + (1-t)y \in K_f$ if and only if $tf(x) + (1-t)f(y) = M$. Because M is the maximum value assumed by f, this is possible if and only if both $f(x)$ and $f(y)$ equal M, so that both x and y are in K_f. □

Remark. In part (c), we may clearly replace maximum by minimum by looking at $-f$. For a closed and bounded polytope in $\mathcal{R}^n$, the optimal values of a linear function on this set must occur at some extreme points since every face contains such points.

Application: Linear Programming. Linear programming seeks to optimize linear functionals on $\mathcal{R}^n$ subject to a set of linear inequalities. The most commonly used method by far is the *simplex algorithm*, which is based directly on the remark above. The set of points satisfying a system of linear inequalities form a convex polytope and the algorithm seeks the optimum by looking at the value of the functional at extreme points.

We will prove the Krein–Milman theorem after giving another formulation of compactness. Suppose we have a collection of closed subsets of a compact set K such that the intersection of any finite subcollection is nonempty. Then, the intersection of the entire collection is nonempty. If this were not true, then the complements of the given family would form an open cover of K that has no finite subcover.

The key to proving the Krein–Milman theorem is the following.

Theorem 7.20. *Let K be a compact subset of a real topological vector space for which the continuous linear functionals separate points. Then, K has an extreme point.*

Proof. The idea is simply to select a minimal closed face K' of K. We claim that K' must consist of a single point. Suppose that, in fact, K' contains two distinct points x and y. Choose a continuous linear functional f for which $f(x) \neq f(y)$. Now, f must assume a maximum value on the set K', which is compact because it is a closed subset of K. But, the face K'_f cannot then include both x and y, contradicting the minimality of K'.

The problem remains, however, to show the existence of a minimal face. In finite dimensions, we can use a dimensional argument, but in general we need to use Zorn's lemma. The set of closed faces is not empty since it conains K. Order the closed faces by *reverse* inclusion. That is, $G \leq F$ if F is a subset of G. Any linearly ordered set in this poset has an upper bound, namely the intersection of all its elements. This is nonempty by the formulation of compactness that we gave before the statement of this theorem and therefore a face by Proposition 7.19(b). We conclude that a maximal element, which will be a minimal face, exists. □

We can now complete the proof of the main theorem.

Proof of Theorem 7.18. Let $J = \overline{\text{co}(E)}$. Then, J is convex by continuity of the operations. Suppose there is an x in K but not in J. By local convexity, there is a convex neighborhood U of x disjoint from J. By the Hahn–Banach separation theorem, there is a hyperplane H separating J and U, and H is equal to $(\text{Re}f)^{-1}(c)$ for some continuous linear functional f and real number c. By replacing f with $-f$ if necessary, we can assume that that for all $y \in J$ we have $\text{Re}f(y) \leq c < \text{Re}f(x)$. Therefore, $K_{\text{Re}f}$ is a face of K that is disjoint from J. But, then, $K_{\text{Re}f}$ is closed since f is continuous and therefore a compact convex subset of K. By Theorem 7.20, $K_{\text{Re}f}$ has an extreme point, and this point, by Proposition 7.19(a), is an extreme point of K, and therefore in J, which is a contradiction. □

Example 7.21. Consider the closed unit ball of ℓ_1 over the reals. We will show that the extreme points are precisely $\pm\delta_n$ for $n = 1, 2, \ldots$

Supppose $\delta_n = tg + (1-t)h$ for some g and h of norm less than or equal to 1 and for some $0 < t < 1$. Then, $1 = tg(n) + (1-t)h(n)$. But, since $g(n)$ and $h(n)$ cannot be bigger than 1, they must both be equal to 1, implying that $g(k) = h(k) = 0$ for all $k \neq n$, which means that $g = h = \delta_n$. Similarly, $-\delta_n$ is extreme.

It is clear that any point with norm less than 1 cannot be extreme, so consider any f of norm 1 other than $\pm\delta_n$. Then, there are least two indices k and n for which $|f(k)|$ and $|f(n)|$ are strictly less than 1. First, suppose that both $f(k)$ and $f(n)$ are of the same sign. Then, define functions g and h that agree with f except that, for some positive r, $g(k) = f(k) + r$, $g(n) = f(n) - r$, $h(k) = f(k) - r$, and $h(n) = f(n) + r$. When $f(k)$ and $f(n)$ are of different signs, we define g and h by $g(k) = f(k) + r$, $g(n) = f(n) + r$, $h(k) = f(k) - r$, and $h(n) = f(n) - r$.

In both cases, we have that $f = (g + h)/2$, and for small enough r, both g and h are unit vectors different from f, so that f is not extreme.

The validity of the Krein–Milman theorem is evident in this example since the convex hull of the extreme points is just those elements in c_{00} of norm less than or equal to 1, and the closure of this set is the closed unit ball in ℓ_1.

There is an interesting application of the Krein–Milman theorem to duality. Given any Banach space, a natural question is to determine whether or not it is the dual of some other space. Our list of examples that we have presented so far shows spaces that do not appear as the dual of another, for example c_0 and L^1. We can now provide reasons why these were excluded.

Theorem 7.22. *Let X be a normed linear space whose closed unit ball has no extreme points. Then, X cannot be a dual space.*

Proof. If X were a dual space, then the space would admit a locally convex, Hausdorff topology such that its closed unit ball is compact, namely the weak* topology. But, this contradicts the Krein–Milman theorem. □

The next question is whether such spaces exist. One example is c_0. Another is certain L^1 spaces. Clearly, not all L^1 spaces have this property as shown by Example 7.21, but the closed unit ball has no extreme points if the measure space is *atomless*, which means that any set of positive measures is the union of two disjoint sets, each of positive measure. An example is any nondegenerate interval of the reals with Lebesgue measure. Verification is left to the exercises.

7.7 OPERATOR TOPOLOGIES

There are some other topologies of interest that we put on the set $\mathcal{B}(X, Y)$, in which X, Y are normed linear spaces.

Definition. The *strong operator topology* is defined by a neighborhood base at 0 consisting of sets $V_{\text{op}}(F;\epsilon)$, in which F is a finite set of unit vectors in X and $\epsilon > 0$. Given F and ϵ, $V_{\text{op}}(F;\epsilon)$ is defined by

$$V_{\text{op}}(F;\epsilon) = \{T \in \mathcal{B}(X, Y) : \| Tx \| < \epsilon \text{ for all } x \in F\}.$$

This topology is just the one of pointwise convergence. A sequence (T_n) converges to T if $(T_n x)$ converges to Tx with respect to the norm for all $x \in X$.

It is clear from the last statement, as well as from the fact that $V_{\text{op}}(F;\epsilon)$ contains the norm 0-neighborhood $B(\epsilon)$, that the strong operator topology is *weaker* than the norm topology on $\mathcal{B}(X, Y)$. One might think from this that the terminology is quite inappropriate, but it is all explained by the fact that the strong operator topology is *stronger* than the *weak operator topology*. This latter is the topology in which (T_n) converges to T if $T_n x$ converges to Tx weakly for all $x \in X$. A neighborhood base is given by the sets $V^*_{\text{op}}(F, G^*;\epsilon)$, in which F is a finite set of unit vectors in X, G^* is a finite set of unit vectors in Y^*, $\epsilon > 0$, and

$$V^*_{\text{op}}(F, G^*;\epsilon) = \{T \in \mathcal{B}(X, Y) : |g(Tx)| < \epsilon \text{ for all } x \in F \text{ and } g \in G^*\}.$$

We will not dwell further on these topologies, except in the exercises.

EXERCISES

7.1 Let K be a compact set in a topological space X.
  (a) Show that any closed subset of K is compact.
  (b) Show that, for any continuous function f on X, $f(K)$ is compact.

7.2 Show that in any topological linear space the closure of a convex set is convex.

7.3 A set A in a topological linear space is said to be *bounded* if for each 0-neighborhood U there is $r > 0$ such that $A \subseteq rU$.
(a) Show that in a normed linear space this agrees with the usual definition.
(b) Show that any compact set is bounded.

7.4 Show that in any linear space, if K is a convex, balanced set with 0 as internal point, then the Minkowski functional p_K is a seminorm.

7.5 Suppose that X and Y are topological linear spaces and that $T : X \to Y$ is linear.
(a) Show that T is continuous if it is continuous at 0.
(b) Suppose that X and Y are normed linear spaces and that T is bounded. Show that T is continuous if we consider each space to have its weak topology.

7.6 Show that convergence with respect to the weak topology is the same as weak convergence defined in Chapter 6.

7.7 Show that in any infinite-dimensional normed linear space any neighborhood of 0 in the weak topology is unbounded with respect to the norm. (You may want to use Lemma 7.15.) Conclude that the weak topology is strictly weaker than the norm topology. Explain why this does not contradict Proposition 6.16.

7.8 Show that the closed unit ball of c_0 has no extreme points.

7.9 Find all the extreme points in the unit ball of ℓ^∞.

7.10 Show that for ℓ^p, $1 < p < \infty$, every unit vector is an extreme point of the closed unit ball.

7.11 Consider $L^1(\Omega)$, in which Ω is atomless. (For simplicity, assume the case of a real space.)
(a) Show that, for any f of norm 1, $\{x : |f(x)| \geq 1/n\}$ has positive measure for some positive integer n.
(b) Show that if f is bounded below on a set A of positive measure there is a function $g \in L^1$ such that $\| f + g \| = 1$.
(c) Conclude that the closed unit ball has no extreme points.

7.12 Show that in the space $C[0, 1]$ over the reals there are only two extreme points of the closed unit ball. Conclude that $C[0, 1]$ cannot be a dual space

7.13 Show that both the strong and weak operator topologies are locally convex and Hausdorff.

7.14 Show that the weak operator topology on $\mathcal{B}(X, Y)$ is weaker than the weak topology on this space.

7.15 Suppose that (S_n) and (T_n) are sequences in $\mathcal{B}(X)$ such that (S_n) and (T_n) converge to S and T, respectively, in the strong operator topology. Show that (S_nT_n) converges strongly to ST.

7.16 In l^2, consider the right- and left-shift operators S and T given by

$$S\delta_n = \delta_{n+1}; \qquad T\delta_n = \delta_{n-1} \quad \text{for } n > 1, \qquad T\delta_1 = 0.$$

(a) Show that for any 0-neighborhood V in the strong operator topology there is some $r > 0$ such that $rS^n \in V$ for all positive integers n.

(b) Show that for any 0-neighborhood V in the strong operator topology and for any $r > 0$ there is a positive integer n such that $T^n/r \in V$.

(c) Conclude that the multiplication mapping from $\mathcal{B}(\ell^2) \times \mathcal{B}(\ell^2)$ to $\mathcal{B}(\ell^2)$ is *not* continuous when $\mathcal{B}(\ell^2)$ has the strong operator topology.

7.17 Let $\mathcal{P}$ be a collection of seminorms on a linear space X. Define subsets $W(F, \epsilon)$ in which $\epsilon > 0$ and F is a finite subset of $\mathcal{P}$ by

$$W(F, \epsilon) = \{x \in X; p(x) < \epsilon \text{ for all } p \in F\}.$$

(a) Show that these sets satisfy the conditions of Theorem 7.5 and define a linear topology on X. Show, moreover, that this results in a locally convex space.

(b) Suppose that given any $x \neq 0$ there is a $p \in \mathcal{P}$ such that $p(x) \neq 0$. Show that the topology is Hausdorff.

(c) Show that all the topological linear spaces that you have encountered in this book so far, with the exception of L^p with $p < 1$, arise in this fashion.

(d) Show that for any locally convex topological linear space there is a collection of seminorms that induce the topology as shown; see Exercise 7.4.

7.18 Let (E_n) be a sequence of projection operators on a Hilbert space such that $E_{n+1} \leq E_n$ for all n. Let $E = \inf E_n$.

(a) Show that (E_n) does *not* converge to E with respect to the norm unless there is some index n_0 such that $E_n = E_{n_0}$ for all $n \geq n_0$.

(b) Show, however, that (E_n) converges to E in the strong operator topology.

8

THE SPECTRUM

8.1 INTRODUCTION

In this chapter, as in Chapter 4, we will deal with the field of complex numbers. The reader may recall from linear algebra that for a finite-dimensional space X a common first step in analyzing an element $T \in \mathcal{L}(X)$ is to compute its *eigenvalues*. An eigenvalue is a scalar λ such that $Tx = \lambda x$ for some nonzero x, called an *eigenvector*. The totality of all eigenvectors for a given λ forms a subspace known as the *eigenspace* for λ.

Every operator on a complex finite-dimensional space has at least one eigenvalue, a fact shown by the fundamental theorem of algebra, which says that every polynomial has some root in the complex numbers.

When the space is infinite-dimensional, there need not exist any eigenvalues. We can, however, generalize the idea involved so it applies to an arbitrary space. For any linear operator T from a space to itself, we can assign a subset of the complex numbers known as the *spectrum* of the operator, which plays a fundamental role in investigating T. In the finite-dimensional case, the spectrum is just the set of eigenvalues. In this chapter, we develop the properties of the spectrum, determine methods of computing the spectrum, and show how it may be used to obtain information about the operator. In addition, we generalize the setting and discuss the spectrum in the context of a Banach algebra. This is an important object of functional analysis which involves a fascinating interplay between analysis and algebra. In this book, we provide a brief introduction to this topic. In Chapters 9 and 11, we will apply some of the results to prove versions of one of the most celebrated results in operator theory, the so-called *spectral theorem*.

The following are two examples to show that eigenvalues need not exist.

Example 8.1. Consider the right-shift operator S on a Hilbert space; see Example 4.29. Suppose that $Sx = \lambda x$, where $x = \sum_{i=1}^{\infty} \alpha_i x_i$ for some orthonormal basis (x_i). We claim that all α_i must be zero. If not, let k be the smallest integer such that

A First Course in Functional Analysis. By S. David Promislow

$\alpha_k \neq 0$. By the definition of the right shift,

$$Sx = \sum_{i=1}^{\infty} \alpha_i x_{i+1},$$

so that the coefficient of x_k in the expansion of Sx is $\alpha_{k-1} = 0$. But, this coefficient also equals $\lambda\alpha_k$. Therefore, $\lambda = 0$, and since S is injective, we must have $x = 0$. We conclude that S has no eigenvalues.

Example 8.2. Here we give an example of a self-adjoint operator without eigenvalues. Let M be the operator of multiplying by t on $L^2[0, 1]$. That is,

$$Mf(t) = tf(t).$$

If $Mf = \lambda f$, then, for t a.e., $(t - \lambda)f(t) = 0$, which implies that $f = 0$ a.e.

We now generalize the concept of an eigenvalue. First, we recall some facts about inverses and clear up a point of possible confusion. In this section, when we say that T is invertible in $\mathcal{B}(X)$, we mean that there exists an operator $T^{-1} \in \mathcal{B}(X)$ such that $TT^{-1} = T^{-1}T = I$. (For a Banach space, this is equivalent to being bijective, as for such an operator, its inverse will be automatically linear and bounded, as we noted in Chapter 3.) Note the key fact here is that T^{-1} is in $\mathcal{B}(X)$, which means that it is defined on all of X. Any injective linear T has an inverse, but if the map is not surjective, this inverse will be defined only on a subspace of X, and we will not consider T as invertible in $\mathcal{B}(X)$. Moreover, when X is not a Banach space, we could even have a bijective operator for which T^{-1} is not bounded, and T will be not be invertible in $\mathcal{B}(X)$.

Definition. The *spectrum* of any $T \in \mathcal{B}(X)$, denoted by $\sigma(T)$, is the set of all scalars λ such that $T - \lambda I$ is not invertible in $\mathcal{B}(X)$.

Note that the spectrum is therefore comprised of the following three sets: (i) all eigenvalues, which are precisely those scalars for which $T - \lambda I$ is not injective; (ii) those scalars for which $T - \lambda I$ is injective but not bounded below; and (iii) those scalars for which $T - \lambda I$ is bounded below [so that $R(T - \lambda I)$ is closed if X is complete by Theorem 3.6] but with $R(T - \lambda I)$ not equal to X. The set of values in (i) is called the *point spectrum*. The subset in (ii) for which $R(T - \lambda I)$ is dense is called the *continuous spectrum*, and the remaining set of values in $\sigma(T)$ is called the *residual spectrum*, which contains all the points in (iii) and possibly some in (ii).

Example 8.3. Consider a multiplication operator M_f on ℓ^2 in which f is some complex-valued function on the positive integers. We will show that its spectrum is the closure of the range of f. All λ in the range of f are eigenvalues since $M_f(\delta_n) = f(n)\,\delta_n$. Any λ not in the range of f but in its closure will not be an eigenvalue but will still be in the spectrum. Indeed, given any such λ and $\epsilon > 0$, there is a $\lambda_0 = f(k)$ for some k such that $|\lambda - \lambda_0| < \epsilon$. Then, $(M_f - \lambda I)\,\delta_k = (\lambda_0 - \lambda)\,\delta_k$, which has norm less than ϵ, thereby showing that $M_f - \lambda I$ is not bounded below.

8.2 BANACH ALGEBRAS

The idea of the spectrum is useful in more structures than spaces of operators. This topic provides us an opportunity to introduce the notion of a general Banach algebra.

We first discuss the general notion of an algebra. Many linear spaces come equipped with a natural multiplication in addition to the linear operations. This is true of $\mathcal{B}(X)$, where we have the operation of composition. It is true, as well, for many spaces of scalar-valued functions, where the multiplication is just the usual pointwise product.

The mathematical object used to model linear spaces with a multiplication is called an *algebra*. This consists of a linear space A equipped with a multiplication operation that assigns to each $x, y \in A$ an element $xy \in A$. The operation must satisfy the following axioms. For all $x, y, z \in A$ and scalars α:

(*i*) (Associative law) $xy(z) = x(yz)$.

(*ii*) (Distributive laws that connect addition and multiplication)

$$x(y+z) = xy + xz, \qquad (y+z)x = yx + zx.$$

(*iii*) (Law connecting multiplication and scalar multiplication)

$$(\alpha x)y = x(\alpha y) = \alpha(xy).$$

An element e in A is called a *unit* if $ex = xe = x$ for all $x \in A$. A unit is clearly unique since if e and f are both units, $e = ef = f$. An algebra that possesses such an element is called *unital*. We will concentrate on unital algebras and will assume that property unless otherwise mentioned.

It is clear that $\mathcal{B}(X)$ is an algebra with composition as the multiplication and with I as a unit. A space of functions with pointwise product as the multiplication is another example. In the latter, the function that is identically 1 is a unit provided that this function is in the specified set of functions, as is true for the bounded functions $B(S)$ or $C(S)$, when S is a compact topological space. An example of a nonunital algebra is $C_0(\mathcal{R})$. More examples of algebras are obtained by looking at subalgebras of $\mathcal{B}(X)$. A subalgebra of an algebra A is just a subspace which contains the product of any two of its elements and contains e if we want it to be unital.

An element x in an algebra is *invertible* if there exists an element x^{-1} such that $xx^{-1} = x^{-1}x = e$. The inverse is unique. Indeed, suppose that an element x has both a right inverse y and a left inverse z. By this we mean that $xy = zx = e$. Then,

$$z = z(xy) = (zx)y = y,$$

and this common element is the inverse of x.

We will make use of some basic relationships between inverses and multiplication. Consider a product $x = x_1x_2 \cdots x_n$. If each factor is invertible, so is x with inverse equal to $x_n^{-1}x_{n-1}^{-1} \cdots x_1^{-1}$. Is the converse true? That is, if x is invertible, is each factor invertible? The answer is yes provided that $x_ix_j = x_jx_i$ for all i, j. Take any factor, say x_1. If y is a right inverse for x, then $x_1[x_2x_3 \cdots x_ny] = 1$, so that $x_2x_3 \cdots x_ny$ is a right inverse for x_1. If z is a left inverse for x, then by rearranging the factors of x, $[zx_2x_3 \cdots x_n]x_1 = 1$, so that $zx_2x_3 \cdots x_n$ is a left inverse for x_1. As shown above, x_1 is invertible.

The notion of a spectrum can be defined in any unital algebra A with precisely the same definition that we used above. Namely, $\sigma(x) = \{\lambda \in C : x - \lambda e$ is not invertible in $A\}$.

Example 8.4. Let $A = C[a, b]$ with pointwise multiplication. We will compute $\sigma(f)$. It is clear that a function f is invertible if and only if $f(t) \neq 0$ for all $t \in [a, b]$. In this case, the inverse is given simply by $f^{-1}(t) = [f(t)]^{-1}$. Clearly, if λ is not in the range of f, then $(f - \lambda e)(t) = f(t) - \lambda \neq 0$ for all t. Hence, $\sigma(f)$ equals the range of f.

After the introduction of normed linear spaces by Banach, it was natural that mathematicians would seek to define axioms connecting multiplication with the norm. This happened quite a bit later, with the development led by Russian mathematicians, chiefly, Israel Gelfand. The resulting structure is known as a *Banach algebra*. A Banach algebra is an algebra A equipped with a complete norm satisfying $\| xy \| \leq \| x \| \| y \|$. This implies that multiplication is continuous. That is, if we have sequences (x_n) and (y_n) converging to x and y, respectively, then $x_ny_n \to xy$. Note also that by induction, given any n elements in a Banach algebra,

$$\| x_1x_2 \cdots x_n \| \leq \| x_1 \| \| x_2 \| \cdots \| x_n \|$$

It is customary to also require that $\| e \| = 1$ in a Banach algebra. Of course, if this does not hold, then we can multiply by a scalar to obtain an equivalent norm in which it does.

Our main examples of algebras given above, $\mathcal{B}(X)$ and $C(S)$ with the sup norm for compact S, are Banach algebras. The multiplication axiom in the first case was verified in Proposition 2.8, and it is an easy exercise in the second. Other examples are furnished by any closed subalgebras of these. (Note that we need the subalgebras to be closed, so that the spaces are complete.)

8.3 GENERAL PROPERTIES OF THE SPECTRUM

In this section, we assume we have a Banach algebra and will develop the main results about spectra.

8.3.1 Algebraic Properties

We start with properties that depend only on the algebraic aspects and that will hold in any algebra, not necessarily one with a norm.

As a motivating example, suppose we know that $\lambda \in \sigma(x)$. What can we say about $\sigma(x^2)$? We might conjecture that $\lambda^2 \in \sigma(x^2)$, and we would be right. By using our facts about inverses above, we deduce that $x^2 - \lambda^2 e = (x - \lambda e)(x + \lambda e)$, which cannot be invertible since the first factor is not and the two factors commute. Conversely, if neither square root of λ^2 is in $\sigma(x)$, both factors are invertible, thereby showing that $\lambda^2 \notin \sigma(x^2)$. This argument extends to any power of x and, more generally, to any polynomial in x. Note that for any one-variable polynomial p, given by $p(t) = \alpha_0 + \alpha_1 t + \cdots + \alpha_n t^n$, we define an element $p(x)$ in the algebra to be $\alpha_0 e + \alpha_1 x + \cdots + \alpha_n x^n$.

Theorem 8.5 (Spectral Mapping Theorem). *$\mu \in \sigma(p(x))$ if and only if $\mu = p(\lambda)$ for some $\lambda \in \sigma(x)\}$.*

Proof. If p is the zero polynomial the conclusion is clear, so assume otherwise. Suppose $\lambda \in \sigma(x)$. The polynomial $p(t) - p(\lambda)$ has λ as a root, so it may be written as $(t - \lambda)q(\lambda)$ for some other polynomial q. In the algebra, this translates to

$$p(x) - p(\lambda e) = (x - \lambda e)q(x).$$

Since the two factors on the right are easily seen to commute and the first is not invertible, our remarks above show that the product is not invertible, thereby implying that $p(\lambda) \in \sigma(p(x))$.

Conversely, suppose that $\mu \in \sigma(p(x))$. Let $\mu_1, \mu_2, \ldots, \mu_n$ be the roots of the polynomial $p(t) - \mu$. In the algebra, this translates to

$$p(x) - \mu e = c(x - \mu_1 e)(x - \mu_2 e) \cdots (x - \mu_n e)$$

for some nonzero scalar c. Since the left side is not invertible and all factors on the right side commute with each other, at least one such factor is not invertible. So, some $\mu_i \in \sigma(x)$ and $p(\mu_i) = \mu$. □

Example 8.6. Following are some useful applications of this theorem. They can be obtained by direct calculation, but it is more elegant to proceed as follows:

(a) Suppose x is an element such that $x^n = 0$ for some n. Such an element is said to be *nilpotent*. Apply the spectral mapping theorem to the polynomial $p(t) = t^n$, noting that the spectrum of the 0 operator clearly contains only 0. Since $p(x) = 0$, we know that if $\lambda \in \sigma(x)$, then $\lambda^n = 0$, so $\lambda = 0$. A nilpotent element then has the property that $x - \lambda e$ is invertible for all scalars $\lambda \neq 0$.

(b) Suppose x is an element such that $x^2 = x$. Such an element is said to be *idempotent*. Apply the theorem to the polynomial $p(t) = t^2 - t$, which has roots 0 and 1. Since $p(x) = 0$, it follows that the spectrum of x must be contained in the set $\{0, 1\}$.

Here is a second algebraic result.

Proposition 8.7. *Suppose x is invertible. Then, $\lambda \in \sigma(x)$ if and only if $\lambda^{-1} \in \sigma(x^{-1})$.*

Proof. Note that the statement makes sense because x invertible means that $0 \notin \sigma(x)$. The theorem follows immediately from our remarks concerning inverses and products and from the identity

$$x^{-1} - \lambda^{-1}e = -\lambda^{-1}(x - \lambda e)x^{-1}. \qquad \square$$

8.3.2 Norm Properties

We now look at properties of the spectrum that depend on the norm.

To assist us in computing spectra, we would like additional criteria for invertibility. What are the invertible elements of A? We know that e is always invertible, so we might conjecture that if an element is "close" to e, it is invertible. That is, we seek $r > 0$ such that if $\| e - x \| < r$, then x is invertible. Now, we cannot take $r > 1$ since our statement would then be satisfied by $x = 0$, which is certainly not invertible, since it follows from the distributive law that $0x = 0$ for all x. But, we can take $r = 1$.

Proposition 8.8. *If $\| e - x \| < 1$, then x is invertible.*

Proof. The trick here is to simply remember the formula for the sum of an infinite geometric series. For a number t satisfying $|1 - t| < 1$, we have $1 + (1 - t) + (1 - t)^2 + \cdots = t^{-1}$. This suggests that a candidate for x^{-1} might be

$$y = e + (e - x) + (e - x)^2 + (e - x)^3 + \cdots.$$

Notice that this makes sense since if $\| e - x \| = s < 1$, then $\| (e - x)^n \| \le s^n < 1$, and the series defining y is absolutely convergent and therefore convergent by our hypothesis of completeness. To show that y is, in fact, the inverse of x, let $y_n = \sum_{k=0}^{n} (e - x)^k$. (We define $x^0 = e$ for all $x \in A$.) Then,

$$y_n x = y_n - y_n(e - x) = e - (e - x)^{n+1},$$

which approaches e, so that $yx = \lim y_n x = e$. Similarly, $xy = e$. $\square$

Here is an important consequence of the above result.

Proposition 8.9. *The set of invertible elements of a Banach algebra is an open set.*

Proof. Take any invertible x. We have to find $r > 0$ (which could depend on x) so that $\| y - x \| < r$ implies that y is invertible. This works with $r = \| x^{-1} \|^{-1}$. Indeed, $\| x - y \| < r$ implies that

$$\| e - x^{-1}y \| = \| x \|^{-1} (x - y) < \| x^{-1} \| \; r = 1.$$

From Proposition 8.8, $x^{-1}y$ is invertible, so that $y = x(x^{-1}y)$ is invertible.
The last two results yield the following. □

Theorem 8.10. *$\sigma(x)$ is a compact subset of $\mathcal{C}$.*

Proof. Suppose that $|\lambda| > \| x \|$. Then,

$$\left\| e - \left(e - \frac{x}{\lambda}\right) \right\| = \frac{\| x \|}{|\lambda|} < 1,$$

so by Proposition 8.8, $e - x/\lambda$ is invertible, and clearly so is $x - \lambda e = -\lambda(e - x/\lambda)$. This shows that $\sigma(x)$ is bounded, being contained in the disc $\bar{B}(\| x \|)$ in $\mathcal{C}$.

If $\lambda \notin \sigma(x)$, then $x - \lambda e$ is invertible. If we choose μ sufficiently close to λ, Proposition 8.9 shows that $x - \mu e$ is invertible. We then know that the complement of the spectrum (known as the *resolvent set*) is open, so the spectrum is closed. □

Definition. The *spectral radius* of an element $x \in A$, denoted by $r(x)$, is defined to be $\sup\{|\lambda| : \lambda \in \sigma(x)\}$. [In a general algebra, we take $r(x) = 0$ if the spectrum is empty, although we show below that this cannot occur in a Banach algebra.]

The proof of Theorem 8.10 shows that $r(x) \leq \| x \|$. By using the spectral mapping theorem, we can improve that bound.

Proposition 8.11. $r(x) \leq \inf \| x^n \|^{1/n}$.

Proof. From the original bound, we know that $r(x^n) \leq \| x^n \|$ for any n. Now the spectral mapping theorem (Theorem 8.5) tells us that $r(x)^n = r(x^n)$. Taking the nth root gives us the conclusion. □

The above immediately gives rise to the following question. Suppose that $\inf \| x^n \|^{1/n} < |\lambda|$. Then, we know that $(x - \lambda e)^{-1}$ exists, and the question is how do we calculate it? Expanding as a formal power series suggests that

$$(x - \lambda e)^{-1} = -\left[\frac{1}{\lambda} + \frac{x}{\lambda^2} + \frac{x^2}{\lambda^3} + \cdots\right],$$

so the question is to determine when this expansion is valid. If $\| x \| < |\lambda|$, then the argument used in Proposition 8.8 shows that this is case, and in fact we can invoke the classical root test for series of nonnegative numbers to cover the more general case for which $\limsup \| x^n \|^{1/n} < \lambda$; see Rudin (1976, Theorem 3.33). We can, however, use the multiplicative property of the norm to cover all cases. It is of interest to look at the general situation.

A sequence of positive reals (a_n) is said to be *submultiplicative* if $a_{n+m} \leq a_n a_m$ for all n, m. The nth roots of such sequences are always convergent as we now show.

Theorem 8.12. *If (a_n) is submultiplicative, then $({a_n}^{1/n})$ converges to $\inf ({a_n}^{1/n})$.*

Proof. Let $a = \inf ({a_n}^{1/n})$. For any $b > a$, we can choose k so that ${a_k}^{1/k} < b$. Given any integer $n > k$, write $n = pk + r$, with p an integer and $0 \leq r \leq k - 1$.

Then, by applying the submultiplicative property inductively, we obtain

$$a_n = a_{k+k+\cdots+k+r} \leq a_k{}^p\, a_r.$$

Now,

$$\frac{p}{n} = \frac{1}{k}\left[1 - \frac{r}{n}\right],$$

so taking the nth root gives

$$a_n{}^{1/n} \leq (a_k{}^{1/k})^{(1-r/n)}\, a_r{}^{1/n} \leq b^{(1-r/n)}\, c^{1/n},$$

in which $c = \max\{a_j : j = 1, 2, \ldots, k-1\}$. Note that $c^{1/n}$ converges to 1, and since $r \leq k-1$, we know that $1 - r/n$ converges to 1. This shows that the right side of the above inequality converges to b. We can conclude that the sequence $(a_n{}^{1/n})$ is Cauchy and, therefore, has a limit a' that is less than b. Since b is an arbitrary number greater than a, we must have $a' \leq a$, but clearly the limit of any sequence can be no smaller than the infimum, so we know that $a' = a$. □

Observe now that the definition of a Banach algebra shows that the sequence (a_n), with $a_n = \| x^n \|$, is submultiplicative. Therefore, $\limsup \| x^n \|^{1/n} = \lim \| x^n \|^{1/n} = \inf \| x^n \|^{1/n}$, and the root test yields the following.

Theorem 8.13. *If* $\inf \| x^n \|^{1/n} < |\lambda|$, *then*

$$(x - \lambda e)^{-1} = -\sum_{k=0}^{\infty} \frac{x^k}{\lambda^{k+1}}.$$

Definition. An element x in a Banach algebra that satisfies $\lim \| x^n \|^{1/n} = 0$ is said to be *quasi-nilpotent.*

This is a generalization of nilpotent elements that we introduced in Example 8.6. The above theorem shows that such an element x has the same property that we demonstrated for nilpotent ones, namely that $x - \lambda e$ is invertible for all nonzero λ.

8.3.3 Deeper Properties

There are two other major facts about the spectrum to demonstrate. We discuss these in a separate section since they are more difficult to show than those above and do not follow directly from the axioms. While it seems to be relatively easy to get criteria for invertibility, it is harder to obtain results which restrict invertibility. We must use results from complex function theory.

The background material that we need can be summarized by the following theorem, which we state without proof. It can be found in any book on complex analysis.

Theorem 8.14. *Let ϕ be a complex-valued function that is defined and differentiable at all points of an open disc $\{z \in C : |z| < r\}$. Then, there is a unique sequence of complex numbers $(a_n), n = 0, 1, \ldots,$ such that the power series $\sum_{n=0}^{\infty} a_n z^n$ converges to $\phi(z)$ at all points of this disc.*

Remark. It is relatively straightforward to verify the uniqueness statement in the above theorem. The only possibility for the sequence is

$$a_n = \frac{f^{(n)}(0)}{n!}$$

[in which (n) denotes the nth derivative]. This follows from elementary calculus. The difficult part is verifying that the series converges at all points of the disk.

We use this theorem to convert the spectral radius inequality developed above into an equality.

Theorem 8.15. *For all elements x in a Banach algebra A, $r(x) = \inf\| x^n\|^{1/n}$.*

Proof. Fix any x, and let $\rho = \inf \| x^n \|^{1/n}$. We will show that

$$|\lambda| > r(x) \text{ implies that} |\lambda| \geq \rho,$$

which together with Proposition 8.11 will complete the proof. It is convenient to work with inverses, so we will show equivalently that

$$|z| < r(x)^{-1} \text{implies that} |z| \leq \rho^{-1}. \qquad (*)$$

[Take $r(x)^{-1} = \infty$ if $r(x) = 0$.]

Let D denote the disc $\{z \in C : |z| < r(x)^{-1}\}$. If $z \in D$ and $z \neq 0$, then $|z^{-1}| > r(x)$, which means that $x - z^{-1}e$ is invertible. This in turn means that $e - zx = -z(x - z^{-1}e)$ is invertible. This is trivially true as well for $z = 0$.

For any bounded linear functional f on A, we can define a complex-valued function ϕ on D by

$$\phi(z) = f[(e - zx)^{-1}].$$

By using the linearity of f and the fact that, for any two commuting elements a, b of an algebra, $a^{-1} - b^{-1} = (b - a)a^{-1}b^{-1}$, we have that

$$\phi(w) - \phi(z) = f[(e - wx)^{-1} - (e - zx)^{-1}] = f[(w - z)x(e - zx)^{-1}(e - wx)^{-1}].$$

Divide by $w - z$ and take limits as $w \to z$ to conclude that ϕ is differentiable for all $z \in D$ with derivative equal to $f[x(e - zx)^{-2}]$. (Note that the calculation of this derivative is not affected by the fact that we have elements from an arbitrary Banach algebra; it simply parallels that of elementary calculus with absolute values replaced by the norm.)

Consider $D' = \{z \in \mathcal{C} : |z| < \| x \|^{-1}\}$, a subdisc of D. From the proof of Proposition 8.8, for $z \in D'$,

$$(e - zx)^{-1} = e + zx + z^2x^2 + \cdots,$$

and by linearity and continuity of f,

$$\phi(z) = f(e) + f(zx) + f(z^2x^2) + \cdots.$$

By the uniqueness of the power series representation given in Theorem 8.14, the above expression for ϕ must hold for all z in the larger disc D. Fix any such z. We must have, then, that the sequence $(f(z^nx^n))$ is bounded. Since this is true for all $f \in A^*$, we can invoke uniform boundedness in the form of Theorem 6.9. We conclude that there is some $K > 0$ such that $\| z^nx^n \| \leq K$ for all positive integers n, which implies that

$$|z| \leq \frac{K^{1/n}}{\| x^n \|^{1/n}}.$$

By taking limits as n goes to infinity and by invoking Theorem 8.12, we obtain $|z| \leq \rho^{-1}$, as required in inequality $(*)$ above. □

We close this section with what might be considered the most basic question when discussing spectra. Is it possible that $x - \lambda e$ is invertible for all λ, so in fact there is nothing in the spectrum? For a Banach algebra the answer is "no."

Theorem 8.16. *For all elements x of a Banach algebra, $\sigma(x) \neq \emptyset$.*

Proof. If x is not invertible, then $0 \in \sigma(x)$. Suppose, then, that x has an inverse y. For any positive integer n, $1 = \| e \| = \| x^ny^n \| \leq \| x^n \| \| y^n \|$, so that

$$\| x^n \|^{1/n} \geq \frac{1}{\| y^n \|^{1/n}} \geq \frac{1}{\| y \|} > 0.$$

From Theorem 8.15, it follows that $r(x) > 0$, which implies the conclusion. □

It is not surprising that complex analysis is required for the last result. Even in the finite-dimensional case, as we pointed out before, the nonemptyness of the spectrum is proved by the fundamental theorem of algebra, and proofs of that theorem invariably use complex analysis.

8.4 NUMERICAL RANGE

All of the properties derived in the preceding section hold in any Banach algebra. We can, however, get more detailed information on the spectrum in the particular case of the algebra $\mathcal{B}(H)$, in which H is a Hilbert space.

Definition. For any $T \in \mathcal{B}(H)$, we define the *numerical range* of T to be the set of all scalars of the form $< Tx, x >$, in which x runs over the unit vectors. We define the *numerical radius* of T, denoted by $w(T)$, to be the supremum of the absolute values of all elements in the numerical range.

From the Cauchy-Schwartz inequality, it is clear that $w(T) \leq \| T \|$.

Proposition 8.17. *$\sigma(T)$ is contained in the closure of the numerical range of T.*

Proof. Let $\lambda \in \sigma(T)$. If λ is a eigenvalue, choose a unit eigenvector x. Then, $< Tx, x > = < \lambda x, x > = \lambda$.

If $T - \lambda I$ is not bounded below, there exists a sequence (x_n) of unit vectors such that $(T - \lambda I)x_n \to 0$. Also,

$$< (T - \lambda I)x_n, x_n > = < Tx_n, x_n > - \lambda,$$

thereby showing that λ is in the closure of the numerical range.

If $T - \lambda I$ is bounded below but the closure of $R(T - \lambda I) \neq H$, then by Proposition 4.27(a), we can find a unit vector y in the null space of $(T - \lambda I)^* = T^* - \bar{\lambda} I$. Then,

$$\bar{\lambda} = < T^* y, y > = < y, Ty > = \overline{< Ty, y >},$$

so that $\lambda = < Ty, y >$ is in the numerical range. □

8.5 SPECTRUM OF A NORMAL OPERATOR

8.5.1 General Results

In this section, we derive more information about the spectra of the classes of operators we looked at in Section 4.6.4. The correspondence with multiplication operators given in that section, together with Example 8.3, suggests what the spectra of these classes should look like. We will now verify these facts.

A basic observation is that if a normal operator is bounded below, we know it is injective and its range is closed, as is true for all operators on a Hilbert space. The significance of normality is that, by the remark following Proposition 4.30, we know the range is dense and it therefore must be all of H. It follows that $\lambda \in \sigma(T)$ if and only if $T - \lambda I$ is not bounded below.

Proposition 8.18. *The spectrum of*

(a) a self-adjoint operator is real;
(b) a positive operator is nonnegative;
(c) a projection is contained in the two-point set $\{0, 1\}$;
(d) a unitary operator is contained in $\{\lambda \in \mathcal{C} : |\lambda| = 1\}$.

Proof. Parts (a) and (b) follow directly from Proposition 8.17, and we demonstrated (c) in Example 8.6. It remains to verify (d). For a unitary operator U and unit vector x, $< Ux, x > \leq \| x \|^2 = 1$, thereby showing that the numerical range of U cannot contain numbers of absolute value strictly greater than 1; therefore $\sigma(U)$ cannot contain any numbers of absolute value strictly greater than 1. But, since $U^{-1} = U^*$ is also unitary, we can invoke Proposition 8.7 to show that the spectrum cannot contain any numbers of absolute value strictly less than 1. □

8.5.2 Special Case of a Self-Adjoint Operator

We now look further at self-adjoint operators. In this case, we can give elementary proofs for the statements of Theorems 8.15 and 8.16.

We first have a preliminary lemma.

Lemma 8.19. *If T is self-adjoint, then for all unit vectors x and y*

$$\text{Re} < Tx, y > \leq w(T).$$

Proof. Apply polarization to the sesquilinear form that maps (x, y) to $< Tx, y >$ (see Section 4.6.2) and use the fact from self-adjointness that $< Ty, x > = \overline{< Tx, y >}$. We then obtain the analogue of equation $(*)$ in the proof of Proposition 4.7, which is

$$\text{Re} < Tx, y > = 1/4[< T(x+y), (x+y) > - < T(x-y), (x-y) >].$$

It follows that for unit vectors x and y

$$\text{Re} < Tx, y > \leq 1/4\, w(T)\, [\, \| x+y \|^2 + \| x-y \|^2\,] = w(T).$$ □

Proposition 8.20. *If T is self-adjoint, then $w(T) = \| T \|$.*

Proof. We observed in the last section that $w(T) \leq \| T \|$ for any T, and we need only show the reverse inequality. Take any unit vector x, and define $y = Tx / \| Tx \|$. Then,

$$\| Tx \| = \| Tx \| \, / \, \| Tx \|^2 = < Tx, y > \leq w(T),$$

where we use Lemma 8.19 for the last inequality. (The above formula holds trivially if $Tx = 0$.) We conclude that $\| T \| \leq w(T)$. □

Proposition 8.21. *For a self-adjoint operator T, either $\| T \| \in \sigma(T)$ or $-\| T \| \in \sigma(T)$.*

Proof. We can assume that $T \neq 0$. Let $w(T) = \| T \| = \lambda$. Since the numerical range of T is real, we know that either λ is the supremum of the numerical range or $-\lambda$ is the infimum. We can assume that λ is the supremum, by replacing T by $-T$ if not, and by using the fact that $\sigma(T) = -\sigma(-T)$.

For any unit vector x, by using the basic identity in Lemma 4.2, we obtain

$$\| (T - \lambda I)x \|^2 = \| Tx \|^2 + \lambda^2 - 2\lambda < Tx, x > .$$

Now, given any $\epsilon > 0$, choose a unit vector x such that $< Tx, x > \; > \lambda - \epsilon/(2\lambda)$. Then,

$$\| (T - \lambda I)x \|^2 \leq \lambda^2 + \lambda^2 - 2\lambda\left(\lambda - \frac{\epsilon}{2\lambda}\right) = \epsilon,$$

thereby showing that $T - \lambda I$ is not bounded below, so that $\lambda \in \sigma(T)$.

From Propositions 8.20 and 8.21, we see that $r(T) = \| T \|$ for a self-adjoint operator T. The importance of this result will be seen in Section 8.6. It is in fact true for all normal operators as we now show.

8.5.3 Spectral Radius of a Normal Operator

Proposition 8.22. *If T is normal, then* $\| T^n \| = \| T \|^n$.

Proof. We will first give the proof when T is self-adjoint. It is clear that the left-hand side is less than or equal to the right, so we must show the other inequality. We start by verifying that if the statement holds for any integer, it will hold for all smaller integers. Suppose it holds for $n + k$. Then,

$$\| T \|^{n+k} = \| T^{n+k} \| \leq \| T^n \| \, \| T^k \| \leq \| T^n \| \, \| T \|^k,$$

and we divide by $\| T \|^k$ to show that the statement holds for n. This, in fact, is true for any operator T. We invoke the self-adjointness to show that if it is true for an integer m, it is true for $2m$. This follows since

$$\| T^{2m} \| = \| T^{*m} T^m \| = \| T^m \|^2 = \| T \|^{2m},$$

in which we use Theorem 4.26(f) for the second equality. Since it is clear for $n = 1$, it is true for all powers of 2 and, by the first statement, true for all positive integers.

Now, take the case for which T is normal. By applying our first result to the self-adjoint operator T^*T, we obtain

$$\| T^*T \|^n = \| (T^*T)^n \| = \| T^{*n} T^n \| \leq \| T^{*n} \| \, \| T^n \| \leq$$

$$\| T^* \|^n \, \| T^n \| = \| T \|^n \, \| T^n \|,$$

in which we use the normality to rearrange in the second equality. From Theorem 4.26(f), by taking nth powers, we get $\| T^*T \|^n = \| T \|^{2n}$, and we divide by $\| T \|^n$ to complete the derivation. □

Combining this result with Theorem 8.15 immediately yields the following.

Theorem 8.23. *$r(T) = \| T \|$ for all normal operators T.*

Remark. Note that this is the first time we have used part (f) of Theorem 4.26, the significance of which was not immediate when introduced.

8.5.4 Eigenvalues of Self-Adjoint Operators

The following elementary results will be useful in subsequent chapters. Suppose that T is a bounded, self-adjoint operator on a Hilbert space H.

Proposition 8.24. *Eigenspaces corresponding to distinct eigenvalues of T are orthogonal.*

Proof. If $Tx = \lambda x$ and $Ty = \mu y$, then

$$\lambda < x, y > = < Tx, y > = < x, Ty > = \mu < x, y >,$$

so that $\lambda \neq \mu$ implies $< x, y > = 0$. □

Proposition 8.25. *If H is separable, the number of distinct eigenvalues of T is either finite or countably infinite.*

Proof. By Proposition 8.24, after dividing all the eigenvectors by their norm, we obtain an orthornormal set, so this result is an immediate consequence of Theorem 4.21. □

These results are in fact true for any normal operator. See Exercise 8.11.

8.6 FUNCTIONS OF OPERATORS

For an operator T from a linear space to itself, any polynomial $p(t) = a_0 + a_1 t + \cdots + a_n t^n$ gives rise to the operator $p(T) = a_0 I + a_1 T + \cdots + a_n T^n$. There are many instances for which it is desirable to define more general functions of an operator. We encountered this in Chapter 4 where we saw that it was useful to know that a positive operator had a square root. We will show here that, in fact, for any bounded, self-adjoint operator on a Hilbert space and for any function f that is continuous on $\sigma(T)$, we can define an operator $f(T)$ so that the usual operations are preserved. This procedure is often known as the *functional calculus*.

Theorem 8.26. *Let T be a bounded, self-adjoint operator on a Hilbert space H. Then, the mapping that sends p to p(T), for all polynomial functions p, can be extended to an isometry from $C(\sigma(T))$, with the sup norm, into $\mathcal{B}(H)$, sending f to $f(T)$ such that the following properties hold: For all f, $g \in C\ (\sigma(T))$ and any $\alpha \in \mathcal{C}$,*

(i) $(f+g)(T) = f(T) + g(T)$ *and* $(\alpha f)(t) = \alpha(f(T))$;

(ii) $\bar{f}(T) = (f(T))^*$;

(iii) $(fg)(T) = f(T)\, g(T)$;

(iv) for any $S \in \mathcal{B}(H)$ *such that* $ST = TS$, *we have* $Sf(T) = f(T)S$.

Proof. The key step is to verify that $\| p \| = \| p(T) \|$, thereby showing that the map on the polynomials is an isometry. By definition, $\| p \| = \sup\{|p(\lambda)| : \lambda \in \sigma(T)\}$. By the spectral mapping theorem (Theorem 8.5), $\| p \|$ is equal to $\sup\{|\mu| : \mu \in \sigma(p(T))\} = r(p(T))$, which by Theorem 8.23 is equal to $\| p(T) \|$.

This argument also shows that the mapping p to $p(T)$ is well defined. This is not obvious from the outset since in the case of a finite spectrum two distinct polynomials p and q could agree on $\sigma(T)$. But, if so, then $\| p - q \| = 0$, thereby showing that $\| p(T) - q(T) \| = 0$, so that $p(T) = q(T)$.

By our basic extension result, Theorem 2.20, we can extend this isometry from the space of polynomial functions to its closure. By the Stone–Weierstrass theorem (Appendix B), this closure is all of $C((\sigma(T))$. The verification of properties (i) to (iv) follow from the fact that they are true for polynomials, from the continuity of the mapping f to $f(T)$, and from the fact that the properties are preserved under limits. We illustrate with property (iii); the other derivations are similar. If we have sequences of polynomials (p_n) and (q_n) converging uniformly to the continuous functions f and g, respectively, then it is not hard to see that $(p_n q_n)$ converges uniformly to fg. It follows that

$$\begin{aligned}(fg)(T) &= \lim\, (p_n q_n)(T) = \lim(p_n(T) q_n(T)) \\ &= \lim p_n(T) \lim q_n(T) = f(T)\, g(T).\end{aligned}$$

□

Remark. Readers might wonder about a function f defined on a domain other than the spectrum of T. We must have the domain include the spectrum, since if not, as the proof of Theorem 8.26 indicates, the mapping from p to $p(T)$ will not be well defined. We certainly can take a domain D larger than $\sigma(T)$ and just define the function of the operator by restriction. That is, let $f(T) = \hat{f}(T)$ where $\hat{f}$ is the restriction of f to $\sigma(T)$. This does not lead to anything new, however, for the same argument shows that if the values of f and g agree on $\sigma(T)$, then $f(T) = g(T)$.

Remark. We have now provided the promised proof for Theorem 4.32 on the existence of square roots for positive operators. Take $T^{1/2} = f(T)$ for the function $f(\lambda) = \lambda^{1/2}$, which makes sense since we know from Theorem 8.18(b) that for positive T we have $\sigma(T) \subseteq [0, \infty)$.

Remark. There are various other important properties that follow from the given ones. Property (ii) implies that, for a real-valued function f, $f(T)$ is self-adjoint. Moreover, if f if nonnegative valued, then from properties (ii) and (iii), $f(T) = (f^{1/2}T)^2$, thereby showing that $f(T)$ is a positive operator. It follows that for two real functions $f \leq g$ we have $g - f \geq 0$ and therefore $f(T) \leq g(T)$ in the ordering given in Section 4.7.

Remark. Theorem 8.26 is also true for all *normal* operators. The proof is more complicated, and we will not give it here. One major difficulty is that we need to consider polynomials in two variables, T and T^*, so a direct use of the Stone–Weierstrass theorem in the form that we have presented in the Appendix C is not possible. For more details see Kadison and Ringrose (1983, Theorem 4.4.5).

Remark. For nonnormal operators, there is a technique using complex analysis to define analytic functions of operators. The Hilbert space structure is not needed here, and one can do this for $\mathcal{B}(X)$, in which X is any Banach space, and even more generally for any Banach algebra. See Rudin (1973, Section 10.21) for a reference. We will not go into details here, but we would like to observe one application. Suppose that an operator T has a finite spectrum $\{\lambda_1, \lambda_2, \ldots, \lambda_n\}$. If T is a self-adjoint operator on a Hilbert space, we know that

$$\prod_{i=1}^{n}(T - \lambda_i I) = 0,$$

since the corresponding polynomial vanishes on $\sigma(T)$. Using the analytic functional calculus, this is true for any operator on a Banch space. This generalizes a well known result of linear algebra, the Cayley–Hamilton theorem.

Sometimes the function of an operator can be described in concrete terms.

Example 8.27. Suppose in n-dimensional space we have a self-adjoint operator T whose matrix with respect to some orthonormal basis is in diagonal form. That is (denoting an operator by its matrix),

$$T = \begin{pmatrix} \lambda_1 & 0 & 0 & \cdots & 0 \\ 0 & \lambda_2 & 0 & \cdots & 0 \\ \vdots & \vdots & \vdots & \ddots & \vdots \\ 0 & 0 & 0 & \cdots & \lambda_n \end{pmatrix}.$$

The spectrum of T is the finite set $\{\lambda_1, \lambda_2, \ldots, \lambda_n\}$. All functions on this set are continuous and, in fact, all given by polynomials. [The last statement is well known and easily demonstrated. Indeed, the polynomial $\prod_{i \neq j}(\lambda - \lambda_i)$ takes all λ_i into 0 except for λ_j. A suitable linear combination of such products will then map the spectrum into any ordered n-tuple of scalars.] From the usual rules for multiplying and adding matrices, we immediately see that

$$f(T) = \begin{pmatrix} f(\lambda_1) & 0 & 0 & \cdots & 0 \\ 0 & f(\lambda_2) & 0 & \cdots & 0 \\ \vdots & \vdots & \vdots & \ddots & \vdots \\ 0 & 0 & 0 & \cdots & f(\lambda_n) \end{pmatrix}.$$

This same result holds for any normal operator.

EXAMPLE 8.28. Suppose that T is the operator of multiplying by some fixed bounded real-valued measurable function g on some L^2 space. That is,

$$(Th)(\omega) = g(\omega)\, h(\omega)$$

for all $h \in L^2$. It is clear that $\sigma(T)$ contains the essential range of g. (In fact, arguing similarly to Example 8.3 shows that the spectrum is the closure of the essential range.) For any continuous f defined on $\sigma(T), f(T)$ is just the multiplication operator that multiplies by the composition $f\ o\ g$. That is,

$$(f(T)\, h)(\omega) = f(g(\omega))\, h(\omega).$$

This is easily seen for the power functions $f(t) = t^n$, and we can pass from there to polynomials and then to any continuous function by taking limits.

Note that Example 8.27 is really a particular case of Example 8.28. Indeed, choosing an orthonormal basis means that we can effectively view our space as the set of functions on n points $\{1, 2, \ldots, n\}$, and the diagonal operator T can be viewed as the multiplication operator that multiplies by the function g defined by $g(i) = \lambda_i$ for $i = 1, 2, \ldots, n$.

8.7 BRIEF INTRODUCTION TO C*-ALGEBRAS

We briefly introduce a type of Banach algebra which has found applications in many branches of mathematics and is a major tool in a deeper study of Hilbert space operators. It is motivated by Section 8.6.

As we noted before, a prime example of a Banach algebra is a closed subalgebra A of $\mathcal{B}(X)$ for some Banach space X. Suppose that X is a Hilbert space and, in addition, A is closed under adjoints. That is, T^* is in A for all $T \in A$. Then, we say that A is a *C^*-algebra*.

A natural question that arises is, then, which Banach algebras are isomorphic to C^*-algebras? Clearly such an algebra must have a mapping to itself, sending x

to x^* that satisfies the properties of Theorem 4.26. It turns out quite remarkably that this is sufficient. Any Banach algebra with such a * mapping that satisfies properties (a) to (d) and (f) of Theorem 4.26 is isometric to a to C^*-algebra. (In modern terminology, such abstract algebras are also called C^*-algebras.) Note that we do not need to specify property (e) of Theorem 4.26 since this is easily derivable from (f).

Another example of such an abstract C^*-algebra is $C(K)$, in which K is a compact Hausdorff space, and the * operation simply assigns to any function f the function $\bar{f}$.

A major technique in the modern approach to operator theory is to study an operator T by looking at $C^*(T)$, the smallest C^*-algebra containing T. This is beyond the scope of this book, but we will look at a particularly tractable case. Suppose that T is self-adjoint. The smallest subalgebra containing T will be the set of all polynomials in T. This will be closed under adjoints, the adjoint of a polynomial in T being obtained by replacing each coefficient with its conjugate. Then, $C^*(T)$ is the closure of the set of polynomials. This closed algebra is also closed under adjoints since property (d) of Propostion 4.26 shows that the * map is an isometry.

Theorem 8.26 can be interpreted as saying that $C^*(T)$ is isometrically *-isomorphic to $C(\sigma(T))$. (Here the * in the word *-isomorphic means that the map preserves the * operation.) One of the main results of C^*-algebra theory states that any unital C^*-algebra that is commutative (that is, $xy = yx$ for all x and y) is isometrically *-isomorphic to $C(K)$ for some compact Hausdorff space K.

EXERCISES

8.1 Let $T \in \mathcal{B}(X)$.

(a) If X is a Banach space, consider the Banach space adjoint T^* as an element of $\mathcal{B}(X^*)$, and show that $\sigma(T^*) = \sigma(T)$.

(b) If X is a Hilbert space, consider the Hilbert space adjoint T^* as an element of $\mathcal{B}(X)$, and show that $\sigma(T^*) = \{\bar{\lambda} : \lambda \in \sigma(T)\}$.

8.2 Let S and T be two bounded linear operators on a Hilbert space.

(a) Show that if λ is a nonzero eigenvalue of ST, then it is also an eigenvalue of TS.

(b) Find an example to show that $\sigma(ST)$ need not be equal to $\sigma(TS)$.

8.3 Let x, y be two elements of an algebra A.

(a) If $e - xy$ has an inverse z, show that $e - yx$ has an inverse $e + yzx$.

(b) Show that the nonzero elements of $\sigma(xy)$ and $\sigma(yx)$ are the same.

(c) (Wielant's theorem) If A is a Banach algebra, show that $xy - yx$ cannot equal e.

8.4 Consider $T \in \mathcal{B}(H)$, in which H is a Hilbert space.
(a) Show that an eigenvalue of T is either an eigenvalue of T^* or in the residual spectrum of T^*.
(b) Show that if λ is in the residual spectrum of T, it is an eigenvalue of T^*.

8.5 Consider the right shift S on a Hilbert space and its adjoint S^*.
(a) Show that $|\lambda| < 1$ implies that λ is an eigenvalue of S^*.
(b) Compute the spectrum of S and of S^*; see Exercise 8.4.

8.6 (Cayley transform) If T is a self-adjoint operator on a Hilbert space, show that $(T - iI)(T + iI)^{-1}$ is a unitary operator.

8.7 Let T be self-adjoint.
(a) By using Theorem 8.26, show that if $\sigma(T) \subset [0, \infty)$, then T is positive.
(b) Suppose that a and b are the minimum and maximum elements, respectively, of $\sigma(T)$. Show that the numerical range is $[a, b]$.

8.8 Let P be an injective positive operator in $\mathcal{B}(H)$ for a Hilbert space H. Let us say that a scalar λ is a P-eigenvalue for an operator T if for some nonzero x (called a P-eigenvector) we have $Tx = \lambda Px$. Suppose that T is self-adjoint and commutes with P. Show that any two distinct P-eigenvectors of T are orthogonal with respect to the inner product $< x, y >_P$ as defined in Exercise 4.10.

8.9 Let T be self-adjoint.
(a) Let f be the function defined on $\sigma(T)$ by $f(t) = e^{it}$. Show that $f(T)$ (usually denoted as just e^{iT}) is a unitary operator.
(b) Fix any real number t_0, and let g be the function defined on $\sigma(T)$ by

$$g(t) = \begin{cases} 1 & \text{if } t \leq t_0, \\ 0 & \text{if } t > t_0. \end{cases}$$

Show that $g(T)$ is a projection.

8.10 For a self-adjoint operator T, extend the spectral mapping theorem from polynomials to all continuous functions. That is, show that, for any $f \in C(\sigma(T))$, $\lambda \in \sigma(f(T))$ if and only if $\lambda = f(\mu)$ for some $\mu \in \sigma(T)$.

8.11 Show that Propositions 8.24 and 8.25 hold for any normal T.

9

COMPACT OPERATORS

9.1 INTRODUCTION AND BASIC DEFINITIONS

We now investigate a significant class of operators on infinite-dimensional spaces. These are the *compact* operators, which are close in behavior to finite-dimensional operators, which are automatically compact. Compact operators are historically important because the integral equations that were studied by early researchers in functional analysis gave rise to this class of operators. More details and applications related to integral equations are given in Chapter 10. In this chapter, we look at basic properties and end with the well-known spectral theorem, which gives us complete information about the class of normal compact operators on a Hilbert space. We now give the relevant definitions.

Definition. A subset K of a normed linear space is said to be *relatively compact* if $\bar{K}$ is compact. Equivalently, K is relatively compact if any sequence is K has a convergent subsequence (not necessarily to a point in K since K need not be closed).

Definition. For normed linear spaces X and Y, an operator $T \in \mathcal{B}(X, Y)$ is said to be *compact* if, for all bounded sets B in X, $T(B)$ is relatively compact. Equivalently, T is compact if given any bounded sequence (x_n) in X the sequence (Tx_n) has a convergent subsequence.

To show that an operator T is compact, it is sufficient to show that $TB(1)$ is relatively compact. We leave this as an exercise.

Since a relatively compact set must be bounded, it is clear that a compact operator is bounded. In finite-dimensional spaces, all bounded sets are relatively compact, so the compact operators simply coincide with the bounded ones, which coincide with all linear operators. We are really concerned, then, with infinite-dimensional spaces. The first question is, are there such operators? We will give more pertinent examples later, but an obvious class is comprised of those operators

A First Course in Functional Analysis. By S. David Promislow

with finite-dimensional range, known as *finite-rank* operators. Could all operators on an infinite-dimensional space be compact? The answer is "no" since the identity takes the closed unit ball into itself, and we know from Theorem 2.34 that this is not relatively compact in an infinite-dimensional space.

9.2 COMPACTNESS CRITERIA IN METRIC SPACES

In order to show that certain operators are compact, we need to be able to show that certain subsets of a normed linear space are relatively compact. The purpose of this section is to identify some criteria for doing so. In infinite-dimensional spaces, the relatively compact sets are precisely the bounded sets. We can get corresponding results in the infinite-dimensional case by generalizing the notion of boundedness. Although we will still deal with normed linear spaces, we want to state conditions for a general metric space X in order to apply them to arbitrary subsets, not necessarily subspaces, of a normed linear space.

Definition. If A and B are subsets of a metric space (X, d) and if $\epsilon \geq 0$, we say that $A \subseteq^{\epsilon} B$ if, given any $a \in A$, there is an element $b \in B$ with $d(a, b) \leq \epsilon$.

We can think of $A \subseteq^{\epsilon} B$ as saying that A is "approximately" contained in B, in which ϵ defines the degree of approximation. Of course, ordinary set inclusion is just the case for which $\epsilon = 0$. The triangle inequality immediately gives us that

$$A \subseteq^{\epsilon} B \quad \text{and} \quad B \subseteq^{\delta} C \quad \text{implies that} \quad A \subseteq^{\epsilon+\delta} C.$$

Definition. A set A in a metric space is said to be *totally bounded* if for any $\epsilon > 0$ there is a finite set F (which can depend on ϵ) such that $A \subseteq^{\epsilon} F$. The set F is often called a ϵ-net for A.

(The reader should not confuse the concept of ϵ-net with the nets mentioned in Section 7.1.2, a completely different concept.)

A totally bounded set A is bounded. (We leave this as an exercise.) In finite-dimensional space, any bounded set is totally bounded. This can easily be visualized in $\mathcal{R}^2$. Draw a square contaning the set, and subdivide this square into smaller squares by drawing vertical and horizontal lines. Given any $\epsilon > 0$, choose this grid fine enough so that all the subsquares have diameter less that ϵ. Then, we can take the vertices of all the subsquares as the finite ϵ-net. See Figure 9.1, which suggests why the word "net" is used. The procedure can immediately be extended to $\mathcal{R}^n$, as well as to $\mathcal{C}^n$, which we can identify with $\mathcal{R}^{2n}$.

To see how the concept of total boundedness ties in with compactness, we have the following.

Proposition 9.1. *A is totally bounded if and only if any sequence (a_n) of points in A has a Cauchy subsequence.*

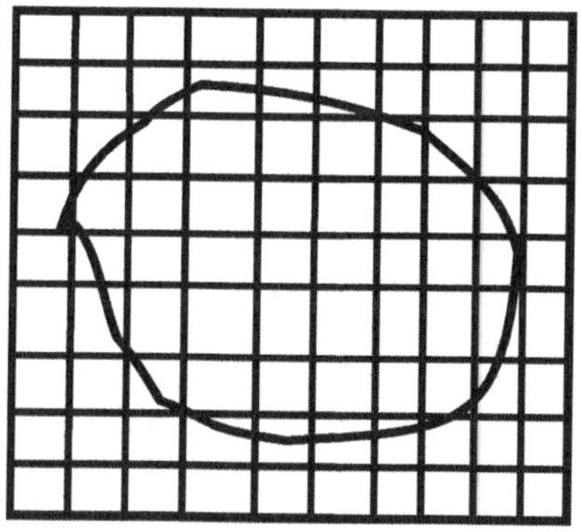

Figure 9.1 Construction of a ϵ-net.

Proof. Suppose that (a_n) is a sequence in the totally bounded set A. We construct a Cauchy subsequence by a diagonal argument. Given $\epsilon > 0$, we can find a finite $(\epsilon/2)$-net F for A. Since there are infinitely many integers and only finitely many points in F, there must be at least one point $y \in F$ such that $d(a_n, y) < \epsilon/2$ for infinitely many values of n. This means we have a subsequence for which any two values are within a distance of ϵ from each other.

We iterate this procedure starting with $\epsilon = 1$ and taking $\epsilon = 1/k$ at the kth stage. Starting with any sequence which we call (a_n^0), we can construct sequences (a_n^k) for $k = 1, 2, \ldots$ satisfying the following: The sequence (a_n^{k+1}) is a subsequence of (a_n^k), and all entries of the sequence (a_n^k) are within $1/k$ of each other. Then, the sequence (a_k^k) is Cauchy since all terms after the Nth are within $1/N$ of each other.

Conversely, if A is not totally bounded, there is an $\epsilon_0 > 0$ such that no finite ϵ_0-net exists for A. We will inductively construct a sequence in A such that any two points are at a distance greater than ϵ_0 from each other, and this cannot possibly have a Cauchy subsequence.

Let a_1 be arbitrary. Having chosen $a_1, a_2, \ldots, a_k$ with this property, we simply choose $a_{k+1} \in A$ such that $d(a_i, a_{k+1}) > \epsilon_0$ for $i = 1, 2, \ldots, k$. Such a point must exist, for if not, $\{a_1, a_2, \ldots, a_k\}$ would be an ϵ_0-net for A. □

As a corollary, any relatively compact set is totally bounded since any sequence has a convergent subsequence. Conversely, in a complete space, any totally bounded set is relatively compact since any sequence has a Cauchy subsequence which necessarily converges.

In Section 2.2.3, we remarked that a compact set could be thought of as a set that is "almost" finite. We see directly from the definition that this is a very apt description of a totally bounded set.

The above result and the relative compactness of bounded sets in finite-dimensional spaces lead to another useful characterization of compactness in a normed linear space.

Proposition 9.2. *A bounded set A of a Banach space X is relatively compact if any only if for any* $\epsilon > 0$ *there is a finite-dimensional subspace Y of X with* $A \subseteq^{\epsilon} Y$.

Proof. Suppose that A has the given property. Given $\epsilon > 0$, choose a finite-dimensional subspace Y such that $A \subseteq^{\epsilon/2} Y$. For each $a \in A$ choose $\acute{a} \in Y$ such that $\| a - a' \| \leq \epsilon/2$. The set A' consisting of all such a' is bounded since diam $(A') \leq \text{diam}(A) + \epsilon$. A' is therefore relatively compact in Y, and we can find a finite set F with $Y \subseteq^{\epsilon/2} F$. It follows that $A \subseteq^{\epsilon} F$, so that A is totally bounded and therefore relatively compact by the completeness assumption.

Conversely, if A is relatively compact, we can find a finite set F with $A \subseteq^{\epsilon} F$ and take $Y = \text{span}(F)$. □

We will use this result to obtain some criteria for compactness in certain Banach spaces. In the case of function spaces, these conditions often have a similar pattern. They state that a certain standard of approximation must hold uniformly for all functions in the set.

Proposition 9.3. *Consider the space B(S) with the sup norm. Suppose that A is a bounded subset satisfying the following: For any* $\epsilon > 0$, *we can partition S into a finite number of pairwise disjoint subsets* $S_1, S_2, \ldots, S_n$ *such that, given any i, any two points s and t in* S_i, *and any* $f \in A$, *we have* $|f(s) - f(t)| \leq \epsilon$. *Then, A is relatively compact.*

Proof. Given $\epsilon > 0$, find the corresponding partition, and let Y be the subspace consisting of all functions that are constant on each set of the partition. Then, Y is finite-dimensional with a basis consisting of $\{1_{S_i} : i = 1, 2, \ldots, n\}$. For any $f \in A$, choose a point, s_i in each S_i, and let g be the element of Y that takes the value $f(s_i)$ on S_i. For all $t \in S$, we have $|f(t) - g(t)| \leq \epsilon$. Therefore, $\| f - g \| \leq \epsilon$. We have shown that $A \subseteq^{\epsilon} Y$, so by Proposition 9.2, A is relatively compact. □

In the particular case that S is a compact metric space, we can look at the space of continuous functions $C(S)$ as a subspace of $B(S)$ and use the above to derive a classical condition for compactness. We first need a definition.

Definition. A set A of functions defined on a metric space S is said to be *equicontinuous at a point* $t_0 \in S$ if given any $\epsilon > 0$ there is a $\delta > 0$ such that $d(t, t_0) < \delta$ implies that $|f(t) - f(t_0)| < \epsilon$ for all $f \in A$. (Note that this is stronger than simply saying that all functions in A are continuous at t_0 since the same δ must hold for all f.) A is said to be ***equicontinuous*** if it is equicontinuous at all points of S.

Example 9.4. Let S be any subset of a normed linear space X, and let A be a bounded subset of X^*. Then, A is equicontinuous on S since, if $\| f \| \leq K$ for all $f \in A$, we can choose $\delta = \epsilon/K$.

Remark. Note that in the above example the same δ could be chosen for any point t_0. This is not always the case, but it is when S is compact. We sketch the derivation here. It is similar to the proof that shows that a continuous function on a compact set is uniformly continuous. For each t, choose δ_t satisfying the definition for $\epsilon/2$.

The collection of open set $B(t_i; \delta_{t_i})$ covers S, and so S is contained in the union of finitely many such balls, $\{B(t_i; \delta_{t_i}) : i = 1, 2, \ldots, n\}$. Take $\delta = \frac{1}{2}\min\{\delta_1, \delta_2, \ldots, \delta_n\}$.

We can now state the following fundamental result.

Theorem 9.5 (Arzela–Ascoli Theorem). *If S is a compact metric space, a subset A of $C(S)$ is relatively compact if any only if it is bounded and equicontinuous.*

Proof. Suppose that A is bounded and equicontinuous. We view $C(S)$ as a subspace of B(s) and invoke Proposition 9.3. Given $\epsilon > 0$, find δ such that $|s - t| < \delta$ implies that $|f(s) - f(t)| < \epsilon$ for all $f \in A$. Choose a finite δ-net $\{t_1, t_2, \ldots, t_n\}$ for S. Then, partition S as follows: Let $S_1 = \{s \in S : d(s, t_1) < \delta\}$ and define inductively $S_k = \{s \in S : s \notin \bigcup_{i=1}^{k-1} S_i \text{ and } d(s, t_k) < \delta\}$. Since the t_i's form a δ-net for S, the union of all the sets S_i is indeed S. The conclusion follows from Proposition 9.3.

We leave the proof of the converse to the reader. (We do not use it in the sequel.) We know that if A is relatively compact, it must be bounded, and it remains to show that it is equicontinuous. □

We can also use Proposition 9.2 to derive a criterion for compactness in ℓ^p.

Proposition 9.6. *Let A be a bounded subset of ℓ^p that has uniformly small tails. That is, for any $\epsilon > 0$, there is a positive integer N such that, for all $f \in A$, $\sum_{i=N}^{\infty} |f(n)|^p < \epsilon$. Then, A is relatively compact.*

Proof. Given $\epsilon > 0$, choose N satisfying the criterion for ϵ^p, and let Y be the set of all f such that $f(k) = 0$ for $k \geq N$. Then, Y is a finite-dimensional subspace with basis $\{\delta_i : i = 1, 2, \ldots, N - 1\}$. For any $f \in A$, the function f' that agrees with f on integers less than N and takes the value 0 for integers greater than or equal to N is in Y and satisfies $\| f - f' \| < \epsilon$. This shows that $A \subseteq^{\epsilon} Y$. □

We can now apply the above results to give examples of compact operators.

Proposition 9.7. *The multiplication operator M_f on ℓ^p is compact if and only if $f(n) \to 0$.*

Proof. Suppose that $f(n) \to 0$. We will show that $M_f B(1)$ has uniformly small tails. For any ϵ, simply pick N so that $|f(n) < \epsilon^{1/p}$ for $n \geq N$. Then, for any $g \in B(1)$, we have

$$\sum_{i=N}^{\infty} |f(i)g(i)|^p \leq \epsilon \sum_{i=N}^{\infty} |g(i)|^p \leq \epsilon \| g \|^p < \epsilon$$

Conversely, if $(f(n))$ does not converge to 0, we can find $\epsilon > 0$ and an infinite subset J of the positive integers such that $|f(n)| \geq \epsilon$ for all, $n \in J$. Then, $M_f B(1)$ contains the points $|f(n)\delta_n$ for all $n \in J$, and the distance between any two such points is clearly greater than ϵ. This means that the set is not totally bounded. This can be verified directly from the definition or seen from Proposition 9.1. □

We now look at an integral operator.

Example 9.8. Let k be a continuous function on the closed unit square of $\mathcal{R}^2$. (This is the set $[0, 1] \times [0, 1]$.) From k, we define an operator K on $C[0,1]$ by

$$Kf(s) = \int_0^1 k(s,t)f(t)\, dt.$$

To get some insight into the operator K, recall the definition for an operator on an n-dimensional space corresponding to an $n \times n$ matrix, which one can think of as a function on the set $\{1,2,\ldots,n\} \times \{1,2,\ldots,n\}$. Let $a(i,j)$ denote the value of the function at (i,j); then, the operator is given by

$$Tf(i) = \sum_{j=1}^{n} a(i,j)f(i).$$

Thus, we can view K as a sort of "continuous" matrix. Since $k(s,t)$ is bounded, it is clear that K is bounded, but more is true.

Proposition 9.9. *The operator K of Example* 9.8 *is compact.*

Proof. We know that $KB(1)$ is bounded; it remains to show that it is equicontinuous, so that we can apply Theorem 9.5. Since the unit square is compact, it is a basic result of real analysis that k is uniformly continuous; see Rudin (1976, Theorem 4.19). Given $\epsilon > 0$, we can choose δ such that if $|s_1 - s_2| < \delta$, then $|k(s_1, t) - k(s_2, t)| < \epsilon$ for all t. Then, for all $f \in B(1)$,

$$|Kf(s_1) - Kf(s_2)| \leq \int_0^1 |k(s_1, t) - k(s_2, t)||f(t)|dt \leq \epsilon \int_0^1 |f(t)|dt < \epsilon \qquad \square$$

9.3 NEW COMPACT OPERATORS FROM OLD

In this section, we look at various ways in which compact operators combine to give new compact operators.

Proposition 9.10 (Compactness and Algebraic Properties). *Suppose we are given normed linear spaces X, Y, and Z, a scalar* α, *operators* S, $T \in \mathcal{B}(X, Y)$, *and* $A \in \mathcal{B}(Y, Z)$. *Then:*

(*a*) *If T is compact, so is* αT.
(*b*) *If S and T are compact, so is* $S + T$.
(*c*) *If T is compact, so is AT.*
(*d*) *If A is compact, so is AT.*

Proof

(a) This is clear from the fact that multiplying a subset by a scalar preserves relative compactness.

(b) Given (x_n) in $B(1)$ we can assume, passing to a subsequence if necessary, that (Tx_n) converges; then, passing to another subsequence, we can assume that (Sx_n) converges, so that $((S+T)(x_n))$ converges.
(c) Given (x_n) in $B(1)$, we can assume that (Tx_n) converges, and since A is continuous, (ATx_n) converges.
(d) Since $TB(1)$ is bounded, $ATB(1)$ is relatively compact. □

Note that parts (a) and (b) of this theorem imply that the compact linear operators form a subspace, while parts (c) and (d) imply that if $X = Y = Z$, then the compact operators form what is called a *two-sided ideal* of the algebra $B(X)$.

Proposition 9.11 (Compactness and Limits). *If $T = \lim T_n$, in which (T_n) is a sequence of compact operators in $\mathcal{B}(X,Z)$, then T is compact.*

Proof. Given $\epsilon > 0$, choose N such that $\|T - T_N\| < \epsilon/2$. Then, by Proposition 9.2, choose a finite-dimensional subspace Y of Z such that $T_N B(1) \subseteq^{\epsilon/2} Y$. Since $TB(1) \subseteq^{\epsilon/2} T_N B(1)$, we have $TB(1) \subseteq^{\epsilon} Y$, which shows that $TB(1)$ is relatively compact by the other direction of Proposition 9.2.

Proposition 9.12 (Compactness and Adjoints). *Let $T \in \mathcal{B}(X,Y)$. If T is compact, then its adjoint T^* is compact. If Y is complete, then compactness of T^* implies that of T.*

Proof. Suppose T is compact. Let (g_n) be a sequence in $B^*(1)$, the unit ball of Y^*. We are going to show that some subsequence of $(T^* g_n)$ converges. Now, $B^*(1)$ is a set of functions on all of Y, but by restriction we can view it as a bounded set of functions on the compact set $\overline{TB(1)}$. Example 9.4 shows that this set of functions is equicontinuous. By invoking Theorem 9.5 and passing to a subsequence if necessary, we can assume that (g_n) converges to some g in $C(\overline{TB(1)})$. Recall that this means uniform convergence, so given any $\epsilon > 0$, we can find N such that for $n \geq N$ and $x \in B(1)$ we have $|g_n(Tx) - g(Tx)| < \epsilon$. Equivalently, $|T^* g_n(x) - T^* g(x)| < \varepsilon$. We conclude that $(T^* g_n)$ converges to $T^* g$.

Conversely, suppose that T^* is compact, so we know by the above argument that T^{**} is compact. Then, letting j denote the natural embedding of X in X^{**}, we have by Exercise 6.5 that $jT(B) = T^{**}(jB)$, thereby showing that $jT(B)$ is totally bounded. Since j is an isometry, it follows that $T(B)$ is totally bounded. If Y is complete, we know from the remarks after Proposition 9.1 that $T(B)$ is relatively compact. □

Remark. From Proposition 9.11, we see that any operator that is the limit of finite-rank operators is compact. An interesting fact is that on a Hilbert space the converse is true, namely, any compact operator is the limit of a finite-rank operator. A proof is outlined in Exercise 9.8. The same fact is true for any Banach space with a Schauder basis; see Giles (2000, Theorem 15.22). A long outstanding question of Banach was whether this was true for any Banach space. Enflo (1973) produces a compact operator that is not the limit of finite-rank operators, and this example provides that counterexample, which we mentioned in Section 2.11, of a Banach space without a Schauder basis.

9.4 SPECTRUM OF A COMPACT OPERATOR

What is in the spectrum of a compact operator $T \in \mathcal{B}(X)$ in which X is a Banach space? In the first place, if X is infinite-dimensional, then 0 belongs. The operator T cannot be invertible, for if so, then $I = TT^{-1}$ would be compact, and we know that it is not. The rest of the spectrum behaves very similarly to the finite-dimensional case. We will show that all nonzero elements of the spectrum are eigenvalues, and each eigenspace is finite-dimensional. Moreover, the eigenvalues either are finite in number or form an infinite sequence tending to 0. In other words, they behave like the eigenvalues of a compact multiplication operator on ℓ_2; see Example 8.3 and Proposition 9.7.

To motivate the analysis, consider again an equation of the form $Sx = y$ in which $S \in \mathcal{B}(X)$ and we have to solve for x. What happens in the finite-dimensional case? There are two possibilities. If S is injective, it must also be surjective, and therefore there is a *unique* solution for all $y \in X$. If S is not injective, it is not surjective, and therefore there are no solutions for certain y and infinitely many solutions for others. Much of the theory of compact operators was developed by looking at certain integral equations for which S was of the form $T - \lambda I$ with T compact and λ a nonzero scalar. We are going to show that such an S is injective if and only if it is surjective, just as in the finite-dimensional case. A major tool will be an algebraic result that does not involve norms or boundedness.

Let S be a linear operator from any linear space to itself. Consider the nested sequences of subspaces:

$$R(S) \supseteq R(S^2) \supseteq R(S^3) \supseteq \cdots,$$
$$N(S) \subseteq N(S^2) \subseteq N(S^3) \subseteq \cdots.$$

Lemma 9.13

(*a*) *Suppose that S is injective but not surjective. Then, all inclusions in the range sequence are strict.*

(*b*) *Suppose that S is surjective but not injective. Then, all inclusions in the null space sequence are strict.*

Proof

(a) Suppose that S is injective. If S is not surjective, we can choose $x \notin R(X)$. Then, Sx is in $R(S)$ but not in $R(S^2)$. If it were, then $Sx = S^2y$ for some y, and we would have by injectivity that $x = Sy$, a contradiction. Similarly, we can show that S^kx is in $R(S^k)$ but not in $R(S^{k+1})$.

(b) Suppose that S is surjective. If S is not injective, we can choose $z \neq 0$ in $N(S)$ and, then, by surjectively choose x such that $Sx = z$. Then, x is in $N(S^2)$ but not in $N(S)$ since $z \neq 0$. Similarly, by choosing x such that $S^k x = z$, we obtain an element in $N(S^{k+1})$ that is not in $N(S^k)$. □

We next prove a rather technical lemma to apply later in conjunction with the previous one.

Lemma 9.14. *Let $T \in \mathcal{B}(X)$ be compact, and let (λ_n) be a sequence of scalars. Suppose we have a strictly increasing sequence of closed subspaces of X:*

$$Y_1 \subset Y_2 \subset Y_3 \subset \cdots$$

such that $(T - \lambda_n I)Y_n \subseteq Y_{n-1}$ for all $n > 1$. Then, $(|\lambda_n|)$ is not bounded below. The same conclusion holds if we have a strictly decreasing sequence of closed subspaces of X:

$$Z_1 \supset Z_2 \supset Z_3 \supset \cdots$$

such that $(T - \lambda_n I)Z_n \supseteq Z_{n+1}$ for all n.

Proof. We give the proof for the increasing sequence. The one for the decreasing sequence is similar. By way of contradiction, suppose the assertion is false, so there exists $r > 0$ such that $|\lambda_n| \geq r$ for all n. Let (y_n) be a sequence of unit vectors such that $y_n \in Y_n$ and $d(y_n, Y_{n-1}) \geq \frac{1}{2}$. This is a sequence of almost perpendiculars, the existence of which was shown by Theorem 2.33. Then, for any positive integers $k < n$, we can write

$$\| Ty_k - Ty_n \| = \| -\lambda_n y_n + [(T - \lambda_k I)y_k - (T - \lambda_n I)y_n + \lambda_k y_k)] \| .$$

Note that by the nested property and our hypothesis the term in square brackets is in Y_{n-1}. This means that

$$\| Ty_k - Ty_n \| \geq |\lambda_n| d(y_n, Y_{n-1}) \geq \tfrac{r}{2}.$$

But, then, no subsequence of (Ty_n) can converge, thereby contradicting the compactness of T. □

Fix a nonzero scalar λ, and let $S = T - \lambda I$ with T compact. According to Lemma 9.14 with $\lambda_n = \lambda$ for all n and $Y_n = N(S^n)$, the inclusions cannot all be strict. Then, by Lemma 9.13, If S is surjective, it is injective. By using the other parts of the same two results, we would like to conclude that if S is injective, it is surjective. However, for this we need to know that the ranges are closed, which we now show. It is sufficient to take $\lambda = 1$.

Proposition 9.15 *If T is compact, then $R(T - I)$ is closed.*

Proof. The first step is to reduce to the injective case. Let S denote $T - I$, and consider the factorization $S = S_1 \pi$ (see Section 2.7.1), in which π is the projection of X onto the quotient space $X/N(S)$ and S_1 is an injective map from $X/N(S)$ to X. We will show that S_1 is bounded below, which by Theorem 3.6 will show that the range of S_1 is closed. Since $R(S) = R(S_1)$ by the surjectivity of π, this will complete the proof.

Suppose, by way of contradiction, that S_1 is not bounded below. Then, we can find unit vectors $z_n \in X/N(S)$ such that $S_1 z_n \to 0$. By the definition of the quotient norm, we can choose $x_n \in X$ such that for each n we have $\pi(x_n) = z_n$ and $\| x_n \| \leq 2$. Then, by passing to a subsequence if necessary, we can assume that there is some x in X such that $Tx_n \to x$. But, also, $Sx_n = S_1 z_n \to 0$. This means that

$$x_n = Ix_n = (T - S)x_n \to x,$$

so that $Sx_n \to Sx$ Therefore, $x \in N(S)$, which means that $\pi(x) = 0$. But, also,

$$z_n = \pi(x_n) \to \pi(x).$$

So, we have that 0 is the limit of a sequence of unit vectors, a contradiction. □

To summarize, we now know from the last three results that $T - \lambda I$ is injective if and only if it surjective, and we can state our main theorem.

Theorem 9.16 *Let T be a compact operator in B(X), in which X is a Banach space. Then, the nonzero elements of the spectrum of T are eigenvalues. There are only countably many eigenvalues, and, in the case of infinitely many, they form a sequence tending to 0. The eigenspaces are all finite-dimensional.*

Proof. If for $\lambda \neq 0$ we have that $T - \lambda I$ is injective, then we know it is surjective and therefore invertible. Hence, if λ is in the spectrum of T, it must be that $T - \lambda I$ is not injective, so that λ is an eigenvalue.

Suppose we are given a sequence of vectors (x_n) of norm less than 1 in the eigenspace for the eigenvalue λ. By compactness, we can assume that the sequence $(Tx_n) = (\lambda x_n)$ converges, so that (x_n) also converges. This shows that the unit ball of the eigenspace is relatively compact, and by Theorem 2.34, the eigenspace is finite-dimensional.

Given any $r > 0$, let $\mathcal{E}_r$ be the set of all eigenvalues of T with absolute value greater than r. We will show that $\mathcal{E}_r$ is finite. Since the set of all $\mathcal{E}$ eigenvalues is the union of countably many sets $\mathcal{E}_{1/n}$, $n = 1, 2, \ldots$, this will show that the number of eigenvalues is countable and, moreover, that the sequence in question converges to 0 when infinite. If the set $\mathcal{E}_r$ were infinite, we could choose a sequence of distinct numbers (λ_k) in $\mathcal{E}_r$. We could then choose an eigenvector x_k for each λ_k. Let $Y_n = \text{span}\{x_1, x_2, \ldots, x_n\}$. By definition, $(T - \lambda_n I)x_n = 0$, so we have that $(T - \lambda_n I)Y_n \subseteq Y_{n-1}$. But, (Y_n) is a strictly increasing sequence of closed subspaces since it is a basic fact of linear algebra that a set of eigenvectors for distinct eigenvalues is linearly independent. Since each $|\lambda_n| \geq r$, we have a contradiction to Lemma 9.14. □

Remark. As mentioned above, 0 is always in the spectrum of a compact operator on an infinite-dimensional space. This occurs in one of two ways. If the operator is not injective, then 0 is an eigenvalue. If it is injective, then it is necessarily not bounded below, and 0 is in the continuous spectrum if $R(T)$ is dense or the residual spectrum if $R(T)$ is not dense.

9.5 COMPACT SELF-ADJOINT OPERATORS ON HILBERT SPACES

In this section, we will derive the spectral theorem for a compact self-adjoint operator on a Hilbert space H.

9.5.1 Finite-Dimensional Case

We first review the finite-dimensional situation. A preliminary version of the spectral theorem is the result from linear algebra that every real symmetric matrix can be diagonalized. For complex matrices, this symmetry must be modified to *Hermitian symmetry,* which means that $a_{i,j}$ is the complex conjugate of $a_{j,i}$. From our remarks about matrices in Section 4.6.3, we recognize this as a statement about self-adjoint operators. It says that, given such an operator T from an n-dimensional Hilbert space to itself, we can choose an orthornormal basis consisting of eigenvectors of T. The matrix of T with respect to such a basis then takes on a particularly simple form. Namely, all entries are zero, except possibly for those on the main diagonal, which by the Hermitian symmetry must be real.

There are equivalent viewpoints that are useful. We can say that any self-adjoint T can be written as a finite linear combination of mutually orthogonal projections. More precisely, if E_λ is the projection onto the eigenspace for the eigenvalue λ, then

$$T = \sum_{\lambda} \lambda E_\lambda,$$

in which λ ranges over the distinct eigenvalues of T.

For example, if the matrix of T for some basis is given by

$$\begin{pmatrix} 1 & 0 & 0 \\ 0 & 2 & 0 \\ 0 & 0 & 2 \end{pmatrix}$$

then

$$T = E_1 + 2E_2,$$

in which (denoting an operator by its matrix)

$$E_1 = \begin{pmatrix} 1 & 0 & 0 \\ 0 & 0 & 0 \\ 0 & 0 & 0 \end{pmatrix} \quad \text{and} \quad E_2 = \begin{pmatrix} 0 & 0 & 0 \\ 0 & 1 & 0 \\ 0 & 0 & 1 \end{pmatrix}$$

Still another way of looking at things is to notice that every self-adjoint operator on a finite-dimensional spae can be viewed as a multiplication operator. Choosing an orthonormal basis means essentially that we view the space as a set of functions on the points $\{1, 2, \ldots, n\}$. In the above example with $n = 3$, our operator can be described by

$$Tf(k) = g(k)f(k),$$

in which $g(1) = 1$ and $g(2) = g(3) = 2$.

An important application is that we immediately obtain an expansion for any vector Tx in terms of the diagonalizing orthonormal basis. To describe this, it is convenient to relabel the eigenvalues to allow repetition. There will be n of them, where one with a k-dimensional eigenspace appears k times. We then have the orthonormal basis $\{e_1, e_2, \ldots, e_n\}$ of eigenvectors with corresponding eigenvalues $\{\mu_1, \mu_2, \ldots, \mu_n\}$. So, in the above example, we have $\mu_1 = 1$, $\mu_2 = 2$, and $\mu_3 = 2$. We then know that for all x

$$x = \sum_{i=1}^{3} < x, e_i > e_i,$$

so that

$$Tx = \sum_{i=1}^{3} \mu_i < x, e_i > e_i,$$

This turns out to be an invaluable device when it comes to solving equations of the form $Tx = y$. (The applications to differential and integral equations are given in Chapter 10.)

We are going to show that the above results hold for compact self-adjoint operators on a possibly infinite-dimensional space, the only difference being that the number of eigenvalues can be infinite. The proof is self-contained and does not use the finite-dimensional result, so we obtain that as a special case.

9.5.2 General Case

Throughout this section, H is a Hilbert space and T is a compact, self-adjoint operator (not equal to 0) in $\mathcal{B}(H)$. We need a preliminary definition and proposition.

Definition. A subspace M of H is said to be *invariant* for T if $TM \subseteq M$.

Note that for any such M the restriction of T to M, denoted by $T|M$., is an operator in $\mathcal{B}(M)$.

Proposition 9.17. *If M is invariant for T, then $M^{\perp}$ is invariant for T. Moreover, the restrictions of T to both M and $M^{\perp}$ are also self-adjoint.*

Proof. For all $x \in M$ and $y \in M^{\perp}$, we have $< Ty, x > = < y, Tx >$, which equals 0 since $Tx \in M$. This shows that $Ty \in M^{\perp}$. The second statement now follows immediately from the identity defining the adjoint in Section 4.6.3.

We are now ready for the main theorem. We showed above that the nonzero elements of the spectrum of a compact operator are eigenvalues (Theorem 9.16). This alone does not show that the operator has any eigenvalues for its spectrum could consist of 0 alone. (An example appears in Section 10.2.2.) We know, however, from Proposition 8.21 that for nonzero self-adjoint operators there are nonzero points in the spectrum. This is the key to proof.

Theorem 9.18 (Spectral Theorem for Compact, Self-Adjoint Operator). *Let T be a compact, self-adjoint operator on a Hilbert space H. There is a sequence (either finite or countably infinite) of mutually orthogonal closed subspaces (M_n) whose closed linear span is all of H. There is a corresponding sequence (λ_n) of real numbers which if countably infinite converges to 0. For all n and $x \in M_n$, we have $Tx = \lambda_n x$. Moreover,* if $\lambda_n \neq 0$, *then M_n is finite-dimensional.*

Proof. Let $\{\lambda_n\}$ be the set of distinct eigenvalues of T. Let M_n be the eigenspace for λ_n, and let K be the closed span of all these eigenspaces. Everything follows from Proposition 8.24 and Theorem 9.16, except for the fact that K is all of H.

Each M_n is invariant under T, so K is invariant under T, and by Proposition 9.17, $K^\perp$ is invariant under T. We will show that $K^\perp = 0$. Suppose not. Let T_1 denote the restriction of T of $K^\perp$. Since a subset of any relatively compact set is relatively compact, a restriction of a compact operator must be compact. Moreover, T_1 is self-adjoint by Proposition 9.17. Observe that T_1 cannot be the zero operator since, if there were a nonzero vector in $N(T)$, it would be in the eigenspace for the zero eingenvalue and therefore in K. By Proposition 8.21, T_1 has a nonzero element in its spectrum, which must be an eigenvalue by Theorem 9.16. This will, of course, also be an eigenvalue for T. But the corresponding eigenvector is then in both K and $K^\perp$, a contradiction. □

Remark. With some minor modifications in the proof, we can derive the same result as above for *normal* compact operators, except that the eigenvalues need not be real. We need Theorem 8.23 and a normal operator version of Proposition 9.17. We leave the details as an exercise.

Theorem 9.18 yields analogous statements to those in the finite-dimensional case, as we now show. For convenience, we state this in the separable case. (There are nonseparable counterparts for the statements below.)

Theorem 9.19. *For T as in Theorem* 9.18,

$$T = \sum_n \lambda_n E_{\lambda_n},$$

in which E_{λ_n} is the projection onto M_n, the eigenspace for λ_n.

Proof. Let $S_n = \sum_{i=1}^n \lambda_n E_{\lambda_n}$, and let $T_n = T - S_n$. We must show that (T_n) converges to 0. Given any $\epsilon > 0$, pick N such that $n \geq N$ implies that $|\lambda_n| < \epsilon$. Then, for any $n \geq N$, T_n is self-adjoint and compact by proposition 9.10, parts (a) and (b). T_n takes the closed linear span of $(M_1, M_2, \ldots, M_n)$ to 0 and acts like T on the orthogonal complement of this space. Consequently, the maximal absolute value of an eigenvalue for T_n is less than ϵ. Therefore $\| T_n \| = r(T_n) \leq \epsilon$. □

As in the finite-dimensional case, we have an orthonormal basis (e_k) For H consisting of eigenvectors of T. We obtain this by choosing an orthonormal basis for each of the spaces M_n of Theorem 9.18 and taking thc union of all these bases. Let μ_k be the eigenvalue corresponding to e_k. [Note that the sequence (μ_k) involves repetition, as in the example given in Section 9.5.1] If the dimension of the eigenspace M_n for the eigenvelue λ_n is equal to r, then λ_n will be repeated r times in the sequence (μ_k). The expression for T given in Theorem 9.19 immediately yields the analogous statement for the expansion of Tx as we observed in the finite-dimensional case. Namely,

$$Tx = \sum_k \mu_k < x, e_k > e_k.$$

More generally, take any function f that is continuous on $\sigma(T)$. (Since 0 is the only limit point of the spectrum, we need only check continuity at 0 to verify this for a given f.) Then, e_n is an eigenvector of $f(T)$ (as defined in Section 8.6) with eigenvalue $f(\mu_n)$. This is verified directly for polynomials and then by a limiting argument for all continuous f. It follows that

$$f(T)x = \sum_k f(\mu_k) < x, e_k > e_k.$$

We can also illustrate the multiplication operator version of the spectral theorem, which we do here more precisely than in Section 9.5.1.

In the following theorem, we use the term "unitary equivalence" in a somewhat more general sense than the definition given in Chapter 4, which referred to two operators on the same space. More generally, we say that an operator S on a Hilbert space K is *unitarily equivalent* to an operator T on the Hilbert space H if there is a bijective isometry U from K to H such that $S = U^{-1}TU$; see Figure 9.2.

Theorem 9.29. *A compact, self-adjoint operator T on a separable Hilbert space is unitarily equivalent to a multiplication operator M_f on ℓ^2.*

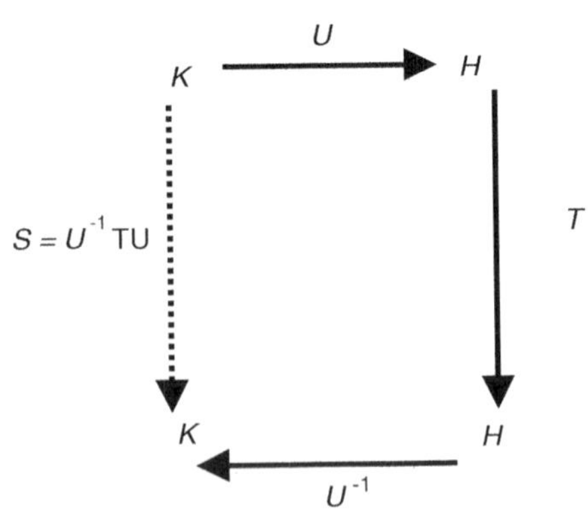

Figure 9.2 Unitary equivalence of S and T.

Proof. Choose an orthonormal basis of eigenvectors (e_n) and corresponding eigenvalues (μ_n) as above. Let U: $\ell^2 \to H$ be the mapping that takes δ_n to e_n. Then, the operator $U^{-1}TU$ maps δ_n to $\mu_n\delta_n$, so it is equal to M_f, in which $f(n) = \mu_n$.

9.6 INVARIANT SUBSPACES

Up to this point, the subject matter in this book has been quite classical. This is, of course, in keeping with the goal to provide an introduction to the topic. In this section, we will deal with somewhat more modern material, which does make connections with present-day research.

We gave the definition of an invariant subspace above. (Although stated for compact, self-adjoint operators on Hilbert spaces, the definition extends to any linear operator on any linear space.) One of the major problems in operator theory over recent years was to determine whether or not any bounded linear operator on a separable Banach space of dimension bigger than 1 has a nontrivial closed invariant subspace. By nontrivial, we mean other than 0 or the whole space, which are both obviously invariant. For Banach space, the problem remained unsolved until 1972 when a counterexample was found by Per Enflo and published some number of years later (Enflo, 1987). Independently, Charles Read published a counterexample (Read 1984) and followed up (Read, 1985) by simplifying the underlying space to ℓ_1. However, for Hilbert spaces, the problem remains unsolved up to the present day. Much of the work has been devoted to showing that certain classes of operators do have invariant subspaces.

A natural class to look at was the compact operators. It was clear that if such an operator has a nonzero element in its spectrum, it would have an eigenvalue and therefore an invariant subspace. That did not, however, handle the case for which the spectrum reduced to zero. The affirmative solution for a compact operator on a Hilbert space was first published by Aronszajn and Smith (1960). The proof was considerably simplified and generalized by Lomonosov (1973), who in fact proved the result for any complex Banach space. We will present a version of Lomonosov's argument. Our proof follows closely the one given by Giles (2000).

We first note that there are various ways of obtaining invariant subspaces, of which we will mention two. First, if T has an eigenvalue, then the eigenspace is obviously invariant, and it will not be trivial (unless T is a scalar multiple of I, in which case any subspace is invariant).

Second, let $\mathcal{P}$ be the set of all elements of $\mathcal{B}(X)$ that can be written as polynomials in T. Choose any nonzero vector x. Then, $\mathcal{P}x = \{Sx : S \in \mathcal{P}\}$ is a subspace and easily seen to be invariant since clearly $TS \in \mathcal{P}$ for all $S \in \mathcal{P}$. By continuity of T, the closure of this space will be a closed invariant subspace. The problem is that it could well be that $\mathcal{P}x$ is dense for all x, so we do not get a proper closed invariant subspace by this method. (Note, however, that this shows why the problem is phrased for separable spaces. The closure of $\mathcal{P}x$ will be separable, spanned by $\{T^n x; n = 0, 1, \ldots\}$, and therefore will necessarily be proper in a nonseparable space.) The trick of the following proof is to show that for compact operators at least one of the two methods above will work. In the following two theorems, X denotes a complex Banach space of dimension greater than 1.

Theorem 9.21. *Any compact $T \in \mathcal{B}(X)$ has a closed proper invariant subspace.*

Proof. We can clearly assume $T \neq 0$. We can also assume that $\mathcal{P}x$ is dense for all nonzero x or we are done, as shown above. Pick a unit vector x_0 so that $Tx_0 \neq 0$. Choose a neighborhood W of Tx_0 such that $\bar{W}$ does not contain 0, and choose a bounded neighborhood G of x_0 such that $T(G) \subseteq W$. We can also ensure that G is small enough so that all elements in G have norm greater than $\frac{1}{2}$. We will show the following:

(∗) There is a positive constant k such that given any positive integer n and any $x \in G$ we can find $S \in \mathcal{P}$ with $\| S \| \leq k^n$ such that $ST^nx \in G$.

Having shown (∗), the proof follows since given any n we take $x = x_0$ and find k and S so that (∗) holds. It follows that

$$\tfrac{1}{2} \leq \| ST^nx_0 \| \leq k^n \| T^n \|,$$

which implies that

$$\| T^n \|^{1/n} \geq \frac{(1/2)^{1/n}}{k} \geq \frac{1}{2k}.$$

From the spectral radius formula (Theorem 8.15), T has a nonzero element in its spectrum, which we know is an eigenvalue from Theorem 9.16.

We first produce the constant k and verify (∗) for $n = 1$. Given any $y \in \overline{T(G)}$, we know that $y \neq 0$, so by our hypothesis, $\mathcal{P}y$ is dense and must intersect the open set G. Thus, we can find $S_y \in \mathcal{P}$ such that $S_yy \in G$. By continuity, we can find a neighborhood V_y of y such that $S_y(V_y) \subseteq G$. By compactness of $\overline{T(G)}$, we can find m points, $y_1, y_2, \ldots, y_m$, so that $\cup_{i=1}^m V_{y_i}$ contains $\overline{T(G)}$. Let $k = \max\{\| S_{yi} \| \; i = 1, 2, \ldots, m\}$.

Now, given $x \in G$, $Tx \in V_{y_i}$ for some i and $S_{y_i}Tx \in G$. We have therefore verified (∗) for $n = 1$ by taking $S = S_{y_i}$.

We proceed by induction by using two obvious facts about $\mathcal{P}$, namely that it is closed under multiplication and that all its elements commute with T. Suppose that (∗) is true for an integer n. Then, given $x \in G$, choose $S_n \in \mathcal{P}$ with $\| S_n \| \leq k^n$ such that $S_nT^nx \in G$. By applying (∗) with $n = 1$, we can find $S_1 \in \mathcal{P}$ with $\| S_1 \| \leq k$ such that $S_1TS_nT^nx = S_1S_nT^{n+1}x \in G$. The condition for $n + 1$ is satisfied with $S_{n+1} = S_1S_n$, which has norm less than or equal to $kk^n = k^{n+1}$. □

We can adapt the proof to obtain a significant generalization, which was part of the work of Lomonosov.

Theorem 9.22. *If $A \in B(X)$ commutes with a nonzero compact operator T, then A has an invariant subspace.*

Proof. We can assume that X is infinite-dimensional because every operator on a finite-dimensional space of dimension greater than 1 has an invariant subspace. Replace $\mathcal{P}$ in the proof of Theorem 9.21 by $\mathcal{A} = \{S \in \mathcal{B}(X) : ST = TS\}$. Since $AS \in \mathcal{A}$ for all $S \in \mathcal{A}$, we can assume as above that $\mathcal{A}x$ is dense for all x. The same proof as above shows that T has a nonzero eigenvalue λ. Let M be the associated eigenspace. Then, $TAx = ATx = \lambda Ax$ all $x \in M$, thereby showing that M is invariant under A. By compactness and the infinite dimensionality of X, T is not a scalar multiple of I and M is proper.

EXERCISES

9.1 Show that a linear operator $T : X \to Y$ is compact if and only $TB(1)$ is relatively compact.

9.2 Show that for a compact operator T on X, if a sequence (x_n) in X converges weakly, then (Tx_n) converges in norm.

9.3 Show that for a compact operator T the space $R(T)$ is separable. (First show that any relatively compact set in a metric space has a countable dense subset.)

9.4 For any operator T let $\widetilde{T}$ be the operator defined on the quotient space $T/N(T)$ as in Section 2.7.1. Show that if T is compact, then $\widetilde{T}$ is compact.

9.5 If T is compact, show that any closed subspace contained in $R(T)$ must be finite-dimensional.

9.6 Let (α_n) be a bounded sequence of scalars, and define an operator T on ℓ^2 by

$$T\delta_n = \alpha_n \delta_{n+1} \text{ for all } n.$$

(T is known as a *weighted shift*.) If (α_n) converges to 0, show that T is compact and find $\sigma(T)$.

9.7 Suppose that (A_n) is a sequence in $\mathcal{B}(X)$, in which X is a Banach space, such that $(A_n x)$ converges to 0 for all $x \in X$. If $T \in \mathcal{B}(X)$ is compact, use Exercises 2.11 and 2.12 to show that $(A_n T)$ converges to 0 in norm.

9.8 Let $T \in \mathcal{B}(H)$ be compact, in which H is a Hilbert space. If (e_n) is an orthonormal basis for $R(T)$, let $M_n = \text{span}(e_n, e_{n+1}, \ldots)$. Apply Exercise 9.7 with $A_n = P_{M_n}$ to show that T is a limit of finite-rank operators.

9.9 If T is a bounded linear operator on a complex Banach space such that T^n is compact for some positive integer n, show that all nonzero points in $\sigma(T)$ are eigenvalues.

9.10 Let P be a projection operator on a Hilbert space. Show that $R(P)$ is invariant under an operator T if and only if $PTP = TP$. Use this to give an alternative derivation of the first statement in Proposition 9.17.

9.11 Adapt the proof of Theorem 9.18 so that it holds for all normal operators.

10

APPLICATION TO INTEGRAL AND DIFFERENTIAL EQUATIONS

10.1 INTRODUCTION

For the most part, this book has concentrated on the fundamentals of functional analysis with only brief allusions to applications. In this chapter, however, we concentrate on one of the major areas in which functional analysis has been applied, namely that of integral and differential equations. In addition to the importance of this subject in all areas of science as well as mathematics, such a survey is of great historical significance as well. The origins of the field of functional analysis can been traced to the work of many researchers beginning early in the twentieth century who were motivated by the study of such equations. This is particularly true of Hilbert spaces. We will see that such key concepts as orthogonal expansions, self-adjointness, and compactness of operators were discovered in this way.

The study of differential equations presents an interesting point. Exercise 2.14 shows that even in the most simple case, operators defined by derivatives are unbounded, presenting an entirely new problem. To study these fully, one needs to develop a theory for such operators, something which is beyond the scope of this book. However, in many cases, as we observed in Example 3.8, the inverses of such differential operators are bounded. We will focus on cases for which this is true and show how our existing theory of bounded operators is sometimes sufficient to obtain information about these unbounded operators.

Our approach here is intended to be self-contained, and no prior knowledge of differential or integral equations is required. For this reason, it is also quite elementary, and readers who are interested in a more advanced treatment of this subject will need to consult other references

10.2 INTEGRAL OPERATORS

10.2.1 Fredholm Integral Operators

In this section, we revisit the integral operator introduced briefly in Example 9.8. It is known as the *Fredholm integral operator*, and it will be a key tool throughout this chapter. We now want to consider this operator as acting on the Hilbert space $H = L^2[a,b]$ rather than on $C[a,b]$ (in which $[a,b]$ is a compact interval on the real line) and investigate it in more detail. As in Chapter 9, choose a continuous function k on the square $S = [a,b] \times [a,b]$, and define the operator K by

$$Kf(s) = \int_a^b k(s,t) f(t)\, dt$$

for all $f \in H$.

We must verify that Kf is in H. Let $M > 0$ be such that $|k(s,t)| \leq M$ for all $s,t \in S$. For fixed $s \in [a,b]$, define the function k_s on $[a,b]$ by $k_s(t) = k(s,t)$. As a function on H, we know that

$$\| k_s \|^2 = \int_a^b |k(s,t)|^2\, dt \leq M^2(b-a),$$

so $\| k_s \| \leq M\sqrt{b-a}$. We can write the operator K in terms of the inner product as

$$Kf(s) = \langle k_s, \bar{f} \rangle.$$

Take a unit vector f. By using the Cauchy–Schwartz inequality and monotonicity of the integral,

$$\| Kf(s) \|^2 = \int_a^b |\langle k_s, \bar{f} \rangle|^2\, dt \leq \int_a^b \| k_s \|^2 \| f \|^2 \; dt \leq M^2(b-a).$$

This shows that K is, indeed, an operator from H to itself with norm less than or equal to $M\sqrt{b-a}$.

The function k is known as the *kernel* of the operator K. (The reader should be careful, as this is another one of those overworked words that appears several times in mathematical literature with completely different meanings.)

Remark. One can define the Fredholm integral operator with a more general kernel. We need only specify that $k \in L^2(S)$, in which S is equipped with two-dimensional Lebesgue measure. We require here some more advanced measure-theoretic tools, chiefly Fubini's theorem (Theorem D.34).

We now want to show that K is compact. While we could adapt our proof of the compactness of K as an operator on continuous functions, it is instructive to present an alternate approach and show that K, in fact, belongs to a subclass of the compact operators known as the *Hilbert–Schmidt operators*.

Given any orthonormal basis $E = (e_1, e_2, \ldots, e_n, \ldots)$ of a separable Hilbert space H and $T \in \mathcal{B}(H)$, let

$$p_E(T) = \sum_{n=1}^{\infty} \| Te_n \|^2,$$

which could take the value of ∞. The first fact to observe about this quantity is as follows.

Lemma 10.1. *$p_E(T)$ is independent of the orthonormal basis E.*

Proof. Let $F = (f_1, f_2, \ldots, f_i, \ldots)$ be any other orthonormal basis. For each n, we have $Te_n = \sum_{i=1}^{\infty} \langle Te_n, f_i \rangle f_i$, so that

$$\| Te_n \|^2 = \sum_{i=1}^{\infty} |\langle Te_n, f_i \rangle|^2,$$

and

$$p_E(T) = \sum_{n,i} |\langle Te_n, f_i \rangle|^2 = \sum_{n,i} |\langle e_n, T^* f_i \rangle|^2 = p_F(T^*).$$

By using the same basis F in both places, we similarly show that $p_F(T^*) = p_F(T^{**}) = p_F(T)$, so that $p_E(T) = p_F(T)$. □

Definitions. $T \in \mathcal{B}(H)$ is said to be a *Hilbert–Schmidt operator* if for some (and therefore any) orthonormal basis E we have $p_E(T) < \infty$. For such an operator, define the *Hilbert–Schmidt norm* of an element $T \in \mathcal{B}(H)$ to be the quantity

$$\| T \|_{\text{HS}} = p_E(T)^{1/2}.$$

The fact that this is a norm is verified in Exercise 10.12. To see how it compares with the operator norm, consider a unit vector $x = \sum_{n=1}^{\infty} \alpha_n e_n$, in which (e_n) is an orthonormal basis. Then,

$$\| Tx \| \leq \sum_{n=1}^{\infty} |\alpha_n| \; \| Te_n \| \leq \left(\sum_{n=1}^{\infty} |\alpha_n|^2 \right)^{1/2} \left(\sum_{n=1}^{\infty} \| Te_n \|^2 \right)^{1/2} = \| T \|_{\text{HS}},$$

in which we use the Cauchy–Schwartz inequality for the second inequality. This shows that the Hilbert–Schmidt norm is greater than or equal to the operator norm.

Example 10.2. Consider the multiplication operator M_f on ℓ^2. By taking $E = (\delta_1, \delta_2, \ldots)$, we have

$$p_E(M_f) = \sum_{n=1}^{\infty} |f(n)|^2,$$

and so M_f is a Hilbert–Schmidt operator if and only if $f \in \ell^2$.

By Proposition 9.7, the Hilbert–Schmidt multiplication operators on ℓ^2 are a subset of the compact multiplication operators. This is true in general.

Theorem 10.3. *Hilbert–Schmidt operators are compact.*

Proof. We show that any Hilbert–Schmidt operator T can be approximated by a finite-rank operator, which proves the assertion. (See the Remark at the end of Section 9.3.) Given $\epsilon > 0$, choose any orthonormal basis (e_n) and a positive integer N so that $\sum_{j=N}^{\infty} \| Te_j \|^2 < \epsilon^2$. For any n, let T_n be the finite-rank operator defined by

$$T_n(e_j) = \begin{cases} Te_j, & j = 1, 2, \ldots, n-1; \\ 0, & j \geq n. \end{cases}$$

We see from the choice of N that $\| T - T_n \|_{\mathrm{HS}} < \epsilon$ for all $n \geq N$. From the comparison result above, it follows that $\| T - T_n \| < \epsilon$. □

Theorem 10.4. *The Fredholm integral operator is a Hilbert–Schmidt operator.*

Proof. Choose any orthonormal basis $E = (e_n)$ of $L^2[a,b]$. Then, for all n, as shown above,

$$\| Ke_n \|^2 = \int_a^b |\langle k_s, \bar{e}_n \rangle|^2 \, ds.$$

It follows that

$$p_E(K) = \sum_{n=1}^{\infty} \int_a^b |\langle k_s, \bar{e}_n \rangle|^2 \, ds = \int_a^b \sum_{n=1}^{\infty} |\langle k_s, \bar{e}_n \rangle|^2 \, ds = \int_a^b \| k_s \|^2 \, ds.$$

The interchange of the sum and the integral above is justified by the series version of the monotone convergence theorem (Theorem D.14). The last step is true since $(\bar{e}_n)$ is easily seen to be an orthonormal basis along with (e_n). Because k is continuous on the compact set $S = [a,b] \times [a,b]$, $|k|$ is bounded by some $M > 0$, from which it follows that $p_E(K) \leq M^2(b-a)$. (In the case of a general L^2 kernel, we do not necessarily have the upper bound on $|k_s|$. In that case, part (a) of Fubini's theorem (Theorem D.34) shows that $k_s \in L^2[a,b]$.) □

Given a Hilbert space operator, it is natural to attempt to determine its adjoint. For the Fredholm integral operator, there is an easy answer, which is suggested immediately when we view our kernel as a continuous matrix, as we outlined in Example 9.8.

Theorem 10.5. *The adjoint of the Fredholm integral operator with kernel k is the integral operator with kernel k^* defined by $k^*(s,t) = \overline{k(t,s)}$.*

Proof. Let K denote the operator with kernel k and J the one with kernel k^*. For any $f, g \in H$,

$$\langle Kf, g\rangle = \int_a^b \left(\int_a^b k(s,t)f(t)\,dt\right)\overline{g(s)}\,ds = \int_S k(s,t)f(t)\,\overline{g(s)}$$

Similarly,

$$\langle Jg, f\rangle = \int_S k^*(s,t)\,g(t)\overline{f(s)} = \int_S k^*(t,s)\,g(s)\overline{f(t)}$$

A direct consequence of the defintion of the integral for complex-valued functions is that for any function h the complex conjugate of $\int h$ equals $\int \overline{h}$. It then follows from the above two expressions that $\langle Kf, g\rangle = \overline{\langle Jg, f\rangle} = \langle f, Jg\rangle$, thereby completing the proof. □

Remark. Above, we used the notation $\int_S$ to denote the integral over the two-dimensional region S. The fact that this is equal to the iterated double integral is a basic fact of analysis for continuous functions. More formally, this is an integral with respect to a product measure (see Section D.11), and for a general L^2 kernel, we need to invoke Fubini's theorem.

Suppose that our kernel k is equal to k^*. Such a kernel is said to be *symmetric*. Then, K is a self-adjoint compact operator, and we can avail ourselves of the spectral theory developed in Chapter 9. We will return to this later.

We next consider the behavior of Fredholm integral operators under multiplication, and the following is another expected result after viewing the kernel as a continuous matrix.

Theorem 10.6. *If K and J are Fredholm integral operators with kernels k and j, respectively, then KJ is a Fredholm integral operator with kernel h given by*

$$h(s,t) = \int_a^b k(s,r)j(r,t)\,dr.$$

Proof. For all $f \in H$ and $s \in [a,b]$,

$$KJf(s) = \int_a^b k(s,r)\,Jf(r)\,dr = \int_a^b \int_a^b k(s,r)j(r,t)f(t)\,dt\,dr = \int_a^b h(s,t)f(t)\,dt,$$

in which the last step follows by reversing the order of integration. □

10.2.2 Volterra Integral Operators

A special case of the Fredholm integral operator is one with a kernel k satisfying $k(s,t) = 0$ for $s < t$. This is known as the *Volterra integral operator*. In this case, we have

$$Kf(s) = \int_a^s k(s,t)f(t)\,dt.$$

A hypothesis of continuity of k on all of S would be overly restrictive in this case as it would force k to be zero on the line $s = t$, which we do not necessarily want. It is sufficient to assume that k is continuous on $\{(s,t) \in S : s \geq t\}$, in order that all the previous arguments of the general Fredholm case go through (without using the measure theory that is needed if we postulate a general L^2 kernel).

Some insight into the behavior of a Volterra integral operator is obtained from the fact that it is analogous to a finite-dimensional operator whose matrix has zero entries above the main diagonal. Actually to get the right analogy, we need to compare it with a matrix such that entries above or *on* the main diagonal are zero. What is the feature of such an operator? If A is its matrix with respect to a basis $\{e_1, e_2, \ldots, e_n\}$, then $Ae_n = 0$. Now Ae_{n-1} will be a multiple of e_n and so $A^2 e_{n-1} = 0$. Proceeding inductively, we see that $A^n = 0$; thus, A is nilpotent. We can show that the general Volterra operator is quasi-nilpotent as defined in Section 8.3.2.

Theorem 10.7. *A Volterra integral operator is quasi-nilpotent.*

Proof. Let K be a Volterra operator with kernel k, and let k_n denote the kernel of the operator K^n. From Theorem 10.6, we know that

$$k_2(s,t) = \int_a^b k(s,r)\,k(r,t)\,dr = \int_t^s k(s,r)\,k(r,t)\,dr$$

for $s \geq t$; 0, otherwise. To verify the limits on the integral, note that for $s < r$ the first term in the product is zero, and for $r < t$ the second term is zero. It follows that K^2 is also a Volterra operator, and by induction, we know that K^n is a Volterra operator with kernel

$$k_n(s,t) = \int_t^s k(s,r)\,k_{n-1}(r,t)\,dr \tag{$*$}$$

for $s \geq t$; 0, otherwise. Let M be the maximum absolute value of k on S. By further induction, it follows that for $s \geq t$

$$|k_n(s,t)| \leq \frac{M^n(s-t)^{n-1}}{(n-1)!}.$$

This is obvious for $n = 1$, and if it is true for $n - 1$, we use $(*)$ to deduce that

$$|k_n(s,t)| \leq \frac{M \cdot M^{n-1}}{(n-2)!} \int_t^s (r-t)^{n-2}\,dr = \frac{M^n(s-t)^{n-1}}{(n-1)!}.$$

Therefore,

$$\| K^n \| \leq \frac{M^n(b-a)^n}{(n-1)!}.$$

The conclusion that $\| K^n \|^{1/n} \to 0$ follows from noting that $((n-1)!)^{1/n}$ approaches ∞ as n approaches ∞. (For example, if n is odd, then $(n-1)! = [1 \cdot (n-1)]$ $[2 \cdot (n-2)] \cdots [((n-1)/2)((n-1)/2+1)]$, and since each product of two terms is greater than or equal to $n-1$, we have that $(n-1)! \geq (n-1)^{(n-1)/2}$.) □

10.3 INTEGRAL EQUATIONS

In this section, we apply the theory of Section 10.2 to some basic integral equations. Suppose we want to find functions f that satisfy

$$\int_a^b k(s,t)f(t)\,dt - \lambda f(s) = g(s) \qquad (*)$$

for all $s \in [a,b]$, in which λ is a nonzero parameter. The first step is to determine the existence and uniqueness of solutions for various values of λ.

We can write this equation in the form

$$(K - \lambda I)f = g,$$

in which K is a Fredholm integral operator. From Theorems 10.3 and 10.4, we know K is compact, and we can invoke the results of Section 9.4. There are then two possibilities:

Alternative 1: $\lambda \notin \sigma(K)$. Then, $K - \lambda I$ is invertible in $\mathcal{B}(H)$, and there is a unique solution to the equation for any g given by

$$f = (K - \lambda I)^{-1} g.$$

Alternative 2: $\lambda \in \sigma(K)$. Then, $N(K - \lambda I)$ is not zero. There is a solution to $(*)$ if and only if $g \in R(K - \lambda I)$, which is equal to $N(K^* - \bar{\lambda} I)^{\perp}$ by Propositions 4.27(b) and 9.15. The solution is never unique, for given any solution we can always add an element in $N(K - \lambda I)$ to get another. So, there are infinitely many solutions, any two of which differ by an element in $N(K - \lambda I)$.

The above dichotomy is sometimes referred to as the *Fredholm* alternative. An important application is that one can show the existence of a solution by proving that if a solution does exist, then it is unique, which is often a much easier task.

Suppose now that k is symmetric so that the operator K is self-ajoint. An expansion in terms of eigenvectors allows us to explicitly write out a solution when alternative 1 holds. Let (e_n) be an orthonormal basis for H consisting of eigenvectors of

K and let (μ_n) be the corresponding sequence of eigenvectors (which can involve repetition). See the discussion following Theorem 9.19. Then, we know that $(K - \lambda I)e_n = (\mu_n - \lambda)e_n$, so that $(K - \lambda I)^{-1}e_n = (\lambda - \mu_n)^{-1}e_n$, and we can write

$$f = \sum_n (\mu_n - \lambda)^{-1}\langle g, e_n\rangle\, e_n.$$

Example 10.8. Consider the equation on $[0, 1]$,

$$\int_0^1 e^{s-t} f(t)\, dt - \lambda f(s) = g(s),$$

in which $\lambda \neq 0$. This equation is of the form $(K - \lambda I)f = g$, in which K has the kernel $k(s, t) = e^{s-t}$.

The kernel of K^2 is then given by

$$k_2(s, t) = \int_0^1 k(s, r)\, k(r, t)\, dr = \int_0^1 e^{s-r}\, e^{r-t}\, dr = e^{s-t} = k(s, t),$$

thereby showing that $K^2 = K$. We know immediately from Example 8.6(b) that the only possible elements that can be in $\sigma(K)$ are 0 and 1. We then consider the following two alternatives:

Alternative 1: $\lambda \neq 1$. We know there is an inverse to $K - \lambda I$. In fact, this can be explicitly calculated as $[\lambda(1 - \lambda)]^{-1}K - \lambda^{-1}I$, which we can verify by direct multiplication. Thus, we get a unique solution to our equation given by

$$f(s) = \frac{1}{\lambda(1 - \lambda)} \int_0^1 e^{s-t}\, g(t)\, dt - \frac{g(s)}{\lambda}.$$

Alternative 2: $\lambda = 1$. The expression

$$\int_0^1 e^{s-t}\, e^t\, dt = e^s$$

shows that 1 is an eigenvalue of K with $f(s) = e^s$ as an eigenvector. Conversely, if $Kf = f$, then we calculate that $f(s) = c\, e^s$, where c is the scalar $\int_0^1 e^{-t} f(t)\, dt$. So, the eigenspace corresponding to this eigenvalue is one-dimensional. Now, we get a solution to our equation if and only if g is orthogonal to $N(K^* - I)$. By Theorem 10.5, the kernel of K^* is the function $k^*(s, t) = e^{t-s}$, and as above, we see then that $N(K^* - I)$ is spanned by the function e^{-s}. Thus, the requirement for g is that

$$0 = \int_0^1 e^{-t}\, g(t)\, dt.$$

The complete set of solutions to the equation for such a g is then given by

$$f(s) = -g(s) + c\,e^s$$

for an arbitrary scalar c, as one can easily verify.

Consider now a general Volterra equation

$$\int_a^s k(s,t)f(t)\,dt - \lambda f = g,$$

which, as above, is of the form $(K - \lambda I)f = g$. We know from the quasi-nilpotency of K that, for all nonzero λ, $(K - \lambda I)^{-1}$ exists and is given by a series expansion. See the discussion after Proposition 8.11. We can immediately write down the unique solution as

$$f = -\left[\frac{g}{\lambda} + \frac{Kg}{\lambda^2} + \frac{K^2 g}{\lambda^3} + \cdots\right].$$

Another form of the Volterra equation that sometimes appears is

$$\lambda \int_a^s k(s,t)f(t)\,dt - f = g, \qquad (*)$$

which is of the form $(\lambda K - I)f = g$. The same argument shows that there is a unique solution for all nonzero λ given by

$$f = -[g + \lambda Kg + \lambda^2 K^2 g + \cdots].$$

Example 10.9. Consider the equation

$$\lambda \int_0^s (s-t)f(t)\,dt + 1 = f(t)$$

for $0 \le s \le 1$. This is of the form of equation $(*)$ with $g = -1$ and-K-the-Volterra operator with kernel $(s-t)_+ = \max\{s-t, 0\}$. A straightforward integration shows that

$$\int_0^s (s-t)\,\frac{t^k}{k!}\,dt = \frac{s^{k+2}}{(k+2)!}\,.$$

Letting p_k denote the function given by $p_k(t) = t^k/k!$, the above formula simply says that $Kp_k = p_{k+2}$. The unique solution is given by

$$f(s) = 1 + \lambda\frac{s^2}{2!} + \lambda^2\frac{s^4}{4!} + \cdots = \frac{e^{\sqrt{\lambda}s} + e^{-\sqrt{\lambda}s}}{2} = \cosh(\sqrt{\lambda}s).$$

10.4 SECOND-ORDER LINEAR DIFFERENTIAL EQUATIONS

In this section, we consider the general second-order linear differential equation. This is of great significance since there are numerous applications to physics. For example, suppose a particle is moving according to certain forces, and we want to determine the position of the particle at a particular point in time, given information regarding its velocity and acceleration. This leads to such an equation.

Let $C^2[a,b]$ denote those functions defined on the interval $[a,b]$ for which the first and second derivatives exist and are continuous. (This applies to the one-sided derivatives at the endpoints.) We consider this as a subspace of $L^2[a,b]$, as justified by our remarks at the end of Section 2.9.6. Given h and j in $C[a,b]$, we define a linear operator D on $C^2[a,b]$ by

$$Df(t) = f''(t) + h(t)f'(t) + j(t)f(t).$$

Note that $R(D)$ will be contained in $C[a,b]$ but we want to view D as an operator from $C[a,b]$ to $L^2[a,b]$ in order to apply Hilbert space techniques.

For a given g in $C[a,b]$, we consider the equation

$$Df = g. \tag{$*$}$$

In the case for which $g = 0$, the equation is said to be *homogeneous*, and the set of all solutions is given by the subspace $N(D)$.

Remark. A more general equation, needed in some applications, involves multiplying the second-derivative term by some other continuous function p. This can cause complications at points t where $p(t) = 0$. We will confine discussion to the simpler version as above.

We now state the major theorem for such an equation. It is motivated by the above-mentioned application. The theorem essentially says that the path of the particle is uniquely determined given the position and velocity at any single point in time. We present a proof which makes use of the Volterra integral operator.

Theorem 10.10. *Given any point u in $[a,b]$ and constants c and d, there exists a unique solution to $(*)$ satisfying $f(u) = c$ and $f'(u) = d$.*

Proof. Preliminary case: We first prove the simple case for which h and $j = 0$. We then have the equation $f'' = g$, which we solve directly by integrating. Suppose f is a solution. Then, for any $s \in [a,b]$,

$$f'(s) = f'(u) + \int_u^s f''(r)\,dr = d + \int_u^s g(r)\,dr. \tag{$\dagger$}$$

We next calculate

$$f(t) = f(u) + \int_u^t f'(s)\,ds = c + d(t-u) + \int_u^t \int_u^s g(r)\,dr\,ds.$$

By reversing the order of integration, the second integral becomes

$$\int_u^t \int_r^t g(r)\,ds\,dr = \int_u^t (t-r)\,g(r)\,dr,$$

and we can explicitly write the solution as

$$f(t) = c + d(t-u) + \int_u^t (t-r)\,g(r)\,dr, \qquad (\ddagger),$$

which verifies the uniqueness. The existence follows simply by differentiating f, as given in (‡), twice and directly verifying that it is indeed a solution to $(*)$ satisfying the given initial conditions.

General case: Now let j and h be arbitrary. Suppose f is a solution to $(*)$, and let $v = f''$. We obtain expressions for f' and f in terms of v from equations (†) and (‡), respectively, with g replaced by v. Now, substitute those expressions in the original equation to get an equation in v:

$$v(t) + dh(t) + h(t)\int_u^t v(r)\,dr + j(t)[c + d(t-u)] + j(t)\int_u^t (t-r)v(r)dr = g(t).$$

We can write this compactly in the form

$$(K + I)v = z, \qquad (**)$$

in which K is the Volterra integral operator with kernel $k(t,r) = h(t) + (t-r)j(t)$ for $t \geq r$ and $z(t) = g(t) - dh(t) - [c + d(t-u)]j(t)$. From the quasi-nilpotency of K and the discussion after Proposition 8.11, it follows that there is a unique function v satisfying equation $(**)$. This immediately gives us the existence of the solution to equation $(*)$ in the general case. Namely, for this function v, find f satisfying the equation $f'' = v$ and the given initial conditions, which exists, as shown in the preliminary case.

Uniqueness follows in a similar fashion for if f_1 and f_2 are two functions satisfying $(*)$ and the intial conditions, then $(K+I)f_1'' = (K+I)f_2''$, which shows that $f_1'' = f_2''$, and by uniqueness in the preliminary case, we know that $f_1 = f_2$. □

What can we say about the null space of D? Take the simplest possible case, namely $Df = f''$. The null space clearly consists of all functions of the form

$a + bt$; therefore, it is two-dimensional, spanned by the constant function and the function $f(t) = t$. This is, in fact, true in general.

Proposition 10.11. *$N(D)$ is two-dimensional.*

Proof. $N(D)$ consists of all solutions to the equation $Df = 0$. Let u be the solution with $u(a) = 1$ and $u'(a) = 0$, and let v be the solution with $v(a) = 0$ and $v'(a) = 1$. Clearly, u and v are linearly independent since neither can be a scalar multiple of the other. Now, given any other solution f, we must have by the uniqueness statement in Theorem 10.10 that $f = f(a)u + f'(a)v$ since the right-hand side is a solution g that satisfies $g(a) = f(a)$ and $g'(a) = f'(a)$. □

From Proposition 10.11, it follows that to find all solutions of $Df = 0$, we need only find any two linearly independent solutions. The following concept is a useful tool for doing so.

Definition. For any two solutions u and v of $Df = 0$, the function $W(u, v) = uv' - vu'$ is known as the *Wronskian* of the two solutions.

It is clear that if one solution is a scalar multiple of the other, then $W(u, v)$ is identically zero. So, if we find a point c at which $W(u, v)$ is not zero, then u and v are a basis for $N(D)$.

We can also show a converse of the above. In fact, a stronger statement holds.

Proposition 10.12. *If $W(u, v)(c) = 0$ for any point c in $[a, b]$, then u and v are linearly dependent.*

Proof. Consider the matrix

$$A = A(u, v)(c) = \begin{pmatrix} u(c) & v(c) \\ u'(c) & v'(c) \end{pmatrix}.$$

Then, $W(u, v)(c)$ is the determinant of A, so if it equals zero, there must exist a nonzero column vector $\mathbf{x} = (\alpha, \beta)$ for which $A\mathbf{x} = 0$. This translates into the statement that the function $f = \alpha u + \beta v$ satisfies $f(c) = 0$ and $f'(c) = 0$. Since the zero function is a solution to $Df = 0$, the uniqueness statement in Theorem 10.10 shows that $f = 0$. □

In the case that the coefficient h of f' is zero, the Wronskian takes a particularly simple form.

Proposition 10.13. *For the operator $Df = f'' + jf$, the Wronskian of any two solutions of $Df = 0$ is constant.*

Proof. For any two solutions u, v of $Df = 0$, let $w = W(u, v) = uv' - vu'$. Then,

$$w' = uv'' - vu'' = -ujv + vju = 0,$$

showing that w is constant. □

When the functions are linearly independent, this constant will be nonzero by Theorem 10.11 and Proposition 10.12. As an example, consider the operator $f'' + f$. One obvious basis for the null space is $u = \cos(t)$ and $v = \sin(t)$, and for these functions, the Wronskian is identically 1.

10.5 STURM–LIOUVILLE PROBLEMS

10.5.1 Introduction

In several applications, instead of providing two initial conditions at one point, we want to impose a single condition at each of two points, normally the endpoints of the interval. These are referred to as *boundary values*.

This may seem like an innocuous change, but the effect is profound for now there need not exist any solutions. A completely new problem is introduced of trying to decide conditions for the existence of a solution, or nonzero solutions in the homogeneous case.

Example 10.14. A classical problem of this type is the following equation on the interval $[0, \pi]$:

$$f'' + \lambda f = 0 \text{ subject to } f(0) = f(\pi) = 0.$$

Because of the linearity of the operator and the boundary conditions, the set of solutions forms a subspace, but it may well be the zero subspace. The first problem is to decide those values of λ for which nontrivial solutions exist.

Suppose first that $\lambda < 0$. One basis for the null space of $Df = f'' + \lambda f$ is given by

$$u(t) = e^{-\sqrt{-\lambda}\,t} \qquad \text{and} \qquad v(t) = e^{\sqrt{-\lambda}\,t},$$

and there is no nontrivial linear combination of these that vanishes at both endpoints. For the case of positive λ, it is more convenient to use the basis

$$u(t) = \sin(\sqrt{\lambda}\,t) \qquad \text{and} \qquad v(t) = \cos(\sqrt{\lambda}\,t),$$

and when $\lambda = n^2$ for some integer n, we satisfy the boundary condition by taking the solution $f(t) = c\sin(nt)$ for any scalar c. One can easily check that no solutions exist for other positive values of λ.

Problems of the above type are known as *Sturm–Liouville* equations after the two mathematicians who developed much of the theory. We will consider certain cases of these equations.

The boundary conditions often involve values of the first derivative as well as the function. In general, we can consider conditions of the following type. For any real constants α and β (not both equal to zero) and real constants γ and δ (not both equal to zero), let X be the subspace of all $f \in C^2[a,b]$ satisfying:

(i) $\alpha f(a) + \beta f'(a) = 0$;
(ii) $\gamma f(b) + \delta f'(b) = 0$.

Throughout this section, we will consider an operator of the form $Df = f'' + jf$ acting on X, where j is real valued. (Note that the coefficient of f' equals zero.) This is known as a *Sturm–Liouville* operator. Observe that we are taking the coefficients in the boundary condition and the function j to be real valued. The significance of this will be made clear later on.

Proposition 10.15. *Either there are no nonzero solutions to $Df = 0$ on X or the solution space is one-dimensional.*

Proof. If there were two linearly independent solutions in X, it would imply that $N(D) \subseteq X$. (We refer here to the null space of D on the whole space $C^2[a,b]$.) This is easily seen to lead to a contradiction in all cases. For example, by Theorem 10.10, we can find $u \in N(D)$ such that $u(a) = 1$ and $u'(a) = 0$; similarly, we can find $v \in N(D)$ such that $v(a) = 0$ and $v'(a) = 1$. If both u and v are in X, then $\alpha = \beta = 0$, a contradiction to the stated condition. □

10.5.2 Injective Case

We now consider a Sturm–Liouville operator D that is injective on X and prove the major theorem for this case. It will allow us to adapt the results of Section 10.3 to analyze D.

Theorem 10.16. *There exists a self-adjoint Fredholm integral operator G on $L^2[a,b]$ such that the restriction of G to $C[a,b]$ is the inverse of the operator D on the space X.*

Proof. We begin by picking a solution $u \in C^2[a,b]$ for the equation $Df = 0$ satisfying condition (i) for the subspace X and a solution $v \in C^2[a,b]$ satisfying condition (ii) for the subspace X. These solutions are easily seen to exist by Theorem 10.10. For example, if $\beta = 0$, we can take u to be any solution with $u(a) = 0$. If $\beta \neq 0$, we can take u to be the solution with $u(a) = 1$ and $u'(a) = -\alpha/\beta$. These solutions must be linearly independent, for if, say, u is a nonzero scalar multiple of v, then u satisfies both conditions and is in the subspace X. However, $Du = 0$, contradicting our hypothesis that D is injective on X.

Proposition 10.13 tells us that the Wronskian of u and v is constant. Therefore, by multiplying by a suitable scalar if necessary, we can assume that $W(u,v)$ is the constant 1.

We now wish to construct a function $k(s,t)$ of two variables, known as the *Green's function* for the given operator D, such that the Fredholm integral operator with kernel k is the inverse of D. This is done as follows: For $s, t \in [a, b]$, define

$$k(s,t) = \begin{cases} u(s)\,v(t) & \text{if } s \leq t, \\ u(t)\,v(s) & \text{if } s > t. \end{cases}$$

Let G be the Fredholm integral operator with kernel k. We know K is compact. Moreover, since the function g and the coefficients in the boundary conditions are real, the construction given in Theorem 10.10 indicates that we can take u and v to be real valued. If follows from Theorem 10.5 that G is self-adjoint. We now verify that the restriction of G to the continuous functions is an inverse to D.

For all $f \in C[a, b]$,

$$Gf(s) = \int_a^b k(s,t)f(t)\,dt = \int_a^s u(t)v(s)f(t)\,dt + \int_s^b u(s)v(t)f(t)\,dt.$$

By differentiating and using the fundamental theorem of calculus, we get

$$(Gf)'(s) = v'(s)\int_a^s u(t)f(t)\,dt + u'(s)\int_s^h v(t)f(t)\,dt.$$

Note that the term $u(s)\,v(s)f(s)$ cancels. Substituting for $s = a$ and b in the above expressions verifies that $Gf \in X$. By differentiating again and now using the fact that u and v are solutions of $Df = 0$, we obtain

$$\begin{aligned}(Gf)''(s) &= -j(s)\,v(s)\int_a^s u(t)f(t)\,dt - j(s)\,u(s)\int_s^b v(t)f(t)\,dt \\ &\quad + [u(s)\,v'(s) - v(s)\,u'(s)]f(s) \\ &= -j(s)\,Gf(s) + W(u,v)(s)f(s) = -j(s)\,Gf(s) + f(s),\end{aligned}$$

which simply says that

$$DG(f) = f.$$

We must also show that $GDf = f$ for all $f \in X$, but this follows immediately since for all $s \in [a, b]$

$$DGDf(s) = Df(s),$$

and from the injectivity of D, we conclude that $GDf(s) = f(s)$. □

Theorem 10.17. *D has countably many eigenvalues μ_n, all real, with $|\mu_n|$ tending to ∞. The eigenspace of each μ_n is one-dimensional. Choosing a unit vector e_n in the eigenspace corresponding to μ_n gives an orthonormal basis of $L^2[a, b]$.*

Proof. This follows by applying Theorem 9.18 to the operator G constructed in Theorem 10.16 together with Proposition 10.15. We also need the easily derived fact that Gf is continuous for all $f \in L^2[a,b]$, which shows that the eigenvectors of G must be continuous. The eigenvectors of D will be the same as those of G, and since G is the inverse of D, the corresponding eigenvalues of D are the reciprocals of those of G. □

Remark. There are some points of interest to note about the conclusions above. We have shown that D is defined on an orthonormal basis of $L^2[a,b]$. Does this mean there is a unique extension of D to the whole space? The answer is no. This would be true if D was bounded, but it is not. In addition, note that all of these basis vectors lie in the subspace X, showing that X must be dense in $L^2[a,b]$.

Theorem 10.17 tells us that the pattern of eigenvalues and eigenvectors observed in Example 10.14 is typical. In fact, by normalizing the eigenvectors in that example to be of norm 1, we see that the set

$$\left\{\sqrt{\frac{2}{\pi}}\sin(nt) : n = 1, 2, \dots\right\}$$

is an orthonormal basis for $L^2[0,\pi]$.

We can go further and use the eigenvector expansion to write explicitly a solution f to the equation

$$(D - \mu I)f = h. \tag{$*$}$$

Let (λ_n) be the sequence of eigenvalues of G and (e_n) the corresponding orthonormal basis of eigenvectors:

Alternative 1. Suppose $\mu \neq \lambda_n^{-1}$ for any n. Let $\lambda = \mu^{-1}$. Multiplying by λG, we see that f is a solution of $(*)$ if and only if f is also a solution of $(G - \lambda I)f = -\lambda Gh$, which means that

$$f = -\lambda(G - \lambda I)^{-1}Gh.$$

For all k, $(G - \lambda I)e_k = (\lambda_k - \lambda)e_k$, from which it follows that

$$-\lambda(G - \lambda I)^{-1}Ge_k = -\lambda\lambda_k(\lambda_k - \lambda)^{-1}e_k = \lambda_k(1 - \mu\lambda_k)^{-1}e_k.$$

There is a unique solution to $(*)$ given by

$$f = \sum_k \lambda_k(1 - \mu\lambda_k)^{-1}\langle h, e_k\rangle\, e_k.$$

Alternative 2. Suppose $\mu = \lambda_n^{-1}$ for some positive integer n. A solution will exist if and only if $\langle h, e_n\rangle = 0$. In this case, it can be verified that all solutions will be of the above form (excluding the term with $k = n$ from the sum) plus any scalar multiple of e_n.

EXAMPLE 10.18. This example illustrates many of the ideas of this section. The thrust of the exposition above is that we can get information about a differential equation by looking at integral equations. This is a two-way street, however, and sometimes it can work in reverse. Suppose we start with the problem of finding eigenvalues and eigenvectors of the Fredholm integral operator K on the interval $[0, \pi]$ with kernel

$$k(s,t) = \begin{cases} s\left(1 - \frac{t}{\pi}\right) & \text{if } s \leq t, \\ t\left(1 - \frac{s}{\pi}\right) & \text{if } s > t. \end{cases}$$

There are different ways to proceed. One approach is to convert the integral equation into a differential equation by differentiation. We start with the equation $Kf = \mu f$, with $\mu \neq 0$, which we write as

$$\left(1 - \frac{s}{\pi}\right) \int_0^s tf(t)\,dt + s \int_s^\pi \left(1 - \frac{t}{\pi}\right) f(t)\,dt = \mu f(s).$$

Differentiate with respect to s and use the fundamental theorem of calculus to obtain

$$\left(1 - \frac{s}{\pi}\right) sf(s) - \frac{1}{\pi} \int_0^s tf(t)\,dt - \left(1 - \frac{s}{\pi}\right) sf(s) + \int_s^\pi \left(1 - \frac{t}{\pi}\right) f(t)\,dt = \mu f'(s).$$

The first and third terms cancel, and by differentiating again, we get

$$-\frac{1}{\pi}\, sf(s) - \left(1 - \frac{s}{\pi}\right) f(s) = \mu f''(s),$$

which simplifies to our familiar equation

$$f'' + \lambda f = 0,$$

with $\lambda = \mu^{-1}$. Moreover, by substituting for $s = 0$ and $s = \pi$ in the first equation, we obtain the boundary conditions

$$f(0) = 0 \qquad \text{and} \qquad f(\pi) = 0.$$

A little reflection will show that we could have discovered this without going through the differentiation. The kernel of the integral operator was in the form of a Green's function with (using our previous notation) $u(t) = t$ and $v(t) = (1 - t/\pi)$. Moreover, the null space of $Df = f''$ is spanned by 1 and t, and we can see that the condition $f(0) = 0$ is satisfied by u and the condition $f(\pi) = 0$ is satisfied by v. This shows that the integral operator we started with is (when restricted to the continuous functions) precisely the inverse of the Sturm–Liouville operator $Df = f''$ on the space $X = \{f : f(0) = f(\pi) = 0\}$.

From Example 10.14, we now know that the eigenvalues of our original integral operator are $-1/n^2$ for all positive integers n with the corresponding eigenvector equal to $\sqrt{2/\pi}\sin(nt)$.

10.5.3 Case for Which *D* Is Not Injective

What happens when D is not injective on X? For example, take the interval to be $[0, \pi]$, let $Df = f''$, and consider the boundary conditions $f'(0) = f'(\pi) = 0$. Any constant c satisfies the boundary conditions, and $Dc = 0$, so D is not injective on X. However, by arguing as in Example 10.14, we can verify that the eigenvalues of D are $-n^2$, where n is a nonnegative integer, with the corresponding one-dimensional eigenspace spanned by $\cos(nt)$. We get the same conclusions as we did in Theorem 10.17. In fact, this always happens.

Theorem 10.19. *The conclusions of Theorem* 10.17 *hold for any Sturm–Liouville operator D, whether injective on X or not.*

Proof. If D is not injective on X, choose any μ_0 that is not an eigenvalue of D, and consider the operator $\tilde{D} = D - \mu_0 I$. Now, $\tilde{D}$ must be injective since $\tilde{D}f = 0$ implies that $Df = \mu_0 f$, which implies that $f = 0$ by the choice of μ_0. It is clear that μ is an eigenvalue of D if and only if $\mu - \mu_0$ is an eigenvalue of $\tilde{D}$. We then apply Theorem 10.16 to $\tilde{D}$, and the conclusion will follow for D. □

Let us now criticize the above "proof." Once again, it is a matter of worrying about that delicate word "choose." Does there really exist such a μ_0, or is it possible that every real number is an eigenvalue of D? This is not immediately clear. Note that D is not bounded and therefore not an element in a Banach algebra, so we cannot apply Theorem 8.10. It will follow, however, once we show that D is *symmetric*. This means that, for all $f, g \in X$, $\langle Df, g\rangle = \langle f, Dg\rangle$. In other words, D satisfies the condition that we defined for a self-adjoint operator in the bounded case. (One can define the concept of an adjoint for an unbounded operator, but there are complications since being symmetric no longer implies self-adjointness. We will not discuss this further as the general theory of unbounded operators is beyond the scope of this text.) We can then apply Proposition 8.25, which is easily seen to apply to any symmetric operator, bounded or not, acting on a linear subspace of a separable Hilbert space, to conclude that the number of eigenvalues is countable. Therefore, we can indeed choose μ_0 as specified.

Theorem 10.20. *The Sturm–Liouville operator D is symmetric.*

Proof. A direct calculation shows that for any $f, g \in X$

$$(Df)g - f(Dg) = f''g - gf'' = W(g,f)'.$$

We also note that since the constants in the boundary conditions for X are real, X is closed under complex conjugation. Moreover, since j is real, we have $\overline{Df} = D\bar{f}$. Therefore, given $f,\ g \in X$,

$$\langle Df, g\rangle - \langle f, Dg\rangle = \int_a^b [(Df)\bar{g} - f(D\bar{g})] = \int_a^b W(\bar{g},f)' = W(\bar{g},f)(b) - W(\bar{g},f)(a).$$

If we multiply the column vectors (α, β) and (γ, δ) on the left by the transpose of the matrices $A(\bar{g},f)(a)$ and $A(\bar{g},f)(b)$, respectively, of Proposition 10.12, we get 0, since both f and $\bar{g}$ are in X. The corresponding determinants then must be 0, and we conclude from the above that $\langle Df, g\rangle = \langle f, Dg\rangle$. □

10.5.4 Further Generalizations

The vanishing of the coefficient of f', which was a defining feature above of our Sturm–Liouville operator, is not as restrictive as might appear. We can always achieve this at the expense of introducing another coefficient into the equation. Given the operator $D(f) = f'' + hf' + jf$, let

$$p(t) = \exp\left(\int_a^t h(s)ds\right),$$

so that $p' = hp$. Then,

$$pD(f) = (pf')' + kf,$$

in which $k = pj$. A more general type of Sturm–Liouville operator is given by the right side of the above equation. We have been looking at the special case for which $p = 1$. When p has no zero in the interval $[a, b]$, this is known as a *regular* Sturm–Liouville operator. The theory goes through much as above, with the conclusions of Theorems 10.16 and 10.17 remaining valid. In the construction of the Green's function, the Wronskian of two solutions is no longer constant; however, $pW(u, v)$ is constant. When p has a zero in the interval, we have what is called a *singular* Sturm–Liouville operator, and the analysis is much more difficult. The singular theory is beyond the scope of this book.

A second generalization involves looking at the equation $Df = \lambda rf$, in which r is a fixed positive continuous function. We once again obtain a sequence of λ's satisfying the criteria of Theorem 10.17. These will not be eigenvalues, but M_r-eigenvalues, as defined in Exercise 8.8, where M_r denotes the multiplication operator. The eigenvectors are orthogonal for the inner product $\langle\cdot,\cdot\rangle_{M_r}$, as defined in Exercise 4.10. See Young (1988, Section 9.3) for more details.

10.6 FIRST-ORDER DIFFERENTIAL EQUATIONS

In this section, we simplify matters in one sense by looking at a first-order, rather than a second-order, equation, but on the other hand, we introduce complications by removing the linearity. A general first-order equation is determined by a continuous function ϕ defined on a open set S in $\mathcal{R}^2$ and is expressed as

$$f'(t) = \phi(t, f(t)). \tag{$*$}$$

Such equations arise naturally whenever we have a quantity f varying over time. The rate of change of f at any instant can depend on both the particular point in time and the value of f at that time.

In order to ensure solutions, we need to impose a certain condition on ϕ.

Definition. The function ϕ is said to satisfy a *Lipschitz condition* in the second variable if there exists a constant $M > 0$ such that given any two points (t, y_1) and (t, y_2) in S we have

$$|\phi(t, y_1) - \phi(t, y_2)| \leq M|y_1 - y_2|.$$

The main result of this section is as follows.

Theorem 10.21. *If ϕ satisfies a Lipschitz condition in the second variable, then given any point $(t_0, y_0) \in S$, there exists a $\delta > 0$ such that $(*)$ has a unique solution f on the interval $(t_0 - \delta, t_0 + \delta)$ satisfying $f(t_0) = y_0$.*

Proof. The trick is to convert our differential equation into an equivalent integral equation. It is immediate that a function f satisfies the given initial condition and satisfies $(*)$ on an open interval I containing t_0 if and only if

$$f(t) = y_0 + \int_{t_0}^{t} \phi(s, f(s))\, ds \tag{$**$}$$

for all $t \in I$.

Viewing this in an operator-theoretic context, we can consider the nonlinear operator T defined by

$$Tf(t) = y_0 + \int_{t_0}^{t} \phi(s, f(s))\, ds.$$

Then, condition $(**)$ simply says that solutions of $(*)$ with $f(t_0) = y_0$ correspond to fixed points of T.

Suppose we can choose closed intervals $I = [t_0 - \delta, t_0 + \delta]$ and $J = [y_0 - \epsilon, y_0 + \epsilon]$ such that the following conditions hold (See Fig. 10.1):

(i) $I \times J \subseteq S$;

(ii) $\delta M < 1$, in which M is the constant in the Lipschitz condition;

(iii) $TX \subseteq X$, in which X is the subset of all $f \in C(I)$ such that $f(t) \in J$ for all $t \in I$.

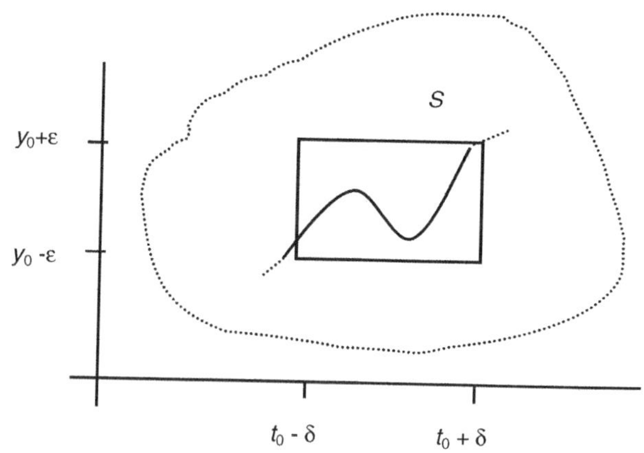

Figure 10.1 Function in the set X of Theorem 10.20.

Let f_1 and f_2 be elements of X. Then, for all $t \in I$ such that $t > t_0$,

$$|Tf_1(t) - Tf_2(t)| \leq \int_{t_0}^{t} |\phi(s,f_1(s)) - \phi(s,f_2(s))|\, ds \leq M \int_{t_0}^{t} |f_1(s) - f_2(s)|\, ds$$
$$\leq M\,(t - t_0)\ \| f_1 - f_2 \| \leq \delta M\ \| f_1 - f_2 \|\ ,$$

in which we use the sup norm on $C(I)$. By arguing similarly for $t \in I$ such that $t < t_0$, we conclude that

$$\| T(f_1) - T(f_2) \| \leq \delta M\ \| f_1 - f_2 \|\ .$$

From condition (ii), T is a contraction mapping, as defined in Section 2.12. By (iii), T maps X into itself. By noticing that X is closed and therefore a complete metric space, we apply Theorem 2.44 to get a unique fixed point of T, which will be the unique solution we seek.

It remains to show that the three conditions above can be satisfied. Since S is open, we can certainly choose δ_1 and ϵ to get intervals I_1 and J satisfying (i) and (ii). Then, we may have to reduce δ_1 to satisfy (iii). Let K be the maximum value of ϕ on $I_1 \times J$, and let δ be the minimum of δ_1 and ϵ/K. Let $I = [t_0 - \delta, t_0 + \delta]$; then, for all $t \in [t_0, t_0 + \delta]$ and $f \in X$, we have

$$|Tf(t) - y_0| \leq \int_{t_0}^{t} |\phi(s,f(s))|\, ds \leq \delta K \leq \epsilon.$$

This, together with a similar argument for $t \in [t_0 - \delta, t_0]$, shows that $Tf \in X$, and we are done.

□

EXERCISES

10.1 Consider the operator T on $L^2[0,1]$ given by $Tf(s) = \int_0^s f(t)\,dt$.

(a) Find the kernel of the operator $T + T^*$. Show that $T + T^*$ is a projection. Find its range.

(b) Find the kernel of the operator T^*T.

10.2 Show that if T is a Hilbert–Schmidt operator and U is a unitary operator, then both UT and TU are Hilbert–Schmidt operators.

10.3 Consider the Fredholm integral operator K with kernel $k(s,t) = s + t$ defined on the unit square.

(a) Show that K has precisely two eigenvalues, the solutions of a certain quadratic equation.

(b) Describe the general solution of the equation $(K - \lambda)f = g$ in terms of λ and g.

10.4 Repeat Exercise 10.2(b) for kernel $k(s,t) = st$, defined on the unit square.

10.5 Solve the Volterra integral equation

$$f(t) = 1 + \lambda \int_0^s (s-t)f(t)\,dt, \qquad 0 \le s \le 1.$$

10.6 Suppose that the Fredholm integral operator K, defined by a continuous kernel k, is positive. Show that $k(s,s) \ge 0$ for all $s \in [a,b]$.

10.7 Find the Green's function, eigenvalues, and corresponding eigenvectors of the operator $Df = f''$ on the subspace of functions f in $C^2[0,\pi]$ such that $f'(0) = 0 = f(\pi)$.

10.8 Find the eigenvalues and eigenvectors for the Fredholm integral operator with kernel $k(s,t) = s$ if $s \le t$, and t if $s > t$, defined on the unit square.

10.9 Consider the following equation on $[0,1]$:

$$f'' + \lambda f = 0, \qquad f'(0) = 0, \qquad f(1) = 0.$$

(a) Find all values of λ for which a solution exists and the corresponding solution.

(b) Find a Green's function for this problem. Compare it with the kernel of T^*T, in which T is the operator in Exercise 10.1.

(c) Show that $\| T \| = 2/\pi$.

10.10 For the interval $[0,\pi]$, consider the Sturm–Liouville operator $Df = f''$ on the subspace of functions $f \in C^2[a,b]$ satisfying $f'(0) = f'(\pi) = 0$.

(a) Show that D is not injective on X.

(b) Find all eigenvalues and associated eigenvectors.

(c) Use your answer to (b), Example 10.14, and Exercise 4.15 to show that the system of Example 4.16(b) is an orthonormal basis for $L^2[-\pi,\pi]$.

10.11 For the interval $[0, \pi]$, consider the operator $Df = f''$ acting on the subspace of functions $f \in C^2[0, \pi]$ satisfying $f(0) = f(\pi) = 0$. Find the Green's function.

10.12 For $T \in \mathcal{B}(H)$, in which H is a separable Hilbert space with an orthonormal basis (e_n), define the scalars

$$a_{i,j} = \langle Te_j, e_i \rangle.$$

(a) Show that

$$\| T \|_{\mathrm{HS}} = \left(\sum_{i,j} |a_{i,j}|^2 \right)^{1/2}.$$

(b) Conclude that $\| \cdot \|_{\mathrm{HS}}$ is a norm on the space of Hilbert–Schmidt operators.

11

SPECTRAL THEOREM FOR BOUNDED, SELF-ADJOINT OPERATORS

11.1 INTRODUCTION AND MOTIVATION

We observed in the last chapter the effectiveness of the spectral theorem in studying compact, self-adjoint operators. This leads naturally to the question of whether there is some sort of generalization to handle an arbitrary bounded, self-adjoint operator. Clearly, in the noncompact case, we cannot hope to recover the exact results that we had for compact operators since Example 8.2 demonstrated that there need not be any eigenvalues. To motivate the procedure, we begin again with the statement that a self-adjoint operator T on a finite-dimensional space is a linear combination of projections; see Section 9.5.1. For our present purposes, it is convenient to express this by summing over *all* real numbers. That is, we write

$$T = \sum_{\lambda \in \mathcal{R}} \lambda E_\lambda,$$

in which E_λ is zero if λ is not an eigenvalue and is the projection onto the eigenspace corresponding to λ if λ is an eigenvalue.

We consider a concrete example. Take the operator T whose matrix with respect to some orthonormal basis is

$$\begin{pmatrix} 2 & 0 & 0 \\ 0 & 3 & 0 \\ 0 & 0 & 5 \end{pmatrix},$$

A First Course in Functional Analysis. By S. David Promislow

in which (denoting an operator by its matrix)

$$E_2 = \begin{pmatrix} 1 & 0 & 0 \\ 0 & 0 & 0 \\ 0 & 0 & 0 \end{pmatrix}, \qquad E_3 = \begin{pmatrix} 0 & 0 & 0 \\ 0 & 1 & 0 \\ 0 & 0 & 0 \end{pmatrix}, \qquad E_5 = \begin{pmatrix} 0 & 0 & 0 \\ 0 & 0 & 0 \\ 0 & 0 & 1 \end{pmatrix}.$$

The key to generalizing this to the infinite-dimensional case is to notice that we can get an alternate representation in the finite-dimensional case by accumulating projections from the left. That is, we define

$$F_\lambda = \sum_{\mu \le \lambda} E_\mu.$$

So, in this example, we have

$$F_\lambda = \begin{cases} 0, & \lambda < 2, \\ E_2, & 2 \le \lambda < 3, \\ E_2 + E_3, & 3 \le \lambda < 5, \\ I, & 5 \le \lambda. \end{cases}$$

We are then faced with the problem of recovering the operator T from $\{F_\lambda\}$ rather than $\{E_\lambda\}$. The obvious answer is that we can easily deduce the collection $\{E_\lambda\}$ from the collection $\{F_\lambda\}$ and then use the original expression for T. Alternatively, we can get a direct expression. Namely, if m and M are the minimal and maximal eigenvalues, respectively, then for any number $m' < m$

$$T = \int_{m'}^{M} \lambda \, dF_\lambda.$$

This expression requires some explanation. The integral is the usual Riemann–Stieljes integral [see, for example, Apostol (1974, Exercise 7.3)], except that the integrator is operator valued rather than numerical valued. We can, however, write down exactly the same definition with norms replacing absolute values. For completeness, we include that here.

The above integral means that given any $\epsilon > 0$ there is a $\delta > 0$ such that for any partition $m' = \lambda_0 < \lambda_1 < \lambda_2 < \cdots < \lambda_n = M$, with $\lambda_{i+1} - \lambda_i < \delta$, and any choice of points u_i in the interval $[\lambda_{i-1}, \lambda_i]$ we have

$$\left\| T - \sum_{i=1}^{n} u_i (F_{\lambda_i} - F_{\lambda_{i-1}}) \right\| < \epsilon.$$

The validity of the integral representation of T can easily be seen in the finite-dimensional case. For a sufficiently fine partition, the terms in the sum given above are all zero, except for those i where the interval $(\lambda_{i-1}, \lambda_i)$ contains a λ for which

$E_\lambda \neq 0$, and the coefficient of u_i is close to λ in that case. Note that we must take m' strictly less than m to ensure that the sum includes E_m.

We have now succeeded in making things much more complicated than they need be in the finite-dimensional setting, but complication is often the first step toward generalization.

For readers familiar with probability theory there is a useful analogy. Suppose we have a discrete random variable X. There are a number of ways to describe the distribution of X. The natural way is through the *probability mass function* f; that is, for any real number t, $f(t)$ is the probability that X takes the value t. We can also describe it through the *cumulative distribution function* F, in which $F(t)$ is the probability that X takes a value less than or equal to t. The advantage of using F is that is applies to *any* random variable, not necessarily discrete.

In our case, E_λ is an analogue of the probability mass function and F_λ is an analogue of the cumulative distribution function. The difference is that rather than assigning numbers, they assign projections. To further the analogy, note that the original formula for T corresponds to the formula for the expectation of a random variable in terms of the probability mass function, and the second integral formula corresponds to the formula for the expectation of a random variable in terms of its cumulative distribution function. For even more insight, suppose we choose any unit vector x. Then, $f(\lambda) = \langle E_\lambda x, x\rangle$ is a probability mass function with corresponding cumulative distribution function $F(\lambda) = \langle F_\lambda x, x\rangle$. It follows from linearity and continuity of the inner product that

$$\langle Tx, x\rangle = \sum_{\lambda} \lambda \langle E_\lambda x, x\rangle = \int_{m'}^{M} \lambda \, d\langle F_\lambda x, x\rangle,$$

in which the last integral is the usual numerical-valued Riemann–Stieljes integral.

This viewpoint appears in quantum mechanics where the vector x represents a certain physical state, the operator T represents an observable, and $\langle Tx, x\rangle$ can be viewed as the expectation of T with respect to the state x.

11.2 SPECTRAL DECOMPOSITION

The family of projections $\{F_\lambda\}$ of the preceding section, corresponding to a finite-dimensional, self-adjoint operator, satisfies the following properties. (We use the order relation on self-adjoint operators given in Section 4.7.)

(i) (Monotonicity) $F_\mu \leq F_\lambda$ for $\mu \leq \lambda$.

(ii) (Right continuity) $F_\lambda = \inf\{F_\mu : \mu > \lambda\}$.

(iii) (Boundedness) There exists real numbers $m \leq M$ such that $F_\lambda = 0$ for $\lambda < m$ and $F_\lambda = I$ for $\lambda \geq M$.

(iv) (Commutativity) $TF_\lambda = F_\lambda T$ for all λ.

(v) The operator T is less than or equal to λI on $R(F_\lambda)$ and greater than or equal to λI on $R(F_\lambda)^\perp$. That is, for all λ,

$$TF_\lambda \le \lambda F_\lambda \qquad \text{and} \qquad T(I - F_\lambda) \ge \lambda(I - F_\lambda).$$

Properties (i) and (ii) follow directly from the definition. For (iii), we can take m as the smallest eigenvalue and M the largest. Property (iv) follows from the fact that the range of each F_λ is an invariant subspace for T. We then have that $F_\lambda TF_\lambda = TF_\lambda$ (see Exercise 9.10); taking adjoints gives (iv). This argument also establishes that TF_λ is self-adjoint, so that the order relations in (v) make sense. The first statement in (v) reduces to showing that for any unit vector $x \in R(F_\lambda)$ we have $\langle Tx, x\rangle \le \lambda$. Any such x can be written as an orthogonal sum $x = x_1 + x_2 + \cdots + x_r$, in which $x_i \in E_{\mu_i}$ for some eigenvalue $\mu_i \le \lambda$. Therefore,

$$Tx = \sum_{i=1}^{r} \mu_i x_i.$$

By orthogonality,

$$\langle Tx, x\rangle = \sum_{i=1}^{r} \mu_i \parallel x_i \parallel^2 \le \lambda \sum_{i=1}^{r} \parallel x_i \parallel^2 = \lambda \parallel x \parallel^2 = \lambda.$$

The second statement follows in a similar way after noting that

$$I - F_\lambda = \sum_{\mu > \lambda} E_\mu.$$

Definition. A family of projections satisfying properties (i) to (v) above is said to be a *bounded decomposition of the identity* for T. (The phrase *resolution of the identity* is also used.)

We have shown, then, that a bounded decomposition of the identity for a bounded, self-adjoint T always exists in the finite-dimensional case. The following is an infinite-dimensional example.

Example **11.1.** Define the multiplication operator T from $L^2[a, b]$ to itself by

$$Tf(s) = sf(s).$$

For all real λ, let e_λ be the function that takes the value 1 on $(-\infty, \lambda]$ and 0 on (λ, ∞), and let F_λ be the operator of multiplication by the restriction of e_λ to $[a, b]$. That is, for $a \le s \le b$,

$$F_\lambda f(s) = \begin{cases} f(s), & s \le \lambda, \\ 0, & s > \lambda. \end{cases}$$

We will show that $\{F_\lambda\}$ is a bounded decomposition of the identity for T.

Since e_λ is real valued, it is clear that F_λ is self-adjoint; moreover, since $F_\lambda^2 = F_\lambda$, we have a family of projections. Monotonicity is clear, as is the fact that $F_\lambda = 0$ for $\lambda < a$ and $F_\lambda = I$ for $\lambda > b$.

Suppose we have a sequence (μ_n) decreasing to a limit of λ. We will show that $F_{\mu_n}(f)$ converges to $F_\lambda(f)$ for all f. (This is *strong* operator convergence, as defined in Section 7.5.) This result will imply that if a function f is in $R(F_{\mu_n})$ for all n, which means that $F_{\mu_n}f = f$, we will necessarily have $F_\lambda f = f$. This will show that $f \in R(F_\lambda)$, verifying (ii). To demonstrate the pointwise convergence, note that for $f \in L^2[a,b]$ the function $1_{(\lambda,\mu_n]}f^2$ decreases to zero with n. By the decreasing version of the monotone convergence theorem (Exercise 2.40),

$$\| F_{\mu_n}f - F_\lambda f \|^2 = \int 1_{(\lambda,u_n]}f^2 \ \downarrow 0.$$

[One can also use a more general convergence theorem for integrals, namely the bounded convergence theorem (Corollary D.20)].

Commutativity is clear since

$$TF_\lambda f(s) = F_\lambda Tf(s) = \begin{cases} sf(s), & s \le \lambda, \\ 0, & s > \lambda. \end{cases}$$

As functions of the operator T, we can write $TF_\lambda = g(T)$, in which $g(t) = te_\lambda(t)$, and $\lambda F_\lambda = \lambda e_\lambda(T)$. The ordering of the two functions implies the ordering of the corresponding operators (see the second set of remarks following the proof of Theorem 8.26), thereby verifying the first statement in (v). In the same fashion, we can demonstrate that the second statement holds.

Application: Systems Theory. Resolutions of the identity were originally developed for the purpose of stating and proving the spectral theorem, as we are about to do. There are, however, other uses. Systems theory is a branch of engineering which analyses inputs and outputs to physical systems at various points of time. In Feintuch and Saeks (1982), resolutions of the identity of operators are used to model *time causality*, that is, the principle that outputs cannot be affected by inputs that occur at a later point of time.

With a little more work, we could adapt the argument in Example 11.1 to show that all bounded multiplication operators on $L^2[a,b]$ have a bounded decomposition of the identity, but more is true. Our main result is that bounded resolutions of the identity exist in general.

Theorem 11.2 (Spectral Theorem). *Let $T \in \mathcal{B}(H)$ be self-adjoint. Then:*

(*a*) *There exists a unique bounded decomposition of the identity $\{F_\lambda\}$ for T.*

(*b*) *For any number $m' < m$, we have*

$$T = \int_{m'}^{M} \lambda \, dF_\lambda.$$

Proof of Part (b). The harder part here is (a), and we require some additional machinery to prove it. We will assume the existence part for the moment and show that (b) follows. Given $\epsilon > 0$, we take $\delta = \epsilon$ and choose any partition $\{\lambda_i\}$ of $[m', M]$ and intermediate points $\{u_i\}$ (that is, $u_i \in [\lambda_{i-1}, \lambda_i]$), as in the definition of the integral given in Section 11.1. Let $F(\Delta_i) = F_{\lambda_i} - F_{\lambda_{i-1}}$, and let $S = \sum_{i=1}^{n} u_i F(\Delta_i)$. For each i, using proposition 4.38 and point (b) on page 95,

$$R(F(\Delta_i)) \subseteq R(F_{\lambda_i}) \cap R(F_{\lambda_{i-1}})^{\perp}.$$

It follows from property (v) of the spectral decomposition that

$$\lambda_{i-1} F(\Delta_i) \leq T F(\Delta_i) \leq \lambda_i F(\Delta_i).$$

Subtracting $u_i F(\Delta_i)$ from each term (which is easily shown to be a valid operation for this ordering) gives

$$(\lambda_{i-1} - u_i) F(\Delta_i) \leq (T - u_i I) F(\Delta_i) \leq (\lambda_i - u_i) F(\Delta_i).$$

Now, sum the above inequalities over all values of i, noting that $\sum_{i=1}^{n} F(\Delta_i) = I$. For the left-hand terms, each $\lambda_{i-1} - u_i$ is greater than $-\epsilon$, so the sum will be greater than or equal to $-\epsilon I$. Similarly, the sum of the right-hand terms will be less than or equal to ϵI. The sum of the middle terms is just $T - S$. The result is that

$$-\epsilon I \leq T - S \leq \epsilon I.$$

From the definition of the ordering relation on self-adjoint operators, it follows that for any unit vector x

$$-\epsilon = -\epsilon \langle x, x \rangle \leq \langle (T - S)x, x \rangle \leq \epsilon \langle x, x \rangle = \epsilon.$$

This means that the numerical radius $w(T) \leq \epsilon$, and by Propostion 8.20, we know that $\| T - S \| \leq \epsilon$. □

This finishes the proof of (b). To get an idea as to how to proceed to prove part (a), we will again return to a finite-dimensional setting and review some basic linear algebra. Given a self-adjoint operator T, the problem is to find the projections $\{F_\lambda\}$ without first finding the projections $\{E_\lambda\}$, which are not going to exist in the general case. For motivation, we consider a particular example.

Example 11.3. Suppose a self-adjoint operator T on $\mathcal{R}^3$ is given in matrix form by

$$T = \begin{pmatrix} \frac{17}{6} & \frac{-5}{6} & \frac{-2}{3} \\ \frac{-5}{6} & \frac{17}{6} & \frac{2}{3} \\ \frac{-2}{3} & \frac{2}{3} & \frac{13}{3} \end{pmatrix}.$$

Find the projection F_3.

We first determine that the eigenvalues are $\{2, 3, 5\}$; therefore, there is an orthonormal basis for which the matrix of T is the one given in Section 11.1. Thus, $F_3 = E_2 + E_3$ has the following matrix with respect to the new basis:

$$\begin{pmatrix} 1 & 0 & 0 \\ 0 & 1 & 0 \\ 0 & 0 & 0 \end{pmatrix}.$$

So, F_3 is a function of the given operator T. In particular, as we showed in Example 8.27, it is that function that maps 2 to 1, 3 to 1, and 5 to 0. In this case, we want a linear combination of the polynomials $(x-3)(x-5)$ and $(x-2)(x-5)$ because the first maps 3 and 5 to 0 and the second maps 2 and 5 to 0. Since the first polynomial maps 2 to 3, and the second maps 3 to -2, we want the polynomial

$$\frac{1}{3}[(x-3)(x-5)] - \frac{1}{2}[(x-2)(x-5)] = \frac{1}{6}[5x - x^2].$$

The matrix for the spectral projection F_3 in the original basis is therefore

$$\frac{5T - T^2}{6} = \begin{pmatrix} \frac{5}{6} & \frac{1}{6} & \frac{2}{6} \\ \frac{1}{6} & \frac{5}{6} & \frac{-2}{6} \\ \frac{2}{6} & \frac{-2}{6} & \frac{2}{6} \end{pmatrix}.$$

We now have a guide as how to proceed in general. In the finite-dimensional case, we have seen that for the self-adjoint operator T the spectral projection F_λ is a function of T. Namely, it is that function of T that maps all numbers less than or equal to λ to 1 and all numbers greater than λ to 0. Indeed, this is a particular case of what we did in Example 8.27. We therefore want to consider the functions e_λ as defined in Example 11.1

A major problem, however, is that we have only defined a function f of a self-adjoint operator in the case for which f is *continuous*, and e_λ is not continuous in general. We could proceed by extending the discussion in Section 8.6 to include a larger class of functions that includes e_λ. These functions are limits of polynomials as the continuous functions are, but they will be only pointwise rather than uniform limits. With the tools we have in hand, however, it is not too much more work to carry out the extension to all bounded Borel functions defined on $\sigma(T)$. This is done in the next section.

11.3 EXTENSION OF FUNCTIONAL CALCULUS

We start with a bounded self-adjoint operator T and recall from Theorem 8.26 that for all $f \in C(\sigma(T))$ we can define $f(T)$ in $\mathcal{B}(H)$ such that the mapping f to $f(T)$ preserves all the relevant operations and is an isometry under the sup norm. We want to extend this mapping in order to be able to define $f(T)$ for any bounded

Borel function. A Borel function is one that is measurable with respect to the σ-algebra of Borel sets as defined in Section 6.1.4. (Recall that this is the smallest σ-algebra containing the open sets.)

We will, of course, want to preserve the properties that we had in the case for which f was continuous.

To this end, we use once again the Riesz representation theorem [Theorem 6.4(b)]. We stated it for continuous functions on an interval, but it will apply similarly when the domain is any compact subset of the reals. (See the remarks at the end of Section D.12.) We will also need the *bounded convergence theorem* for integrals, Theorem D.20. The reader is also referred to Section D.9.3 for facts about complex measures and to Section D.12 for more details on the Riesz representation theorem.

From the Riesz representation theorem, given any bounded linear functional ϕ on $C(\sigma(T))$, there is a unique complex measure μ on the Borel sets of $\sigma(T)$ such that for all $h \in C(\sigma(T))$

$$\phi(h) = \int h\, d\mu.$$

Let $\hat{\phi}$ denote the resulting extension to a linear functional on the bounded Borel functions. That is,

$$\hat{\phi}(f) = \int f\, d\mu$$

for all bounded Borel functions f on $\sigma(T)$.

In particular, for any two vectors $x, y \in H$, define a linear functional $\phi_{x,y}$ on $C(\sigma(T))$ by

$$\phi_{x,y}(h) = \langle h(T)x, y\rangle.$$

Now, fix any bounded measurable function f, and define a sesquilinear form by

$$(x, y) \to \hat{\phi}_{x,y}(f).$$

If $\mu_{x,y}$ is the measure corresponding to $\phi_{x,y}$, then

$$|\mu_{x,y}(\sigma(T))| = \left|\int 1_{\sigma(T)}\, d\mu_{x,y}\right| = |\phi_{x,y}(1_{\sigma(T)})| = |\langle x, y\rangle| \leq \| x \| \| y \|,$$

so that

$$|\hat{\phi}_{x,y}(f)| = \left|\int f\, d\mu_{x,y}\right| \leq M \| x \| \| y \|,$$

showing that $|f|$ is bounded by M.

This argument shows that the sesquilinear form is bounded, so as shown in Theorem 4.24, it defines a bounded linear operator $f(T)$. To summarize, for any bounded Borel function f on the spectrum, we define $f(T) \in \mathcal{B}(H)$ such that for all $x, y \in H$

$$\langle f(T)x, y\rangle = \int f\, d\mu_{x,y} = \hat{\phi}_{x,y}(f).$$

We state the properties of this correspondence in the following theorem. The first four parallel those of Theorem 8.26. (The reader should note in the following that we are talking about a particular σ-algebra, specifically the Borel sets, but we do not have a fixed measure on this set. We cannot, then, talk about concepts that depend on a particular measure, such as "essential supremum" or "almost everywhere.")

Theorem 11.4. *Given any bounded Borel functions f, g on $\sigma(T)$ and any $\alpha \in \mathbb{C}$:*

(i) $(f+g)(T) = f(T) + g(T)$ *and* $(\alpha f)(T) = \alpha(f(T))$.
(ii) $\bar{f}(T) = (f(T))^*$.
(iii) $(fg)(T) = f(T)\, g(T)$.
(iv) *For any $S \in \mathcal{B}(H)$ such that $ST = TS$, we have $Sf(T) = f(T)S$.*
(v) *Suppose we are given a sequence (f_n) of bounded Borel functions that converges pointwise to a bounded Borel function f. Suppose there is a constant M such that $|f_n| \le M$ for all n. Then, $(f_n(T)x)$ converges to $f(T)x$ for all $x \in H$; that is, we have convergence in the strong operator topology.*

Proof

(i) The proof of this statement is straightforward and will be left to the reader.
(ii) For any $h \in C(\sigma(T))$, we have

$$\overline{\int h\, d\mu_{y,x}} = \overline{\langle h(T)y, x\rangle} = \langle (h(T))^* x, y\rangle = \langle \bar{h}(T)x, y\rangle = \int \bar{h}\, d\mu_{x,y},$$

in which for the third equality we use Theorem 8.26(ii). The definition (as given in Section D.9.3) of integration with respect to a complex measure $\mu = \nu + i\rho$ (in which ν and ρ are signed measures) tells us that

$$\overline{\int h\, d\mu} = \int \bar{h}\, d\bar{\mu},$$

in which $\bar{\mu} = \nu - i\rho$. By applying this expression to the identity above, after conjugating both sides, we obtain, by the uniqueness statement in the Riesz representation theorem, that $\mu_{y,x} = \overline{\mu_{x,y}}$, so that the chain of equalities holds when h is replaced by any bounded Borel function f.

(v) We can assume that f_n and f are real valued by applying the result to the real and imaginary parts separately. Let $g_n = (f_n - f)^2$, which converges pointwise to to 0. Since $g_n(T)$ is self-adjoint by (ii), it follows that

$$\| f_n(T)x - f(T)x \|^2 = \hat{\phi}_{x,x}(g_n) = \int g_n \, d\mu_{x,x},$$

which converges to 0 by the bounded convergence theorem (Corollary D.20).

(iv) By Theorem 8.26(iv), for any continuous f and $x, y \in H$, we have

$$\phi_{Sx,y}(f) = \langle f(T)Sx, y\rangle = \langle Sf(T)x, y\rangle = \langle f(T)x, S^*y\rangle = \phi_{x,S^*y}(f).$$

This shows that $\phi_{Sx,y} = \phi_{x,S^*y}$. By extending to bounded Borel functions, we obtain $\hat{\phi}_{Sx,y}(f) = \hat{\phi}_{x,S^*y}(f)$ for general f, and the above chain of equalities persists and shows the result.

(iii) We leave the proof of this part for the last because it is a little more delicate. We first establish a general fact relating to the extension given by the Riesz representation theorem. For a bounded linear functional ϕ on $C(\sigma(T))$ and continuous function g, let ϕ_g denote the functional defined by $\phi_g(h) = \phi(gh)$ for $h \in C(\sigma(T))$. For the corresponding measure μ, we have

$$\phi_g(h) = \int hg \, d\mu = \int h \, d\mu_g$$

(using the notation of Section 6.1.3), so that the measure corresponding to ϕ_g is precisely μ_g. Then, for a bounded Borel function f,

$$\hat{\phi}_g(f) = \hat{\phi}(gf) = \int gf \, d\mu = \int f \, d\mu_g = \widehat{\phi_g}(f).$$

In other words, we have shown that

$$\hat{\phi}_g = \widehat{\phi_g}. \qquad (*)$$

We now return to verifying the multiplication property. Fix $x, y \in H$. We want to show that for all bounded measurable f and g we have

$$\langle f(T)g(T)x, y\rangle = \langle (fg)(T)x, y\rangle.$$

For any fixed bounded Borel function g, let ψ_g denote the linear functional defined on the bounded Borel functions by

$$\psi_g(f) = \langle f(T)g(T)x, y\rangle,$$

and let ϕ denote $\phi_{x,y}$. Then, the condition that we must verify translates to the statement that

$$\psi_g(f) = \hat{\phi}_g(f) \qquad (**)$$

for all bounded Borel functions f. Equation $(**)$ is true by Theorem 8.26 (iii) when both g and f are continuous. By extending it to the bounded Borel functions and by invoking $(*)$, it is true for all f when g is continuous. From (iv), $f(T)$ and $g(T)$ commute, so by symmetry, it is true for all g when f is continuous. By extension again, it is true in general. □

Continuation of Proof of Theorem 11.2. We now provide the existence proof of part (a) of Theorem 11.2. We consider the functions e_λ, as defined in Example 11.1, restricted to $\sigma(T)$. These are clearly bounded Borel functions. Therefore, for each real λ, we can define the operator

$$F_\lambda = e_\lambda(T).$$

We next verify that $\{F_\lambda\}$ is a decomposition of the identity for T. Since e_λ is real valued and equal to its square, parts (ii) and (iii) of Theorem 11.4 ensure that F_λ is a projection. Let m and M be the minimum and maximum values, respectively, of $\sigma(T)$. If $\lambda < m$, then e_λ is the zero function, so that $F_\lambda = 0$. If $\lambda \geq M$, then e_λ is identically equal to 1, and $F_\lambda = I$. Thus, boundedness holds.

Monontonicity follows from the same argument as given in the second set of remarks following Theorem 8.26. We argue in the same way to prove property (v) of a decomposition of the identity by noting that the function f given by $f(t) = (\lambda - t)e_\lambda(t)$ satisfies $f \geq 0$; we infer that $(\lambda I - T)F_\lambda \geq 0$. Similarly, the function g given by $g(t) = (t - \lambda)(1 - e_\lambda(t))$ satisfies $g \geq 0$; which implies that $(T - \lambda I)(I - F_\lambda) \geq 0$.

Commutativity follows directly from Theorem 11.4(iv). In fact, we get the stronger statement as given by the latter result.

Right continuity follows since given any sequence (λ_n) decreasing to λ we know that the sequence (e_{λ_n}) decreases to e_λ and is bounded above by the function with a constant value of 1. Therefore, (F_{λ_n}) converges strongly to F_λ by Theorem 11.4(v). For any $x \in \bigcap_n R(F_{\lambda_n})$, we have $F_{\lambda_n}x = x$ and, therefore, $F_\lambda x = x$, showing that $x \in R(F_\lambda)$.

This completes the proof of the existence of a decomposition of the identity. We will show the uniqueness after first proving another important fact. □

As well as representing T as an integral, we can similarly express continuous functions of T as an integral.

Theorem 11.5. *For any* $f \in C(\sigma(T))$,

$$f(T) = \int_{m'}^{M} f(\lambda)\, dF_\lambda.$$

Proof. Let S be an approximating sum for the integral representation of T, as given in the proof of Theorem 11.2(b). That is,

$$S = \sum_{i=1}^{n} u_i F(\Delta_i).$$

Then, for any positive integer k,

$$S^k = \sum_{i=1}^{n} u_i^k F(\Delta_i)$$

because, for $i \neq j$, $F(\Delta_i)$ and $F(\Delta_j)$ are orthogonal projections and multiply to zero.

By refining the partition, we can make S^k arbitrarily close to T^k, and the right-hand side of the above expression arbitrarily close to $\int_{m'}^{M} \lambda^k \, dF_\lambda$, which proves the conclusion for functions of the form $f(t) = t^k$. By linearity, it is then true for all polynomials. The conclusion is verified for any continuous f by finding a sequence of polynomials that converges uniformly to f. We leave details to the reader. □

Continuation of Proof of Theorem 11.2. From Theorem 11.5, we can verify the uniqueness of the spectral decomposition. Suppose $\{G_\lambda\}$ is another such family of projections. We want to show that for any $x, y \in H$ and for all λ

$$\langle F_\lambda x, y \rangle = \langle G_\lambda x, y \rangle.$$

Fix any $x, y \in H$, and let $v(\lambda) = \langle F_\lambda x, y \rangle - \langle G_\lambda x, y \rangle$. Then, for all continuous functions f, we have $\int_{m'}^{M} f \, dv = 0$ by Theorem 11.5. But, v is right continuous, and $v(m') = v(M) = 0$. It is a well-known fact of Riemann–Stieljes integrals that this implies that $v = 0$. □

Remark. There are easier proofs of the existence of the decomposition of the identity. One such method is based on the result of Exercise 11.5. Our approach here is more complicated, but it is more natural. Most importantly, it highlights the extension of the functional calculus, which by itself is an important tool in operator theory.

11.4 MULTIPLICATION OPERATORS

We saw in Chapter 9 that another interpretation of the spectral theorem for compact operators is that every compact, self-adjoint operator is unitarily equivalent to a multiplication operator on l^2. An analogue of this statement is true for all bounded self-adjoint operators and justifies our concentration from the beginning of the book on multiplication operators.

Theorem 11.6. *If T is a bounded, self-adjoint operator on a Hilbert space H, there is a finite measure space $(\Omega, \mathcal{S}, \mu)$, a bijective isometry U from $L^2(\Omega)$ to H, and a fixed real-valued bounded measurable function g on Ω such that*

$$U^{-1}TUf(t) = g(t)f(t)$$

for all $f \in L^2(\Omega)$ and $t \in \Omega$.

Proof. The key is to modify the mapping of Section 8.6 that sends f to $f(T)$. Suppose we pick a vector x in H. We can define a mapping U from $C(\sigma(T))$ to the space H by

$$Uf = f(T)x.$$

While our Chapter 8 map was an isometry with respect to the sup norm, this is now an isometry when we equip $C(\sigma(T))$ with the L^2-norm determined by the measure $\mu_{x,x}$. This follows since

$$\begin{aligned}\| f(T)x \|^2 - \langle f(T)x, f(T)x\rangle &= \langle (f(T))^* f(T)x, x\rangle \\ &= \langle \bar{f}(T)f(T)x, x\rangle = \langle |f|^2(T)x, x\rangle,\end{aligned}$$

and from from the construction given in Section 11.3

$$\langle |f|^2(T)x, x\rangle = \int |f(t)|^2 \, d\mu_{x,x} = \| f \|_2{}^2.$$

Since $C(\sigma(T))$ is dense in $L^2(\sigma(T))$, we can by Theorem 2.20 extend U to an isometry on $L^2(\sigma(T))$. Now, in general, this will not be surjective. Suppose, however, we can find a vector x such that the set $\{p(T)x\}$ as p ranges over the polynomials is dense in H. (Such an x is known as a *cyclic vector* for the operator T.) Then, the range of U must be all of H.

Now, consider the map $U^{-1}TU$. U maps the function f to $f(T)x$. This is then multiplied by T to get the vector $Tf(T)x$, which is the image under U of the function h given by $h(t) = tf(t)$. So, the conclusion holds with $g(t) = t$.

Consider the case when no cyclic vector exists, which we will sketch quickly. One can proceed as follows. Choose any nonzero vector x_1, and let M_1 be the closure of $\{p(T)x_1\}$. It will be a closed subspace of H. If M_1 is all of H, we are in the situation above. If not, pick a nonzero vector x_2 in $M_1^{\perp}$. Then, the closure of $\{p(T)x_2\}$ is a closed subspace M_2, and by the self-adjointness of T, it will be in $M_1^{\perp}$. (See Proposition 9.17.) Continue until we have a maximal family of vectors $\{x_\lambda : \lambda \in \Lambda\}$ such that $\{p(T)x_\lambda : \lambda \in \Lambda\}$ is a mutually orthogonal family of closed subspaces. Now, of course, for an infinite-dimensional space, we have to show that the above process will terminate, and the tool for doing so is Zorn's lemma. By maximality, the closed linear span of all of these subspaces is H.

Then, form a big measure space by taking the union of disjoint copies of $C(\sigma(T))$, indexed by $\lambda \in \Lambda$ with the measure $\mu_{x_\lambda, x_\lambda}$ on the appropriate copy. One can then piece together the isometries U^{-1} from all the subspaces to give us an isometry on all of H, and the proof is completed as above. □

Example 11.7. We illustrate the proof of Theorem 11.6 with a finite-dimensional example. Any such example is necessarily rather trivial, but it will illustrate the idea. Let T be the operator on $\mathcal{R}^3$ with matrix

$$\begin{pmatrix} 1 & 0 & 0 \\ 0 & 2 & 0 \\ 0 & 0 & 2 \end{pmatrix}$$

with respect to some orthonormal basis. It is clear from the outset that T is essentially a multiplication operator, but we want to demonstrate the construction outlined above for the general case.

Due to the repeated value of 2, there is no cyclic vector, as can be verified. Suppose we start with the vector $x_1 = (1, 1, 1)$. Then, $\{p(T)x_1\}$ is the subspace M_1 consisting of all vectors of the form (a, b, b). Next, pick a vector in $M_1^{\perp}$, say, $x_2 = (0, 1, -1)$. Then, $\{p(T)x_2\}$ is the subspace M_2 consisting of all vectors of the form $(0, c, -c)$. The whole space, namely $\mathcal{R}^3$, is the span of M_1 and M_2, and we have completed the first part of the process.

The measure space corresponding to x_1 consists of the two points in $\sigma(T)$, namely 1 and 2. Let e_1 be the function on $\sigma(T)$ that takes the value of 1 at 1 and 0 at 2. Then, the matrix of $e_1(T)$ is just

$$\begin{pmatrix} 1 & 0 & 0 \\ 0 & 0 & 0 \\ 0 & 0 & 0 \end{pmatrix},$$

so that the measure of the point 1 is $\langle e_1(T)x_1, x_1\rangle = 1$. Similarly , we can show that the measure of the point 2 is 2. The isometry $U_1 : C(\sigma(T)) \to \mathcal{R}^3$ is as follows. If f is the function with $f(1) = a$ and $f(2) = b$, then $U_1 f = (a, b, b)$. One can verify that $\| U_1 f \| = \| f \| = \sqrt{a + 2b^2}$.

For x_2, let the two points in the measure space be denoted by $1'$ and $2'$. Calculating as above, we find that the measure of $1'$ is 0 and the measure of $2'$ is 2. The isometry U_2 in this case maps the function f with $f(1') = a$ and $f(2') = b$ to $(0, b, -b)$. One can verify that $\| U_2 f \| = \| f \| = \sqrt{b}$.

Putting these together, we have the big measure space Ω consisting of three points $\{1, 2, 2'\}$ with measures of $(1, 2, 2)$, respectively. (We do not need to include the point $1'$ since it has measure 0.) The isometry U^{-1} on $\mathcal{R}^3$ is then determined as follows. Take any vector $(a, b, c) \in \mathcal{R}^3$; then,

$$(a, b, c) = \left(a, \tfrac{1}{2}(b + c), \tfrac{1}{2}(b + c)\right) + \left(0, \tfrac{1}{2}(b - c), -\tfrac{1}{2}(b - c)\right)$$

is the decomposition of (a, b, c) as a sum of an element in M_1 and one in M_2. The mapping U_1^{-1} takes the first vector to the function f_1 given by $f_1(1) = a$ and $f_1(2) = (b + c)/2$. The mapping U_2^{-1} takes the second vector to the function f_2

given by $f_2(2') = (b - c)/2$. Therefore, U^{-1} maps the vector (a, b, c) to the function f given by $f(1) = a$, $f(2) = (b + c)/2$, and $f(2') = (b - c)/2$.

The mapping U takes the function f with $f(1) = a$, $f(2) = b$, and $f(2') = c$ to the vector $(a, b + c, b - c)$. It follows that for such f the function $h = U^{-1}TUf$ satisfies

$$h(1) = a, \qquad h(2) = 2b, \qquad h(2') = 2c,$$

so that $U^{-1}TU$ is indeed the multiplication operator M_g, in which g is given by $g(1) = 1$, $g(2) = 2$, and $g(2') = 2$.

EXERCISES

11.1 Suppose E, F, and G are mutually orthogonal projections in a Hilbert space. Suppose a self-adjoint operator T has a decomposition of the identity $\{F_\lambda\}$, in which F_λ is defined by

$$F_\lambda = \begin{cases} 0, & \lambda < -2, \\ E, & -2 \leq \lambda < 1.5, \\ E + F, & \leq \lambda < 2, \\ E + F + G, & 2 \leq \lambda < 2.4, \\ I, & 2.4 \leq \lambda. \end{cases}$$

Write T as a linear combination of projections.

11.2 Let $\{F_\lambda\}$ be the decomposition of Example 11.1, in which $a = 0$ and $b = 2$. Consider the decomposition $\{F'_\lambda\}$ given by

$$F'_\lambda = \begin{cases} F_\lambda, & \lambda < 1, \\ I, & 1 \leq \lambda. \end{cases}$$

(a) If T is a self-adjoint operator corresponding to $\{F'\}_\lambda$ then find all eigenvalues of T.

(b) Find a continuous function g so that the operator M_g on $L^2[0, 2]$ has $\{F'_\lambda\}$ as its spectral decomposition.

11.3 If g is a continuous real-valued function on $[a, b]$, describe the spectral resolution of the operator T on $L^2[a, b]$ given by

$$Tf(t) = g(t)f(t) \quad \text{for all} f \in L^2[a, b].$$

11.4 Show that a self-adjoint operator on a finite-dimensional space has a cyclic vector if and only if all eigenspaces are of dimension 1.

11.5 For any bounded, self-adjoint operator T, let E be the projection onto $N(T - (T^2)^{1/2})$. Show that $TE \geq 0$ and $T(1 - E) \leq 0$.

APPENDIX A

ZORN'S LEMMA

There are many applications in analysis in which one wants to show the existence of maximal elements in partially ordered sets. The major tool for doing this is Zorn's lemma. We will state this result, without proof, after some preliminary definitions.

Definition. A *partially ordered* set, often called a *poset*, consists of a set S together with a binary relation $\leq$, known as a *partial order*, that satisfies the following properties:

(i) (*Transitivity*) $x \leq y$ and $y \leq z$ imply that $x \leq z$.
(ii) (*Reflexivity*) For all $x \in S$, we have $x \leq x$.
(iii) (*Antisymmetry*) $x \leq y$ and $y \leq x$ imply that $x = y$.

Definition. An element x in a poset is said to be *maximal* if there does not exist a "bigger" element. That is, $x \leq y$ implies $x = y$. A maximal element should be distinguished from a *maximum* element x, namely one that that satisfies $z \leq x$ for all $z \in S$.

A maximum element is the unique maximal element of the set if it exists, but in general, there can be several maximal elements, none of which is a maximum. For example, if $S = \{a, b, c\}$ and the only related distinct pairs are $a \leq b$ and $a \leq c$, then both b and c are maximal, but neither of them is the maximum.

Any subset of a poset is also a poset with the same relation.

Definition. Given a subset A of the poset S, an element $x \in S$ is said to be an *upper bound* for A if $a \leq x$ for all $a \in A$. An upper bound that is also in A is the maximum element of the poset A.

Example A.1. Let X be any set, and let 2^X denote the power set of X, that is, the collection of all subsets of X. Then, 2^X becomes a poset by defining $C \leq D$ if

A First Course in Functional Analysis. By S. David Promislow

$C \subseteq D$. The set X is a maximum element of this poset. Suppose we consider the subset A of 2^X consisting of all *proper* subsets of X. Then, A has many maximal elements, namely the complements of the sets consisting of a single point.

The word "partial" comes from the fact that certain pairs may be incomparable.

Definition. A poset A is said to be *linearly ordered* (or *totally ordered*) if there are no incomparable pairs. That is, for all x and y in A, either $x \leq y$ or $y \leq x$. (A linearly ordered poset is also called a *chain*.)

The major result we want is as follows.

Theorem A.2 (Zorn's Lemma). *If every linearly ordered subset of a nonempty poset S has an upper bound, then S has a maximal element.*

For further discussion and a proof of Zorn's lemma, we refer the reader to Kelley (1975).

Applications of Zorn's lemma often involve a variation on Example A.1. We begin with a set X, and we consider the poset A consisting of all subsets of X that have some fixed property that we are interested in. (In Example A.1, the fixed property was that of being proper, but it could be any property.) We want to show that there is a maximal set with the given property. We then take a general linearly ordered subset C of A and need to produce an upper bound. An obvious candidate for this upper bound is the union of all the sets in C. The thrust of the proof, then, is showing that this union does have the given property, so that it is in the poset A and therefore an upper bound to C. Generally, an arbitrary union of sets in A does not have the property we want, so we must make use of the linear ordering in the proof. The following example illustrates this situation.

Example A.3. Let X be the positive integers and A the subset of 2^X consisting of all sets with the property that any two elements in the set have a common divisor (other than 1). So, for example, the set $\{2,4,6\}$ is in A, and the set $\{3,6\}$ is in A. Their union $\{2,3,4,6\}$ is not in A since 2 and 3 have no common divisor. However, suppose we take a linearly ordered subset C of A, and let W be the union of all the sets in C. Take any y, z in W. Now, y is in some set $Y \in C$ and z is in some set $Z \in C$. From the linear ordering, either $Y \subseteq Z$ or $Z \subseteq Y$. Assume the former. Then, y and z both belong to Z. Hence, they have a common divisor, from which it follows that W is an upper bound for C. Zorn's lemma tells us there is a maximal set in A. Of course, we don't need Zorn's lemma in this case because we can easily identify several maximal elements of A, such as the set of all even integers. However, in many cases there is no way that one can explicity specify maximal elements, and Zorn's lemma is needed to show the existence of such.

APPENDIX B

STONE–WEIERSTRASS THEOREM

B.1 BASIC THEOREM

One of the nicest class of functions defined on subsets of the real line is the polynomials. Values are obtained immediately. Polynomials can be differentiated or integrated as many times as you like with ease. By contrast, a general continuous function can be difficult to deal with. It need not possess a derivative at any point. More often than not, definite integrals can be obtained only through a numerical approximation. For this reason, one of the most basic and useful theorems of real analysis is due to Karl Weierstrass, who showed that any real-valued continuous function defined on a compact interval of the reals can be uniformly approximated as closely as one likes by a polynomial.

This result naturally led mathematicians to attempt to generalize the result to functions defined on arbitrary compact spaces. Of course, the first step here is to formulate the right question since a polynomial function is not defined when the domain is a general set. We can view the problem as follows. If P is a subset of the space of continuous functions on a compact Hausdorff space T, what conditions are needed on P in order that it be dense with respect to the sup norm?

The insight was provided by M. H. Stone, a leading twentieth-century figure in functional analysis and topology. There is one obvious condition. It is a well known result in topology that the set of all real-valued continuous functions on a compact Hausdorff space separates points. That is, given $s \neq t$ in T, there is a real-valued continuous function f on T with $f(s) \neq f(t)$. The same condition must hold for any dense set, for if $p(s) = p(t)$ for all p in some subset P, it follows that $f(s) = f(t)$ for all f in the closure of P. Stone noticed that, in the case of real-valued functions, the only other condition needed for density was that P be a unital algebra (as defined in Chapter 8). These conditions hold for the set of polynomial functions, which clearly is a unital algebra, and the single function $p(t) = t$, by itself, separates points. So, our basic theorem, which we state after some preliminary notation, yields the original result of Weierstrass as a corollary.

A First Course in Functional Analysis. By S. David Promislow

We are concerned here with the linear space $C^r(T)$ of all real-valued continuous functions on a compact space T equipped with the sup norm. This is a unital algebra under pointwise multiplication, with the unit e being the function that has a value of 1 on all points. Note that this is linear space over the *reals*, as indicated by the superscript of r.

Theorem B.1 (Stone–Weierstrass Theorem). *If P is a unital subalgebra of $C^r(T)$ that separates points, then $\bar{P} = C^r(T)$.*

We will break the proof up into several steps. We first recall a classical criterion for uniform convergence.

Lemma B.2 (Dini's Theorem). *Let (f_n) be a monotone (either increasing or decreasing) sequence of real-valued continuous functions converging pointwise to a function f on a compact space T. Then, (f_n) converges to f uniformly.*

Proof. By considering either $(f_n - f)$ or $(f - f_n)$, we can reduce to the case in which (f_n) decreases to 0. Let $\epsilon > 0$ be given. For each $t \in T$, choose an integer n_t such that $f_{n_t}(t) < \epsilon/2$. Then, by continuity, choose a neighborhood V_t of t such that $f_{n_t}(s) < \epsilon$ for all $s \in V_t$. By the monotonicity of (f_n), we have $f_n(s) < \epsilon$ for all $n \geq n_t$ and $s \in V_t$. By compactness, choose a finite set F of points so that the collection of neighborhoods $\{V_s : s \in F\}$ covers T. Let $N = \max\{n_s : s \in F\}$, and suppose $n \geq N$. Any $t \in T$ is in V_s for some $s \in F$, and since $n \geq N \geq n_s$, we have $f_n(t) < \epsilon$. This shows that $\| f_n \|_{\sup} < \epsilon$. □

The next step essentially proves some special cases of the original Weierstrass theorem.

Lemma B.3

(*a*) *There is a sequence of polynomials that converges uniformly to the function $f(t) = 1 - (1-t)^{1/2}$ on the interval $[0, 1]$.*

(*b*) *There is a sequence of polynomials that converges uniformly to the function $f(t) = |t|$ on the interval $[-1, 1]$.*

Proof.

(a) We define a sequence of polynomials inductively by $p_0 = 1$ and $p_n(t) = \frac{1}{2}(t + p_{n-1}^2(t))$ for $n = 1, 2, \ldots$. It is clear by induction that $0 \leq p_n \leq 1$ on $[0, 1]$ for all n. It, then, only remains to show that $p_n \geq p_{n+1}$ on $[0, 1]$ for all n, from which it follows that $p(t) = \lim p_n(t)$ exists for all $t \in [0, 1]$. After that, we take limits in the recursion formula to produce

$$p(t) = \tfrac{1}{2}(t + p^2(t)),$$

which we rearrange to obtain

$$(1 - p(t))^2 = 1 - t.$$

By using the upper bound we established, we take positive square roots of both sides to establish that $p(t) = 1 - (1-t)^{1/2}$ and, finally, apply Dini's theorem to show the uniform convergence.

To show the monotonicity of (p_n), subtract the equation $2p_{n+1}(t) = t + {p_n}^2(t)$ from the equation $2p_n(t) = t + {p_{n-1}}^2(t)$ to see that

$$2(p_n - p_{n+1}) = p_{n-1}^2 - p_n^2 = (p_n + p_{n-1})(p_{n-1} - p_n).$$

Monotonocity follows by an obvious induction using the lower bound of 0 for p_n that we established above and the fact that $p_0 \geq p_1$.

(b) Define the sequence (q_n) of functions on $[-1, 1]$ by $q_n(t) = 1 - p_n(1 - t^2)$, which from (a) decreases to $(t^2)^{1/2} = |t|$. We again invoke Dini's theorem to obtain uniform convergence of (q_n). □

We next obtain some information about certain subsets of $C^r(T)$.

Lemma B.4. *If A is a closed, unital subalgebra of $C^r(T)$, then for any f and g in A the functions $f \vee g$ and $f \wedge g$ are in A, in which*

$$(f \vee g)(t) = \max\{(f(t), g(t)\} \qquad \text{and} \qquad (f \wedge g)(t) = \min\{f(t), g(t)\}.$$

Proof. It is sufficient to show that for all $h \in A$ the function $|h| \in A$, in which $|h|(t) = |h(t)|$. This is shown by the fact that

$$f \vee g = \tfrac{1}{2}(f + g + |f - g|) \qquad \text{and} \qquad f \wedge g = \tfrac{1}{2}(f + g - |f - g|).$$

We can clearly restrict to the case for which $\| h \|_{\sup} \leq 1$.

Note first that since A is a unital algebra, a composition of the form $q \circ h \in A$ for any polynomial q. Given $\epsilon > 0$, by Lemma B.3(b), choose a polynomial q such that $|q(s) - |s|| < \epsilon$ for all $s \in [-1, 1]$. Then, for all $t \in T$, we have $|q \circ h(t) - |h|(t)| = |q(h(t)) - |h(t)|| < \epsilon$, showing that $\| q \circ h - |h| \|_{\sup} < \epsilon$. We conclude that $|h| \in \bar{A} = A$. □

Lemma B.5. *Let A be a linear subspace of $C^r(T)$ that separates points and that contains the unit e. Then, for any $s, t \in T$ with $s \neq t$ and for any $a, b \in \mathcal{R}$, there is $f \in A$ such that $f(s) = a$ and $f(t) = b$.*

Proof. Choose any $g \in A$ such that $g(s) \neq g(t)$. Then, we define $f \in A$ by

$$f = \frac{a - b}{g(s) - g(t)} g - \frac{g(t)a - g(s)b}{g(s) - g(t)} e. \qquad \square$$

We now conclude the proof of Theorem B.1 by using two strategies that are worth noting. The first is to invoke compactness in a familiar manner, by going from a local property to a global one, aided by the existence of the finite covering that compactness gives us. The second is that we break up the absolute value inequality into two statements and prove them separately, which is often a useful idea.

Conclusion of Proof of Theorem B.1. Let $f \in C^r(T)$, and let $\epsilon > 0$ be given. We will produce $p \in \bar{P}$ such that $\| f - p \|_{\sup} \leq \epsilon$. This is equivalent to showing that for all $r \in T$:

(i) $p(r) \leq f(r) + \epsilon$;
(ii) $p(r) \geq f(r) - \epsilon$.

This will show that f is in the closure of $\bar{P}$, which is of course just $\bar{P}$ itself. Since f is arbitrary, $\bar{P} = C^r(T)$, thereby completing the proof.

We can clearly assume that $\| f \|_{\sup} \leq 1$. Note that by induction we deduce from Lemma B.4 that the maximum or minimum of any finite number of functions in $\bar{P}$ is in $\bar{P}$. Let D be the set of all elements in $\bar{P}$ satisfying (i) for all $r \in T$. Suppose we show that for each $t \in T$ we can find a neighborhood V_t of t and a function $p_t \in D$ that satisfies (ii) for all $r \in V_t$. Then, we are done. Indeed, choose such a neighborhood for each point in T, and pick a finite subset F of T such that $T = \bigcup_{s \in F} V_s$. We then simply take p to be the maximum of p_s over $s \in F$. This will clearly be in D, and given any $t \in T$, we have that $t \in V_s$ for some $s \in F$, so that $p(t) \geq p_s(t) \geq f(t) - \epsilon$.

To satisfy the above condition, we need another application of compactness in order to ensure that we can pick the functions p_t to be in D. Fix t, and for all $s \in T$, choose, by Lemma B.5, $p_{t,s} \in P$ such that $p_{t,s}(s) = f(s)$ and $p_{t,s}(t) = f(t)$. Let $U_{t,s} = \{r \in T : p_{t,s}(r) \leq f(r) + \epsilon\}$. By continuity and the fact that $s \in U_{t,s}$, these sets form an open cover of T. We can find a finite subset E such that $T = \bigcup_{s \in E} U_{t,s}$. Let p_t be the minimum of $p_{t,s}$ over $s \in E$. Then, any point $r \in T$ is in $U_{t,s}$ for some $s \in E$, so that

$$p_t(r) \leq p_{t,s}(r) \leq f(r) + \epsilon.$$

This shows that $p_t \in D$.

Now, if we let $V_t = \{r \in T : p_{t,s}(r) \geq f(r) - \epsilon \text{ for all } s \in E\}$, then V_t is a neighborhood of t, and p_t satisfies (ii) on V_t. □

B.2 NONUNITAL ALGEBRAS

If A is a nonunital algebra, the conclusion to Theorem B.1 need not hold. For example, let t_0 be any point of T, and let $A = \{f \in C^r(T) : f(t_0) = 0\}$. Then A is an algebra that separates points, but $\bar{A} = A$ is certainly not equal to all of $C^r(T)$ since clearly $C^r(T)$ contains functions that do not vanish at t_0. However, other than the whole space, such an algebra is the only other possibility for the closure of a point-separating algebra. To show this, we start by deriving another version of

Lemma B.5, which in the case of an algebra involves a weaker condition than being unital.

Lemma B.6. *The conclusion of Lemma B.5 holds under the hypothesis that A is an algebra (not necessarily unital) that separates points and has the property that for any $t \in T$ there is an $f \in A$ such that $f(t) \neq 0$.*

Proof. It is enough to show that given any $s, t \in T$ we can find $f \in A$ such that $f(s) \neq 0$ and $f(t) = 0$. By taking a suitable linear combination of this function and a function that is zero on s and nonzero on t, we achieve the desired result.

We start by finding a function g such that $g(s) \neq 0$ and $g(t) \neq g(s)$. To this end, first pick any function h with $h(s) \neq h(t)$. If $h(s) \neq 0$, we set $g = h$. If $h(s) = 0$, choose another function k with $k(s) \neq 0$. We then set $g = k$ if $k(s) \neq k(t)$ or $g = h + k$ if $k(s) = k(t)$.

Having chosen g, we set $f = g$ if $g(t) = 0$; otherwise, we can assume, multiplying by a scalar if necessary, that $g(t) = 1$ and then set $f = g - g^2$. Since $g(s) \neq 0$ or 1, we must have $f(s) \neq 0$. □

We can now prove a generalization of Theorem B.1.

Theorem B.7 *If A is a subalgebra (not necessarily unital) of $C^r(T)$ that separates points, then either $\bar{A}$ is all of $C^r(T)$ or there exists some $t_0 \in T$ such that $\bar{A}$ is the subalgebra consisting of all functions in $C^r(T)$ that vanish at t_0.*

Proof. First, suppose that for every $s \in T$ there is an $f \in T$ with $f(s) \neq 0$. Then, from Lemma B.6, the conclusion of Lemma B.5 holds. Moreover, the conclusion of Lemma B.4 holds. This is true since we can show inductively that, for the polynomials defined in Lemma B.3, $p_n(1) = 1$ and so $q_n(0) = 0$. Therefore, any polynomial q involved in the proof of Lemma B.4 satisfies $q(0) = 0$. This means that q has a zero constant term; therefore, $q \circ h \in A$ for all $h \in A$. Our proof of Theorem B.1 goes through unchanged, and we conclude that $\bar{A} = C^r(T)$.

Next, suppose the hypothesis above is not true. This means that there exists $t_0 \in T$ such that $f(t_0) = 0$ for all $f \in A$. If B is the subalgebra of all functions vanishing at t_0, then it is easily seen that B is closed, so $\bar{A} \subseteq B$. For the other direction, consider the algebra D consisting of functions of the form $ce + f$, in which c is a constant and $f \in A$. Our original result, Theorem B.1, shows that D is dense $C^r(T)$. Now, take any $g \in B$, and let any $\epsilon > 0$ be given. We can find some constant c and some f in A such that $\| g - f - ce \|_{\sup} \leq \epsilon/2$. In particular,

$$|c| = |(g - f - ce)(t_0)| \leq \| g - f - ce \|_{\sup} \leq \tfrac{1}{2}\epsilon,$$

implying that

$$\| g - f \|_{\sup} \leq \| g - f - ce \|_{\sup} + |c| \leq \epsilon,$$

thereby showing that $B \subseteq \bar{A}$. □

B.3 COMPLEX ALGEBRAS

Our arguments above involve the order properties of the real numbers and do not apply to the algebra of continuous *complex* functions. Indeed, the corresponding result is false in that case. As an example for those familiar with some complex analysis, let A be the algebra of all complex-valued continuous funtions defined on the closed unit disc in the complex plane that are analytic in the interior. Then, A is a closed, unital algebra that separates points but does not include the function $f(z) = \bar{z}$. This example suggests a natural additional hypothesis, which is sufficient to recover the theorem. We let $C^c(T)$ denote the algebra of complex-valued continuous functions on T, and we consider this a complex algebra. We can then consider $C^r(T)$ as a subset of $C^c(T)$.

Theorem B.8 (Complex form of Stone–Weierstrass Theorem). *Let A be a unital subalgebra of $C^c(T)$ that separates points and is closed under conjugation. That is, $f \in A$ implies that $\bar{f} \in A$. Then, $\bar{A} = C^c(T)$.*

Proof. Let A^r consist of all functions in A that are real valued. It is clear that A^r is a real unital algebra. Given any $x \neq y$, we can find $f \in A$ such that $f(x) \neq f(y)$. Therefore, we must have that either $\operatorname{Re} f(x) \neq \operatorname{Re} f(y)$ or $\operatorname{Im} f(x) \neq \operatorname{Im} f(y)$. But, closure under conjugation means that $\operatorname{Re} f = \frac{1}{2}(f + \bar{f})$ and $\operatorname{Im} f = (1/(2i))(f - \bar{f})$ are in A and, therefore, in A^r. This establishes that A^r separates points, so by our original real-valued version of the theorem, we know that any function in $C^r(T)$ can be approximated as closely as we wish by a function in A^r. So, take any $f \in C^c(T)$ and any $\epsilon > 0$. Choose $g, h \in A^r$ such that $\| \operatorname{Re} f - g \|_{\sup} \leq \epsilon/2$ and $\| \operatorname{Im} f - h \|_{\sup} \leq \epsilon/2$. Then,

$$\| f - (g + ih) \|_{\sup} \leq \| \operatorname{Re} f - g \|_{\sup} + \| \operatorname{Im} f - h \|_{\sup} \leq \epsilon. \qquad \square$$

APPENDIX C

EXTENDED REAL NUMBERS AND LIMIT POINTS OF SEQUENCES

C.1 EXTENDED REALS

The extended real number system, denoted by $\hat{\mathcal{R}}$, consists of the real numbers $\mathcal{R}$ together with two additional points, ∞ and $-\infty$. Normal operations of arithmetic are extended by defining:

$$
\begin{aligned}
&x + \infty = \infty, \qquad x + -\infty = -\infty \qquad \text{for all } x \in \mathcal{R};\\
&x \cdot \infty = \infty; \qquad x \cdot -\infty = -\infty \qquad \text{for all } x > 0;\\
&x \cdot \infty = -\infty; \qquad x \cdot -\infty = \infty \qquad \text{for all } x < 0;\\
&0 \cdot \infty = 0 \cdot -\infty = 0;\\
&\infty + -\infty \text{ is undefined.}
\end{aligned}
$$

(The justification for the second to the last statment is simply that it works, mainly in measure and integration theory. Theorems involving multiplication of numbers turn out to be true with this interpretation.)

The ordering on the reals is extended in the natural way to

$$-\infty < r < \infty$$

for all $r \in \mathcal{R}$.

With this ordering $\hat{\mathcal{R}}$ becomes a linearly ordered space; see Appendix A. Moreover, any subset $A \subseteq \hat{\mathcal{R}}$ will then have a *least upper bound*, also called a *supremum*, denoted by $\sup(A)$. The notation $x = \sup(A)$ means that x is an upper bound for A, and if y is another upper bound for A, then $x \leq y$, Similarly, any subset $A \subseteq \hat{\mathcal{R}}$ will then have a

A First Course in Functional Analysis. By S. David Promislow

greatest lower bound, also called a *infinum*, denoted by $\inf(A)$. Here, $x = \inf(A)$ means that x is an lower bound for A, and if y is another lower bound for A, then $x \geq y$,

Note that $\sup(A) = \infty$ if and only if A is unbounded above, and $\inf(A) = -\infty$ if and only if A is unbounded below. If $A = \emptyset$, then it follows logically from the defintion that $\sup(A) = -\infty$ and $\inf(A) = \infty$.

C.2 LIMIT POINTS OF SEQUENCES

A *basic neighborhood* of a point $a \in \hat{\mathcal{R}}$ is an open interval containing a if $a \in \mathcal{R}$, an interval of the form $(b, \infty]$ if $a = \infty$, or an interval of the form $[-\infty, b)$ if $a = -\infty$.

Definition. Let (a_n) be a sequence in $\hat{\mathcal{R}}$. An element $a \in \hat{\mathcal{R}}$ is said to be a *limit point* of this sequence if given any basic neighborhood I of a and any positive integer N there exists $n \geq N$ such that $a_n \in I$. As an equivalent formulation, a is a limit point of (a_n) if and only if there is some subsequence of (a_n) that converges to a.

Definitions. Suppose E is the set of all limit points of a given sequence (a_n) in $\hat{\mathcal{R}}$. Then, $\sup(E)$ is known as the *upper limit* point of the sequence and is denoted by $\limsup a_n$, and $\inf(E)$ is known as the *lower limit* point of the sequence and denoted by $\liminf a_n$.

We know that E is nonempty. Indeed, if E is bounded, this follows from the fact that E is relatively compact. If E is unbounded, then either ∞ or $-\infty$ or both are limit points. We can therefore say that

$$\liminf a_n \leq \limsup a_n.$$

A sequence (a_n) converges to a if and only if $a = \liminf a_n = \limsup a_n$ is the unique limit point of the sequence.

In proving facts about lim sup or lim inf, we need confine ourselves to only one of the two quantities, since it is clear that

$$\limsup a_n = -\liminf(-a_n).$$

Example C.1. Consider the sequence

$$\left(100, -100,\ 1,\ \frac{1}{2},\ 1,\ \frac{1}{3},\ 1, \frac{1}{4}, \ldots\right).$$

The lim sup of the sequence is 1, and the lim inf is 0. There are no other limit points. Note that the upper and lower limits should not be confused with the supremum and infimum of the entries in the sequence, which in this example are 100 and -100, respectively.

When the set E is infinite, it is often important to note the following.

Proposition C.2. $\limsup a_n$ *and* $\liminf a_n$ *are limit points of* (a_n).

Proof. We will show this for the lim sup, which we will denote by a. The statement is clear if $a = \infty$, which occurs only if the sequence is unbounded, and is clear if $a = -\infty$, which occurs only if (a_n) converges to $-\infty$. Assume, then, that $a \in \mathcal{R}$, and let an open interval I containing a and a positive integer N be given. Then, I must contain a point $b \in E$. But, then, since I is also a basic neighborhood of the limit point b, I must contain a_n for some $n \geq N$. Therefore $a \in E$. □

The following proposition is sometimes given as the definition of the upper and lower limits. We prefer to present the definition as above and view the proposition as a recipe for obtaining these points.

Proposition C.3. *If* (a_n) *is a sequence in* $\hat{\mathcal{R}}$, *then*

$$\limsup\ a_n = \inf_n \sup_{k \geq n} a_k \qquad \textit{and} \qquad \liminf\ a_n = \sup_n \inf_{k \geq n}\ a_k.$$

Proof. We will show the first statement. Let $b_n = \sup_{k>n} a_k$, so that $b_n \downarrow \inf_n \sup_{k>n} a_k$; denote this limit by a. If $a = \infty$, it is clear that the sequence is unbounded above, so $\limsup(a_n) = \infty$.

Suppose, then, that $a \neq \infty$. Let I be a basic neighborhood of a and N a positive integer. Then, for some $n > N$, we must have $b_n \in I$. But, then, for some $k \geq n > N$, we must have $a_k \in I$. This shows that a is a limit point of (a_n).

Now take $c \in \mathcal{R}$ with $c > a$. Then, for some N, we have $b_N < c$. We can therefore find an open interval I containing c such that b_N lies to the left of I. Given any $n \geq N$, $a_n \leq b_N$, so that $a_n \notin I$. This shows that c is not a limit point of (a_n), thereby verifying that a is the largest limit point. □

To see how this works in a concrete case, look again at Example C.1. We have $\inf_{k \geq n} a_k = -100$ for $n = 1$, 2 and 0 for $n \geq 3$ taking the supremum given the lim inf as 0.

Another equivalent way of formulating a limit point a of a sequence (a_n) is to require that every basic neighborhood of a contains points a_n for infinitely many values of n. This suggests another characterization for the upper and lower limits, which ties in well with the usual definition of convergence. We will leave the proof to the reader. This is stated for the lim sup, but there is an obvious analogous formulation for the lim inf.

Proposition C.4. $a = \limsup a_n$ *is the unique element of* $\hat{\mathcal{R}}$ *with the property that, given any* $b < a$, *we have* $a_n > b$ *for infinitely many values of* n *and, given any* $c > a$, *we have* $a_n < c$ *for all but a finite number of values of* n. □

One must be careful when dealing with upper and lower limit points of a sum of sequences, as illustrated by the following example.

Example C.5. Let $a_n = (-1)^n$ and $b_n = (-1)^{n+1}$. Then,

$$\liminf a_n = \liminf b_n = -1, \qquad \liminf(a_n + b_n) = 0.$$

Thus, we see that the lower limit of the sum is not the sum of the lower limits.

The inequality always goes the way of the above example, however. The following is the most general statment we can make concerning sums. We leave the proof to the reader.

Proposition C.6

$$\liminf a_n + \liminf b_n \leq \liminf(a_n + b_n) \leq \limsup a_n + \liminf b_n$$

provided the right and left sides are defined. □

The parallel statement can be formulated for the lim sup of a sum. There are analogous inequalities for a product of two sequences, which we do not state here.

APPENDIX D

MEASURE AND INTEGRATION

D.1 INTRODUCTION AND NOTATION

The purpose of this appendix is to provide detailed proofs for most of the assertions that we summarized in Sections 2.9, 6.1, and 10.1, dealing with measure and integration. We consider throughout a measure space $(\Omega, \mathcal{S}, \mu)$, in which $\mathcal{S}$ is a σ-algebra of subsets of Ω and μ is a measure on $\mathcal{S}$ as defined in Section 2.9.2. We assume at the beginning that μ is a positive measure (Later, we consider real-valued and complex-valued measures.)

We wish to consider extended real-valued functions $f : \Omega \to \hat{\mathcal{R}}$, in which $\hat{\mathcal{R}}$ denotes the extended real numbers, as outlined in Section C.1. We restate the definition of a *measurable function* for this setting

Definition. A function $f : \Omega \to \hat{\mathcal{R}}$ is *measurable* if $f^{-1}[-\infty, r] \in \mathcal{S}$ for all $r \in \mathcal{R}$.

The term nonnegative will indicate the interval $[0, \infty]$. Let L denote the set of all extended real-valued measurable functions, let L^+ denote the set of nonnegative functions in L, and let L_0^+ denote the set of simple functions in L^+ as defined in Chapter 2. As in that chapter, the values of simple functions are restricted to $\mathcal{R}$.

For a sequence of functions (f_n), we say that $f_n \uparrow f$ means that, for all $n, f_n \leq f_{n+1}$ and that $f = \sup f_n$. A similar notion is used for a sequence of numbers (a_n). For a sequence of sets (A_n), we say that $A_n \uparrow A$ means that, for all n, $A_n \subseteq A_{n+1}$ and $A = \bigcup_n A_n$. We similarly use $\downarrow$ for decreasing sequences of functions, numbers, or sets. For sets A and B, we write the complement of A as A^c and write $A - B$ to denote $A \cap B^c$. (The context will allow the reader to avoid confusing the symbol "$-$" with subtraction in linear spaces.)

For any $A \in \mathcal{S}$, a *partition of* A means a collection of pairwise disjoint sets in $\mathcal{S}$ whose union is A. (Note that while this term is sometimes used more generally, we will always mean that all sets in the partition are measurable.)

For any $A \in \mathcal{S}$, we let $\int_A f$ denote $\int f 1_A$.

A First Course in Functional Analysis. By S. David Promislow

D.2 BASIC PROPERTIES OF MEASURES

In this section, we develop some simple properties of a measure that follow from the definition.

Proposition D.1. *Any measure μ satisfies the following*:

(a) (Monotonicity) *For sets $A, B \in \mathcal{S}$ with $B \subseteq A$, we have $\mu(B) \leq \mu(A)$. Additionally, if $\mu(B) < \infty$, then $\mu(A - B) = \mu(A) - \mu(B)$.*

(b) (Countable subadditivity) *For any finite or countably infinite sequence of sets (A_n) in $\mathcal{S}$ with union A, we have*

$$\mu(A) \leq \sum_n \mu(A_n).$$

(c) (Continuity) *For a sequence of sets in (b), $A_n \uparrow A$ implies that $\mu(A_n) \uparrow \mu(A)$, and if $\mu(A_1) < \infty$, then $A_n \downarrow A$ implies that $\mu(A_n) \downarrow \mu(A)$.*

Proof

(a) $\mu(A) = \mu(B) + \mu(A - B) \geq \mu(B)$. For the second statement, subtract $\mu(B)$ from each of the first two terms.

(b) Let $B_1 = A_1$, and for $n > 1$, define B_n inductively by $B_n = A_n - \bigcup_{k=1}^{n-1} B_k$. Then, (B_n) is a countable, pairwise disjoint sequence of sets with union A, and each $B_n \subseteq A_n$. By countable additivity and monotonicity,

$$\mu(A) = \sum_n \mu(B_n) \leq \sum_n \mu(A_n).$$

(c) For the increasing sequence, we assume that all sets have finite measure, as the result is immediate if not. We define sets B_n as in (b) and note that, for $n > 1$, B_n is just $A_n - A_{n-1}$, which, by the proof of (a) and the finiteness assumption, has measure equal to $\mu(A_n) - \mu(A_{n-1})$. Let $A_0 = \emptyset$. Then, by countable additivity,

$$\mu(A) = \sum_{k=1}^{\infty} \mu(B_k) = \sum_{k=1}^{\infty} [\mu(A_k) - \mu(A_{k-1})] = \sup_n \sum_{k=1}^{n} [\mu(A_k) - \mu(A_{k-1})] = \sup_n A_n.$$

In the case of a decreasing sequence, we apply the increasing case to the sequence $(A_1 - A_n)$, which increases to $A_1 - A$. We then use the finitenesss assumption to subtract $\mu(A_1)$ from the resulting equality. □

Note that the decreasing result in (c) fails without the finiteness hypothesis. For example, in a Lebesgue measure on the real line, let $A_n = [n, \infty)$, which has measure equal to ∞. (See Section D.7.) However, $A_n \downarrow \emptyset$, which has measure 0.

We conclude this section with a result that is often useful in verifying that a given set function is a measure.

Proposition D.2. *A nonnegative function μ defined on $\mathcal{S}$ is countably additive provided that both of the following assertions hold:*

(*a*) *For any pair of disjoint sets $A, B \in \mathcal{S}$, $\mu(A \cup B) \geq \mu(A) + \mu(B)$.*
(*b*) *μ is countably subadditive.*

Proof. From (b), equality holds in (a), and by induction, we have additivity for any finite number of summands. Then, given (A_n), a countable partition of a set $A \in \mathcal{S}$, it follows that

$$\mu(A) \geq \mu\left(\bigcup_{n=1}^{N} A_n\right) = \sum_{n=1}^{N} \mu(A_n)$$

for all N. By taking a limit as $N \to \infty$, we conclude $\mu(A) \geq \sum_{n=1}^{\infty} \mu(A_n)$, and countable subadditivity gives the other direction. □

D.3 PROPERTIES OF MEASURABLE FUNCTIONS

In this section we show how measurable functions can be combined to produce other measurable functions. In proving that functions are measurable, we repeatedly make use of the fact that $\mathcal{S}$ is a σ-algebra and therefore closed under countable unions and intersections as well as complements. As an example, we can get an equivalent definition of measurability by replacing the interval $[-\infty, r]$ by $[-\infty, r)$ since

$$f^{-1}[-\infty, r) = \bigcup_{n=1}^{\infty} f^{-1}\left[-\infty, r - \frac{1}{n}\right] \quad \text{and} \quad f^{-1}[-\infty, r] = \bigcap_{n=1}^{\infty} f^{-1}\left[-\infty, r + \frac{1}{n}\right).$$

By using complements, we can also equivalently use the interval $[r, \infty]$ or $(r, \infty]$. Similar arguments show that if f is measurable, then

$$f^{-1}(I) \in \mathcal{S} \text{ for any interval } I \subseteq \hat{\mathcal{R}}.$$

Proposition D.3 (Algebraic Properties). *Suppose that f and g are (not-extended) real-valued measurable functions and $\alpha \in \mathcal{R}$. Then, each of the following is measurable:*

(*a*) αf; (*b*) $f + g$; *and* (*c*)fg.

Proof. We leave part (a) as an exercise.

(b) Observe that

$$(f+g)^{-1}(-\infty,r)=\bigcup_q\left\{f^{-1}(-\infty,q)\bigcap g^{-1}(-\infty,r-q)\right\}$$

as q runs over the countable set of rational numbers. Indeed, the inclusion of the left-hand side in the right follows since if $f(x)+g(x)<r$, there is some rational number in the interval $(f(x),r-g(x))$. The other inclusion is clear.

(c) For all $r>0$,

$$(f^2)^{-1}(-\infty,r]=f^{-1}[-\sqrt{r},\sqrt{r}].$$

This shows that f^2 is measurable. Now, from parts (a) and (b)

$$fg=\frac{1}{4}[(f+g)^2-(f-g)^2]$$

is measurable. □

Remark. Analogous results hold for the more general case of extended real-valued functions. Some modification is necessary in the statement and proof of (b) since $f+g$ might not be defined at certain points.

Proposition D.4. *Given a countable collection* $\{f_n : n=1,2,\ldots\}$ *of functions in* L, $\sup\, f_n$ *and* $\inf\, f_n \in L$.

Proof. Since $\inf f_n=-\sup(-f_n)$, we need only prove the statement concerning the supremum. If $f=\sup f_n$, then for any $r\in\mathcal{R}$

$$f^{-1}(-\infty,r]=\bigcap_{n=1}^{\infty}{f_n}^{-1}(-\infty,r],$$

so that f is measurable. □

From Propositions D.4 and C.3, we immediately obtain that $\liminf\, f_n=\sup_n \inf_{k\geq n} f_k$ is measurable. Similarly, $\limsup\, f_n$ is measurable, so $\lim\, f_n$, if it exists, is measurable.

Applying Proposition D.4 to a finite union gives us the following.

Proposition D.5. *If* $f\in L$ *so are* f^+, f^-, *and* $|f|$.

Proof

$$f^+=\sup\{f,0\},\qquad f^-=-\inf\{f,0\},\qquad |f|=f^++f^-.$$ □

Remark. Note that if both f^+ and f^- are measurable, so is $f=f^+-f^-$. However, the measurability of $|f|$ clearly does not imply that of f. For example, take a set A that is not in $\mathcal{S}$, and let f take the value of 1 on A and -1 on A^c.

D.4 INTEGRAL OF A NONNEGATIVE FUNCTION

We will now state the precise definition of the integral of a nonnegative function and prove the six properties given in Section 2.9.2 (although in a different order). We define, for any $f \in L^+$,

$$\int f = \sup\left\{\int s : s \in L_0^+ \text{and } s \leq f\right\},$$

in which the integral of s is as defined in Section 2.9.2.

Proposition D.6 (Property 3). *For $f, g \in L^+$, $f \leq g$ implies $\int f \leq \int g$.*

Proof. We first prove this monotonicity of the integral if f and g are simple. In fact, it is essential to do this in order to show that the above definition of the integral for a general measurable function is consistent with the one given for simple functions. Let $c_1, c_2, \ldots, c_n$ be the distinct values of f and let $E_i = f^{-1}(c_i)$. Let $d_1, d_2, \ldots, d_m$ be the distinct values of g, and let $F_j = g^{-1}(d_j)$. Then, $\bigcup_i E_i = \bigcup_j F_j = \Omega$. It follows that

$$\mu(E_i) = \sum_{j=1}^{m} \mu(E_i \cap F_j) \qquad \text{and} \qquad \mu(F_j) = \sum_{i=1}^{n} \mu(E_i \cap F_j).$$

For $x \in E_i \cap F_j$, we have $c_i = f(x) \leq g(x) = d_j$, thereby implying that for all i, j

$$c_i\mu(E_i \cap F_j) \leq d_j\mu(E_i \cap F_j),$$

since both sides equal zero when $E_i \cap F_j = \emptyset$. This observation leads to

$$\int f = \sum_{i=1}^{n} c_i\mu(E_i) = \sum_{i,j} c_i\mu(E_i \cap F_j) \leq \sum_{i,j} d_j\mu(E_i \cap F_j) = \sum_{j=1}^{m} d_j\mu(F_j) = \int g.$$

The general result follows immediately since for any simple function $s \leq f$ we also have $s \leq g$, so $\int f$ will be the supremum over a smaller set than the one used to define $\int g$. □

This monotonicity property enables a simple demonstration of three more of our six stated properties.

Proposition D.7 (Property 2). *For any $\alpha \geq 0$ and $f \in L^+$, we have $\int \alpha f = \alpha \int f$.*

Proof. The statement follows immediately from the fact that it holds for simple functions, and for $\alpha > 0$ and $s \in L_0^+$, we have have $s \leq \alpha f$ if and only if $s/\alpha \leq f$. □

Proposition D.8 (Property 4). *$\int f = 0$ if and only if $f = 0$ a.e.*

Proof. Suppose there is a set A of measure 0 such that for all $x \notin A$ we have $f(x) = 0$. We will show that $\int f = 0$ by showing that $\int s = 0$ for any nonnegative simple function $s \le f$. Take any such $s = \sum_i c_i 1_{A_i}$ in which each $c_i > 0$ and $A_i \in \mathcal{S}$. For any i and $x \in A_i$, we have $0 < c_i = s(x) \le f(x)$, which implies that $x \in A$. This shows that $A_i \subseteq A$, so that $\mu(A_i) = 0$ for all i. Therefore, $\int s = \sum_i c_i \mu(A_i) = 0$.

Conversely, suppose that f is not equal to zero a.e. Let $B_n = \{x : f(x) > 1/n\}$ for $n = 1, 2, \ldots$. Then, $\bigcup_{n=1}^{\infty} B_n = \{x : f(x) > 0\}$, a set whose measure is not zero. By countable subadditivity, there must be some k for which $\mu(B_k) > 0$. Now, $f \ge (1/k) 1_{B_k}$, so by Propositions D.6 and D.7, we have

$$\int f \ge \frac{1}{k} \int 1_{B_k} = \frac{1}{k}\, \mu(B_k) > 0. \qquad \square$$

Proposition D.9 (Property 5). *$\int f < \infty$ implies $f < \infty$ a.e.*

Proof. Let $A = \{x : f(x) = \infty\}$. Then, for any $\alpha \in [0, \infty)$, we have $\alpha 1_A < f$, so from Propositions D.6 and D.7, it follows that $\int f \ge \alpha \mu(A)$. If $\mu(A) > 0$, we must then have $\int f = \infty$, since we can take α arbitrarily large. $\square$

We need to work more to prove the other two properties. As an introduction, it is worthwhile to note that proofs about nonnegative measurable functions often proceed in three stages. We first prove a result about measures, then use this to prove a result about simple functions, and then in stage 3 move from the simple function result to the general case. We have already covered the stage 1 arguments that we need in Proposition D.1.

We next prove one of the major results of integration theory.

Theorem D.10 (Monotone Convergence Theorem). *If (f_n) is a sequence of nonnegative measurable functions such that $f_n \uparrow f$, then $\int f_n \uparrow \int f$.*

Proof. We can assume that $\int f_n < \infty$ for all n because, if not, the result is clear from the monotonicity of the integral.

The stage 1 version of this theorem is Proposition D.1(c). The stage 2 result is as follows. Suppose we are given a sequence of measurable sets (A_n) such that $A_n \uparrow A$. For a function s in L_0^+ with distinct values $c_1, c_2, \ldots, c_m$ and $F_i = s^{-1}(c_i)$, we have

$$\int s 1_{A_n} = \sum_{i=1}^{m} c_i\, \mu(F_i \cap A_n),$$

and since $F_i \cap A_n \uparrow F_i \cap A$,

$$\int_{A_n} s \uparrow \int_A s$$

by invoking the result for sets, Proposition D.1(c).

For the final stage, let $M = \sup \int f_n$, which is less than or equal to $\int f$, by Proposition D.6. To show the other direction, we use a *multiplicative* relaxation argument. That is, we relax the inequality by showing that $\int f \leq cM$ for all $c > 1$. Since c is arbitrary, we will be done. We have to show, then, that for any nonnegative simple function $s \leq f$ we have that

$$\int s \leq cM. \tag{$*$}$$

For any such s, let $A_n = \{x : s(x) \leq cf_n(x)\}$. Then,

$$\int_{A_n} s \leq \int cf_n = c \int f_n. \tag{$**$}$$

The sets A_n necessarily increase along with the functions f_n. Also, for any x, we know that $s(x)/c < f(x)$, so for n sufficiently large, we must have $s(x)/c < f_n(x)$, showing that $x \in A_n$. It follows that $A_n \uparrow \Omega$, so that by the above stage 2 result we have $\int_{A_n} s \uparrow \int s$. By using this result and taking a supremum of both sides of $(**)$, we obtain $(*)$. □

Fatou's lemma is now an immediate consequence of the monotone convergence theorem. We provide a somewhat more general statement, which does not require the sequence (f_n) to converge.

Proposition D.11 (Property 6, Fatou's Lemma). *Given a sequence (f_n) in L^+,*

$$\int \liminf f_n \leq \liminf \int f_n.$$

Proof. We use the characterization of lim inf given in Proposition C.3. Let $g_k = \inf_{n \geq k} f_n$. By Proposition D.6,

$$\int g_k \leq \inf_{n \geq k} \int f_n.$$

If $f = \liminf f_n$, then $f = \sup_k g_k$. It follows from the monotone convergence theorem that

$$\int f = \sup_k \int g_k \leq \sup_k \inf_{n \geq k} \int f_n = \liminf \int f_n.$$ □

The monotone convergence theorem, or some variant of it, is often the key to the stage 3 proofs, as we will demonstrate. In order to carry out the argument, it is essential to know that there is a sufficient supply of simple functions that are less than a given nonnegative measurable function. The precise statement is as follows.

Proposition D.12. *Given any $f \in L^+$, there is a sequence (s_n) in L_0^+ such that $s_n \uparrow f$. Moreover, if f is bounded, then (s_n) converges uniformly to f.*

Proof. For $n = 1, 2, \ldots$, define a simple function $s_n \in L_0^+$ by

$$s_n(x) = \begin{cases} n, & f(x) > n \\ k/2^{n+1}, & f(x) \leq n, \end{cases}$$

where in the latter case k is the integer such that $k/2^{n+1} \leq f(x) < k/2^n$. Since f is measurable, it is clear that each s_n is as well. We leave it to the reader to verify that $s_n \uparrow f$, and the uniform convergence when f is bounded. □

Proposition D.13 (Property 1). *For $f, g \in L^+$, we have $\int (f + g) = \int f + \int g$.*

Proof. Stage 1 is simply the additivity axiom for measures. We next carry out stage 2 by proving additivity for simple functions. By using the notation in the proof of Proposition D.6, it follows that $f + g$ assumes the value of $c_i + d_j$ on the set $E_i \cap F_j$ (which could be empty); therefore,

$$\int (f + g) = \sum_{i,j} (c_i + d_j)\mu(E_i \cap F_j) = \sum_i c_i\mu(E_i) + \sum_j d_j\mu(F_j) = \int f + \int g.$$

For general nonnegative measurable functions, we choose sequences (s_n) and (t_n) of simple functions increasing to f and g, respectively. Then, $(s_n + t_n) \uparrow (f + g)$, and the conclusion follows by applying the monotone convergence theorem to all three sequences. □

Having established additivity, we can derive a series version of Theorem D.10.

Theorem D.14 (Series Version of Monotone Convergence Theorem). *Given a sequence $(f_n) \in L^+$, let h be the pointwise sum of all the sequence entries. That is,*

$$h(x) = \sum_{n=1}^{\infty} f_n(x), \qquad x \in \Omega.$$

Then,

$$\int h = \sum_{n=1}^{\infty} \int f_n.$$

Proof. From additivity of the integral, induction shows that

$$\int \sum_{k=1}^{n} f_k = \sum_{k=1}^{n} \int f_k.$$

Apply the monotone convergence theorem to complete the proof. □

Remark. An important application of the last theorem is concerned with the set function

$$\mu_f(A) = \int 1_A f \, d\mu,$$

where $f \in L^+$. The fact that this is a measure follows immediately from Theorem D.14 by noting that given a countable partition $\{A_n\}$ of a set $A \in \mathcal{S}$ we have that

$$1_A f = \sum_n 1_{A_n} f$$

D.5 INTEGRAL OF AN EXTENDED REAL-VALUED FUNCTION

For an arbitrary $f \in L$ with values in $\hat{\mathcal{R}}$ we define

$$\int f = \int f^+ - \int f^-,$$

provided that at least one of these integrals is finite. If both $\int f^+$ and $\int f^-$ are equal to ∞, then $\int f$ is undefined.

We will first extend the order and linearity properties to this general case. We assume in the following that all functions are measurable. For simplicity, we also assume that all integrals are finite. In view of Propositions D.8 and D.9 applied to both positive and negative parts, we can then assume that all functions are real valued.

Proposition D.15 (Integral Triangle Inequality)

$$\left| \int f \right| \leq \int |f|.$$

Proof. If $\int f^+ = a$ and $\int f^- = b$, then by using the additivity that we established for nonnegative functions we obtain

$$\left| \int f \right| = |a - b| \leq a + b = \int (f^+ + f^-) = \int |f|.$$

□

We leave the proof of the next proposition as an exercise for the reader.

Proposition D.16. *For any* $c \in \mathcal{R}$, $\int cf = c \int f$.

Proposition D.17. $\int (f + g) = \int f + \int g$.

Proof. We can appeal to the previous nonnegative result if both f and g are nonnegative or both are nonpositive. In the case for which $f \geq 0$ and $g \leq 0$, let

$h = f + g$. Then, $f + g = h^+ - h^-$, so that $f + h^- = h^+ - g$. From this expression, we obtain

$$\int f + \int h^- = \int h^+ + \int (-g) = \int h^+ - \int g,$$

and the conclusion follows by rearranging. Finally, for general functions f and g, we apply the above case to the two functions $f^+ + g^+ \geq 0$ and $-(f^- + g^-) \leq 0$ and rearrange as necessary. □

Remark. We will not prove it here, but the additivity remains true for general extended real-valued functions with possibly infinite integrals, provided that the right-hand side is defined, that is, not of the form $\infty + -\infty$.

Proposition D.18. *If $f \leq g$, then $\int f \leq \int g$.*

Proof. By using the corresponding property for nonnegative functions and the linearity of the integral, we obtain

$$0 \leq \int (g + (-f)) = \int g + \int (-f) = \int g - \int f. \qquad \square$$

It is possible to prove versions of Fatou's lemma and the monotone convergence theorem for general real-valued functions, but the most useful form of such a convergence theorem is the following.

Theorem D.19 (Dominated Convergence Theorem). *Suppose (f_n) is a sequence of measurable functions converging a.e. to a function f. Suppose there exists a measurable function $g \geq 0$ with $\int g < \infty$ such that $|f_n| \leq g$ for all n. Then,*

$$\int f = \lim_n \int f_n.$$

Proof. By hypothesis, $(g + f_n)$ and $(g - f_n)$ are sequences of nonnegative functions that converge a.e. to $g + f$ and $g - f$, respectively. We apply Proposition D.11(Fatou's lemma), linearity, and standard facts involving upper and lower limits (see Appendix C) to get

$$\int g + \int f = \int (g + f) \leq \liminf \int (g + f_n) = \int g + \liminf \int f_n$$

and

$$\int g - \int f = \int (g - f) \leq \liminf \int (g - f_n) = \int g - \limsup \int f_n.$$

By subtracting $\int g$ from both sides in both equations and multiplying the second equation by -1, we get

$$\int f \leq \liminf \int f_n \leq \limsup \int f_n \leq \int f,$$

and the conclusion follows. □

It is instructive to refer back to the "tall skinny" and "short fat" examples given in discussing Property 6 of the integral in Section 2.9, for which the conclusion of Theorem D.19 fails. In both cases, we see that any function g that satisfies the hypothesis of dominating all f_n in absolute value necessarily has an infinite integral.

There is a commonly used corollary of the dominated convergence theorem.

Corollary D.20 (Bounded Convergence Theorem). *If (f_n) converges to f a.e. on a finite measure space, and if there is a constant $M > 0$ such that $|f_n| \leq M$ for all n, then the conclusion of Theorem D.19 holds.*

Proof. Take g to be the function that has a constant value of M. □

D.6 INTEGRAL OF A COMPLEX-VALUED FUNCTION

We complete the definition of the integral by defining it for complex-valued functions. For such a function, we define

$$\int f = \int \operatorname{Re} f + i \int \operatorname{Im} f,$$

provided that both integrals are finite.

Since $|\operatorname{Re} f|$ and $|\operatorname{Im} f|$ are both less that $|f|$, both integrals are finite if and only if $\int |f| < \infty$, and in such case the conclusion of Proposition D.15 holds by an argument similar to the one used in the proof of Theorem 5.3. Indeed, if $\int f = re^{i\Theta}$ in polar form, then $r = \int (e^{-i\Theta} f)$ is real, so that

$$r = \operatorname{Re} \int (e^{-i\Theta} f) = \int \operatorname{Re}(e^{-i\Theta} f) \leq \int |e^{-i\Theta} f| = \int |f|.$$

Moreover, Proposition D.16 (for all complex scalars) and Proposition D.17 hold, as is easily checked from the definitions. The dominated and bounded convergence theorems hold and can be proved by applying the real-valued versions to the real and imaginary parts of the functions.

D.7 CONSTRUCTION OF LEBESGUE MEASURE ON $\mathcal{R}$

Up to this point in this appendix, we have not given any example of a measure. In fact, it is not easy to do so, except for what might be called trivial cases, obtained as follows. Take an arbitrary set Ω and any σ-algebra $\mathcal{S}$ of subsets. For any nonnegative function f on Ω we can define a measure

$$\mu(A) = \sum_{x \in A} f(x)$$

for all $A \in \mathcal{S}$. (See the end of Section 4.4.3, where we discuss sums over possibly uncountable index sets.) We can in fact extend such a measure to the σ-algebra consisting of all subsets of Ω. If Ω is countable, then all measures can be obtained in this way.

One special case of this is given by $f(x) = 1$ for all x, which leads to the so-called *counting measure*. Another is the one for which f takes a value of 0 everywhere, except at a single point x_0 where it takes the value 1. This is known as the *unit mass* at x_0, and the measure of a set is either 1 or 0, according to whether it does or does not contain x_0.

However, the whole apparatus of measure theory is not needed to deal with this type of measure, which we may term as trivial. Any example of a nontrivial measure necessarily involves a reasonably complicated construction. In this section, we present the original motivating example, that of the Lebesgue measure on the real line.

The underlying technique goes back to ancient Greek mathematics and is more easily visualized in the plane. Suppose we wish to calculate the area of an arbitrary region. We could cover it by small rectangles (not necessarily disjoint) and, as an approximation, add up the areas of all the rectangles. We would, of course, get an answer that is too big, due to the portions of the rectangles that lay outside the region as well as the overlapping. But, the smallest number we could get in this way should give a good approximation to what we want. This suggests the procedure for the real line, which we now state precisely.

Given any $A \subseteq \mathcal{R}$, consider a covering of A by a countable collection of bounded open intervals. This is a collection $\{I_n = (a_n, b_n)\}$ such that $A \subseteq \bigcup_n I_n$. An initial attempt to obtain a measure is to define

$$\mu^*(A) = \inf\left\{\sum_n (b_n - a_n) \text{ over all possible coverings}\right\}.$$

It to common to refer to μ^* as an *outer measure*.

It is clear that $\mu^*(x) = 0$ for a singleton $\{x\}$ and that $\mu^*(\emptyset) = 0$. Moreover, $A \subseteq B$ implies that $\mu^*(A) \leq \mu^*(B)$. The next fact seems intuitively clear, but the proof is more involved than what one might expect.

Proposition D.21. *For any interval I with endpoints $a \leq b$, we have $\mu^*(I) = b - a$. (This will be ∞ if $b = \infty$ or $a = -\infty$.)*

Proof. We will first prove that $\mu^*(I) = b - a$ for a closed and bounded interval $[a, b]$. For any $\epsilon > 0$, the single set $(a - \epsilon, b + \epsilon)$ is a cover of $[a, b]$, so

$\mu^*([a,b]) \leq b - a - \epsilon$. It follows that $\mu^*([a,b]) \leq b - a$. We must now rule out the fact that some strange covering could actually yield a number strictly less than $b - a$. This is not trivial as indicated by the fact that we must use the Heine–Borel theorem (which states that a closed and bounded interval is compact) to reduce the problem to one involving finite coverings. Given any such finite subcovering, we inductively select a subcollection as follows (renumbering as indicated). Choose $I_1 = (a_1, b_1)$ containing a. Having chosen $I_n = (a_n, b_n)$, choose $I_{n+1} = (a_{n+1}, b_{n+1})$ containing b_n. We stop when we reach an interval I_N containing b, which must occur by the finiteness of the collection. We then know that $a_1 \leq a$, $a_{n+1} \leq b_n$ for $n = 1, 2, \ldots, N-1$, and $b \leq b_N$. It follows that for our original cover the sum of the lengths of all intervals is bounded below by

$$\sum_{n=1}^{N}(b_n - a_n) = (b_1 - a_1) + (b_2 - a_2) + \cdots + (b_N - a_N)$$
$$\geq (a_2 - a) + (a_3 - a_2) + \cdots + (b - a_N) = b - a.$$

We conclude that $\mu^*([a,b]) \geq b - a$.

The conclusion now follows for an arbitrary bounded interval I since for any sufficiently small $\epsilon > 0$

$$[a + \epsilon, b - \epsilon] \subseteq I \subseteq [a, b].$$

Finally, when $b = \infty$, choose any real number $c > a$. Then, for all $r > 0$, we have $[c, c + r] \subseteq I$, so that $\mu^*(I) \geq r$. It follows that $\mu^*(I) = \infty$. Similarly, we deduce that $\mu^*(I) = \infty$ if $a = -\infty$. □

Proposition D.22. *μ^* is countably subaddtive.*

Proof. Take any countable collection of sets A_i with union A. Let $\epsilon > 0$ be given. For each i, choose a collection of open intervals $\{(a_{i,j}, b_{i,j})\}$ covering A_i such that

$$\mu^*(A_i) \geq \sum_j (b_{i,j} - a_{i,j}) - \frac{\epsilon}{2^i}.$$

The totality of all such open intervals for all i will cover A. Therefore, since $\sum_i \epsilon/2^i = \epsilon$,

$$\mu^*(A) \leq \sum_{i,j}(b_{i,j} - a_{i,j}) \leq \sum_i \mu^*(A_i) + \epsilon.$$

Since ϵ, is arbitrary, the proof is complete. □

If we could improve the above result to countable additivity, then μ^* would be a measure defined on all subsets, but unfortunately this cannot be done. We seek, instead, a reduced collection of sets on which μ^* is additive. To proceed, start with the following question. Given two disjoint sets A and B, when is it true that

$$\mu^*(A \cup B) \geq \mu^*(A) + \mu^*(B), \tag{†}$$

which is the sufficient condition we need from Proposition D.2?

Lemma D.23. *Condition* (†) *holds in either of the following cases*:

(*a*) $\mu^*(B) = 0$;
(*b*) *there is an* $r \in \mathcal{R}$ *such that* $A \subseteq (-\infty, r]$ *and* $B \subseteq (r, \infty]$.

Proof

(a) From the monotonicity of μ^*, which follows directly from the definition, we obtain

$$\mu^*(A \cup B) \geq \mu^*(A) = \mu^*(A) + \mu^*(B).$$

(b) It follows from part (a) that the value of μ^* is unchanged if we remove a single point from a set. By removing $\{r\}$ from A if necessary, we can assume that $A \subseteq (-\infty, r)$. Let $\epsilon > 0$ be arbitrary. Choose a covering $\{(a_n, b_n)\}$ of $A \cup B$ such that

$$\sum_n (b_n - a_n) \leq \mu^*(A \cup B) + \epsilon.$$

The collection $\{(a_n, r)\}$ covers A, and the collection $\{(r, b_n)\}$ covers B, so that

$$\mu^*(A) + \mu^*(B) \leq \sum_n (r - a_n) + \sum_n (b_n - r) = \sum_n (b_n - a_n) \leq \mu^*(A \cup B) + \epsilon.$$

□

While there are many methods for constructing the σ-algebra on which μ^* is additive, the most ingenious is a method due to Carathéodory. He observed that we want to select those sets that have the property that we showed for the interval $(-\infty, r]$ above.

Definitions. The class $\mathcal{M}$ of *Lebesgue measurable sets* on the real line consists of all sets E with the property that for any $A \subseteq E$ and $B \subseteq E^c$ we have $\mu^*(A \cup B) = \mu^*(A) + \mu^*(B)$. The restriction of μ^* to $\mathcal{M}$ is known as *Lebesgue measure*.

In order to justify the terminology, we must prove the following theorem.

Theorem D.24. *$\mathcal{M}$ is a σ-algebra, and μ^* is a measure on $\mathcal{M}$.*

We will prove Theorem D.24 after some preliminary steps.

Lemma D.25. *For all $E, F \in \mathcal{M}$, we have $E \cup F \in \mathcal{M}$.*

Proof. Take any $A \subseteq E \cup F$ and $B \subseteq (E \cup F)^c$. Let $A_1 = A \cap E$, $A_2 = A - E$. Then,

$$\begin{aligned}\mu^*(A \cup B) &= \mu^*(A_1 \cup (A_2 \cup B)) = \mu^*(A_1) + \mu^*(A_2 \cup B)\\ &= \mu^*(A_1) + \mu^*(A_2) + \mu^*(B)\\ &= \mu^*(A) + \mu^*(B).\end{aligned}$$

In the first line, we use the fact that $E \in \mathcal{M}$, $A_1 \subseteq E$, and $A_2 \cup B \subseteq E^c$. In the second line, we use the fact that $F \in \mathcal{M}$, $A_2 \subseteq F$, and $B \subseteq F^c$. In the third line, we use the fact that $E \in \mathcal{M}$, $A_1 \in E$, and $A_2 \in E^c$. □

We now show that the "splitting" property of a set in $\mathcal{M}$ can be extended from two to any finite number of sets. The precise statement is as follows.

Lemma D.26. *Given any n pairwise disjoint sets $\{E_1, E_2, \ldots, E_n\} \in \mathcal{M}$ and sets $A_i \subseteq E_i$ for $i = 1, 2, \ldots, n$, we have*

$$\mu^*\left(\bigcup_{i=1}^{n} A_i\right) = \sum_{i=1}^{n} \mu^*(A_i).$$

Proof. We show this equality first for $n = 2$. If $E_1 \cup E_2 = \Omega$, it holds by the definition of $\mathcal{M}$. If not, replace E_1 by $E_1 \cup (E_1 \cup E_2)^c$. Lemma D.25 and the fact that $\mathcal{M}$ is, by definition, closed under complements imply that the latter set is in $\mathcal{M}$. The general result follows by a straightforward induction, again using Lemma D.25. □

We can now prove the main theorem.

Proof of Theorem D.24. It follows from Lemma D.26 that given any disjoint $E, F \in \mathcal{M}$ we have $\mu^*(E \cup F) = \mu^*(E) + \mu^*(F)$. It will then follow from Propositions D.22 and D.2 that μ^* is a measure on $\mathcal{M}$ once we have shown that $\mathcal{M}$ is a σ-algebra.

By definition, $\mathcal{M}$ is closed under complements. It is trivial that Ω is in $\mathcal{M}$. It remains to show that for a countable collection of sets in $\mathcal{M}$ their union is in $\mathcal{M}$. This is true from Lemma D.25 for finite unions, so we need only deal with the countably infinite case. By using the "disjointification" procedure found in the proof of Proposition D.1(b) and by invoking the finite union result, we can assume that we have a pairwise disjoint famlily $\{E_n\}$ of sets in $\mathcal{M}$, and we want to show that their union $E \in \mathcal{M}$. Take $A \subseteq E$ and $B \subseteq E^c$. Let $A_N = \bigcup_{n=1}^{N}(A \cap E_n)$. We know that $A_N \in \mathcal{M}$ because $\mathcal{M}$ is closed under finite unions and intersections. Then, for any N,

$$\mu^*(A \cup B) \geq \mu^*(A_N \cup B) = \mu^*(A_N) + \mu^*(B) = \sum_{n=1}^{N} \mu^*(A \cap E_n) + \mu^*(B),$$

in which the first inequality holds since $B \subseteq (E_N)^c$ and the last equality follows from Lemma D.26. We now take limits to see that

$$\mu^*(A \cup B) \geq \sum_{n=1}^{\infty} \mu^*(A \cap E_n) + \mu^*(B).$$

From the countably subadditivity of μ^*, we obtain

$$\mu^*(A \cup B) \geq \mu^*(A) + \mu^*(B),$$

thereby completing the proof. □

Theorem D.24 will be of significance only if we can show that $\mathcal{M}$ contains a large number of sets, including the ones that we want to be in there. This is indeed true. Take any s, r with $-\infty \leq s < r < \infty$. From Lemma D.23(b), $\mathcal{M}$ contains all intervals of the form $(-\infty, r]$ and, therefore, any interval of the form $(s, r] = (-\infty, r] - (\infty, s]$. Then, as a σ-algebra, it must contain the open intervals $(s, r) = \bigcup_n (s, r - 1/n]$. Moreover, the intervals $(r, \infty) = (-\infty, r]^c$ are in $\mathcal{M}$. Because any open set is a countable union of open intervals, we know that all open sets are in $\mathcal{M}$, and closure under complementation tells us that all closed sets are in $\mathcal{M}$. In fact, $\mathcal{M}$ contains all the *Borel sets*, those in the σ-algebra generated by the open sets. In addition, Lemma D.23(a) shows that any set of outer measure zero is in $\mathcal{M}$.

Is it possible that, in fact, $\mathcal{M}$ contains all subsets of the reals. Although it is not easy to find a counterexample, the answer is "no." We present one example of a nonmeasurable set. This requires some knowledge of basic group theory.

Example D.27 (Nonmeasurable Subset of $\mathcal{R}$). We first notice an important property of Lebesgue measure, namely *translation invariance*: For all subsets A and real numbers c, we see that $\mu^*(A + c) = \mu^*(A)$. This is obviously true when A is an open interval; therefore, from the definition of μ^*, it holds in general. It also follows easily that if $A \in \mathcal{M}$, then $A + c \in \mathcal{M}$.

Define the operation $\circ$ on the interval $[0, 1)$ as addition modulo 1. That is, $a \circ b$ equals $a + b$ if $a + b < 1$ and $a + b - 1$ if $a + b \geq 1$. Under this operation $[0, 1)$ is a group. Take the subgroup H consisting of all rational numbers in $[0, 1)$, and choose a transversal T for the quotient group $[0, 1]/H$. That is, T is a set formed by selecting one element from each coset. (Readers familiar with set theory will note that the existence of T depends on the axiom of choice.) From the construction of T, it follows that

$$[0, 1) = \bigcup_{q \in H} (T + q),$$

a countable union of pairwise disjoint sets.

This expression implies that T cannot be measurable. If so, let $r = \mu(T)$. (μ denotes the restriction of μ^* to $\mathcal{M}$.) From the translation invariance and countable additivity of μ, we would have an infinite series with each entry equal to, say, r summing to $\mu([0, 1)) = 1$. But, this is impossible for such a sum can only be ∞ if $r > 0$ or 0 if $r = 0$.

D.8 COMPLETENESS OF MEASURES

Definition. A measure space $(\Omega, \mathcal{S}, \mu)$ is said to be *complete* if given any $A \in \mathcal{S}$ with $\mu(A) = 0$, then any subset B of A is also in $\mathcal{S}$. (It is common terminology to just say that the measure μ is complete.) Of course, by monotonicity, we will necessarily have that $\mu(B) = 0$.

Lebesgue measure is complete because we showed that any subset of $\mathcal{R}$ of outer measure 0 is in $\mathcal{M}$ [Lemma D.23(a)]. A useful consequence of completeness is as follows.

Proposition D.28. *If f is a measurable function with respect to a complete measure μ, and if g is any function that is equal to f a.e., then g is measurable.*

Proof. Let $A \in \mathcal{S}$ be a set of measure 0 such that $f(w) = g(w)$ for all $w \in A^c$. Given the interval $I = [-\infty, r]$, we have

$$g^{-1}(I) = [g^{-1}(I) \cap A^c] \cup [g^{-1}(I) \cap A)].$$

The first set on the right is measurable since it is equal to $f^{-1}(I) \cap A^c$, while the second set is measurable since it is a subset of A. $\square$

What happens in the above situation when μ is not complete? Then, g need not be measurable, but we can make it so by *completing* the measure, which can always be done. We sketch the process, leaving the details to the reader. Given an arbitrary measure μ, we consider the collection of all sets $\hat{\mathcal{S}}$ of the form $A \cup N$ in which $A \in \mathcal{S}$ and N is a subset of a set in $\mathcal{S}$ of measure 0. Define a set function $\hat{\mu}$ on $\hat{\mathcal{S}}$ by $\hat{\mu}(A \cup N) = \mu(A)$. First, verify that this is well defined because a given set can have several representations of this form. It is clear that $\hat{\mu}$ extends μ since one can take $N = \emptyset$ if A is in $\mathcal{S}$. Next, verify that $\mathcal{S}$ is a σ-algebra and that $\hat{\mu}$ is a measure on $\hat{\mathcal{S}}$. It is clear from the definition of $\mathcal{S}$ that $\hat{\mu}$ is complete.

Working with complete measures is useful in the following frequently arising situation. We have a function f which is only defined on a measurable subset A^c of Ω such that $\mu(A) = 0$ and we want to compute $\int f$. If f is measurable on A^c, then we can arbitrarily define it on A. When μ is complete, the extended function will be measurable by the same argument as in the proof of Proposition D.28. Moreover, the integral of this extended function is independent of the particular extension.

D.9 SIGNED AND COMPLEX MEASURES

D.9.1 Introduction

Throughout this section, we will take Ω and $\mathcal{S}$ as before but consider variations in the definition of μ, in that we allow it to take on negative or possibly complex values. We, first, consider the real-valued case and give a somewhat more general definition than the one in Section 6.1.3.

Definition. A *signed measure* is a function ν defined on $\mathcal{S}$ that is countably additive, sends the empty set to 0, and takes values in either $[-\infty, \infty)$ or $(-\infty, \infty]$.

We do not allow both ∞ and $-\infty$ in the range of μ because there is no consistent way to define the sum of these two values. The basic example of a signed measure is to take two measures μ_1 and μ_2, one of which is *finite*, and set $\mu = \mu_1 - \mu_2$. Our main result about signed measures will be to show that *all of them* arise in this way, thereby justifying the emphasis from the beginning on nonnegative set functions. This result is clear when Ω is countable since the problem then reduces to the decomposition of a function into its positive and negative parts.

D.9.2 Hahn and Jordan Decompositions

Definitions. Given a signed measure ν on $(\Omega, \mathcal{S})$, a set A in $\mathcal{S}$ is said to be *positive* if $\nu(B) \geq 0$ for *all* measurable $B \subseteq A$ and *negative* if $\nu(B) \leq 0$ for *all* measurable $B \subseteq A$.

Definition. A set that is both positive and negative, so that all measurable subsets have measure 0, is called a *null set.*

Remark. The above definition agrees with our previous definition of null set for a positive measure. It is important to note that for signed measures a set of measure 0 is not necessarily a null set.

Our main conclusions follow from the fact that any set of positive finite measure contains a positive set of positive measure. The proof of this fact depends on the following strategy. Remove a set of negative measure from the given set if it is is not positive, then remove a further set of negative measure and continue the process. The problem is that it might never end. For example, consider the signed measure on the nonnegative integers given by $\nu\{0\} = 1$ and $\nu\{k\} = -1/2^k$ for $k \geq 1$. We might remove $\{2\}$, then $\{3\}$, then $\{4\}$, and so on, but if we never remove $\{1\}$, we will never get to the only positive set, which is just $\{0\}$. The trick to avoid this is to remove at each stage a set whose negative measure is (in absolute value) close to being as large as possible.

Proposition D.29. *Any measurable set of finite positive measure contains a positive set of positive measure.*

Proof. Let A_1 be a nonpositive set of positive finite measure. We inductively constuct a sequence of subsets (A_n) and (B_n) as follows. Let k_n be the smallest positive integer such that A_n contains a set B_n with $\nu(B_n) \leq -1/k_n$, and let $A_{n+1} = A_n - B_n$. Let $B = \bigcup_n B_n$, and let $A = A_1 - B$. Then, $\nu(A_1) = \nu(A) + \nu(B)$.

We cannot have $\nu(B) = \sum_n \nu(B_n) = -\infty$, for that would contradict the fact that $\nu(A_1) > 0$. This shows as well that the series $\sum_n (1/k_n)$ converges, so that (k_n) converges to ∞. Since $\nu(B) < 0$, we know that $\nu(A) > \nu(A_1) > 0$. It remains to show that A is positive. Suppose, to the contrary, that there is a subset D of A with $\nu(D) < 0$, and let k be an integer such that $\nu(D) < -1/k$. Choose n such that $k_n > k$. Since $D \subseteq A$ and it is clear that $A \subseteq A_n$, this contradicts the definition of k_n. □

Proposition D.30. *Let* $\alpha = \sup\{\nu(P) : P \text{ positive}\}$. *Then, there exists a positive set* P_0 *such that* $\nu(P_0) = \alpha$.

Proof. Choose a sequence of positive sets (P_n) such that $\nu(P_n) \to \alpha$, and let $P_0 = \bigcup_n P_n$. It is clear that P_0 is positive. From the definition of α, it follows that $\nu(P_0) \leq \alpha$, and we must show the other inequality.

For any n, $P_0 - P_n$ is a subset of the positive set P_0, so that

$$\nu(P_0) = \nu(P_n) + \nu(P_0 - P_n) \geq \nu(P_n).$$

Thus, necessarily, $\nu(P_0) \geq \alpha$. □

Theorem D.31 (Hahn Decomposition). *There exist a positive set* P *and a negative set* N *such that* $\Omega = P \cup N$.

Proof. We can assume that ν does not take the value ∞. (If it does, we obtain the decomposition corresponding to $-\nu$ and can reverse the sets.) From Proposition D.30, choose a positive set P such that $\nu(P) = \sup\{\nu(Q) : Q \text{ positive}\}$, and let $N = \Omega - P$. Suppose that N is not negative. Then, N contains a set of positive measure, and by Proposition D.29, it contains a positive set A with $\nu(A) > 0$. We now have a positive set $P \cup A$ that, since $\nu(P) < \infty$, has measure greater than P, a contradiction. □

Definition. Two measures μ_1 and μ_2 on the same measure space are said to be *singular* if there exists a partition (A, B) of Ω such that $\mu_1(A) = 0$ and $\mu_2(B) = 0$.

An example of a pair of singular measures on the real line is to take μ_1 to be Lebesgue measure and μ_2 to be a unit mass at some point $x \in \mathcal{R}$. We take the set A of the partition to be $\{x\}$.

We are now able to state and prove the fundamental result mentioned in the introduction above. Recall the following definition (made in Chapter 6).

Definition. For any measure μ on $\mathcal{S}$ and $f \in L^+$ we define the measure μ_f by

$$\mu_f(A) = \int f \, d\mu$$

Theorem D.32 (Jordan Decomposition). *For any signed measure ν, there is a unique pair of singular measures ν^+ and ν^- such that $\nu = \nu^+ - \nu^-$.*

Proof. Let (P, N) be a Hahn decomposition for $(\Omega, \mathcal{S}, \nu)$, and let $\nu^+ = \mu_{1_P}$ and $\nu^- = -\nu_{1_N}$, which clearly satisfy the conditions in the statement of the theorem. To prove their uniqueness, suppose that $\nu = \rho^+ - \rho^-$ for another pair of singular measures, and let (A, B) be a partition of Ω such that $\rho^+(B) = \rho^-(A) = 0$. For any measurable set $E \subseteq P \cap A$, we have $\nu^+(E) = \nu(E) = \rho^+(E)$. If $C \subseteq P \cap B$, then $\nu(C) = \nu^+(C) \geq 0$, but also $\nu(C) = -\rho^-(C) \leq 0$. It must be that $\nu^+(C) = 0$. We have shown that $\nu^+ = \rho^+$; similarly, $\nu^- = \rho^-$. □

D.9.3 Complex Measures

Definition. A *complex measure* is a countably additive set function on $\mathcal{S}$ that takes values in $\mathcal{C}$. (Note that this means we exclude a value of either ∞ or $-\infty$.)

A complex measure ρ can be formed from any two finite signed measures μ and ν by setting $\rho = \mu + i\nu$. Conversely, any complex measure arises in this way since both Re ρ and Im ρ are finite signed measures. For a complex measure ρ, we use a.e. to mean "except for a set that is a null set with respect to both Re ρ and Im ρ." Integration with respect to a complex measure ρ is defined in a natural way as

$$\int f \, d\rho = \int f \, d(\text{Re}\, \rho) + i \int f \, d(\text{Im}\, \rho).$$

D.10 RADON–NIKODYM DERIVATIVES

We follow the notation and terminology of Sections 6.1.2 and 6.1.3. We will give a proof of the main theorem in its simplest version. An interesting aspect of the derivation is that although the theorem is stated for positive measures, we use the theory of signed measures in the proof because such an object appears naturally as the difference of two measures.

Theorem D.33 (Radon–Nikodym). *Suppose that μ and v are finite measures on $\mathcal{S}$ such that v is absolutely continuous with respect to μ. Then, there exists a function $f \in L^1(\mu)$ such that $v = \mu_f$. Moreover, f is unique a.e. $[\mu]$.*

Proof. Uniqueness follows since given two such functions f and g, let $A = \{\omega \in \Omega : f(\omega) > g(\omega)\}$. Then, $\int 1_A(f - g)\, d\mu = v(A) - v(A) = 0$, which implies by Proposition D.8 that $1_A(f - g) = 0$ a.e. $[\mu]$. Since $f - g > 0$ on A, this can only happen if $1_A = 0$ a.e. $[\mu]$, which means that $\mu(A) = 0$. Similarly, the set of points for which g is strictly bigger than f has μ-measure 0.

Let $K = \{f \in L^+ : \mu_f(A) \leq v(A)$for all $A \in \mathcal{S}\}$. Note that K is not empty since it at least contains 0. (The idea of the proof is to select in some sense the "biggest" element of K, which we hope will result in equality.)

We note first that for $f, g \in K$ the maximum $f \vee g \in K$. This holds since, given any $A \in \mathcal{S}$, let $B = \{x \in A : f(x) \geq g(x)\}$ and $C = \{x \in A : f(x) < g(x)\}$. Then,

$$\begin{aligned}\mu_{f\vee g}(A) &= \int_B (f \vee g)\, d\mu + \int_C (f \vee g)\, d\mu = \int_B f\, d\mu + \int_C g\, d\mu \\ &= \mu_f(B) + \mu_g(C) \leq v(B) + v(C) = v(A).\end{aligned}$$

Let $\alpha = \sup\{\int f\, d\mu : f \in K\}$, and choose a sequence (f_n) in K such that $\int f_n\, d\mu \to \alpha$. We know that for $f \in K$,

$$\int f\, d\mu = \mu_f(\Omega) \leq v(\Omega),$$

thereby showing that $\alpha \leq v(\Omega) < \infty$.

By replacing f_n with $f_1 \vee f_2 \vee \cdots \vee f_n$, which we know is in K by induction from the $n = 2$ result, we can assume that the sequence (f_n) is nondecreasing. Then, if $f = \sup f_n$, we can apply the monotone convergence theorem to conclude that $f \in K$ and $\int f\, d\mu = \alpha$.

We want to show that the measure $v - \mu_f = 0$ for this choice of f. For each positive integer n, define a signed measure

$$\rho_n = v - \mu_f - \frac{\mu}{n}.$$

Let (P_n, N_n) be a Hahn decomposition for ρ_n. Let $P = \bigcup_n P_n$ and $N = \bigcap_n N_n$. We will complete the proof by showing that the measure $v - \mu_f$ is 0 on both N and P, which form a partition of the whole space.

For each n, N_n is a negative set for the measure ρ_n, and since $N \subseteq N_n$ we must have $\rho_n(N) \leq 0$. This in turn implies that $(v - \mu_f)(N) \leq \mu(N)/n$. Since this is true for all n, it must be that $(v - \mu_f)(N) = 0$.

Consider now, for any n, the function $g_n = f + (1/n)1_{P_n}$. For any $A \in \mathcal{S}$,

$$\mu_{g_n}(A) = \mu_f(A) + \frac{\mu(A \cap P_n)}{n} = v(A \cap P_n) - \rho_n(A \cap P_n) \leq v(A \cap P_n) \leq v(A).$$

We conclude that $g_n \in K$. By taking $A = \Omega$ in the above, we obtain

$$\alpha \geq \int g_n \, d\mu = \alpha + \frac{\mu(P_n)}{n}.$$

Since α is finite, $\mu(P_n) = 0$, showing by [Proposition 4.1(b)] that $\mu(P) = 0$. By absolute continuity of ν with respect to μ, we have $\nu(P) = 0$, thereby completing the proof that $(\nu - \mu_f)(P) = 0$. □

There are many extensions. We can take μ to be σ-finite as indicated in Chapter 6. We cannot, however, completely remove the finiteness restriction from μ, as seen by the case for which μ is counting measure on the interval $[0, 1]$ and ν is Lebesgue measure. There is clearly no Radon–Nikodym derivative since given any nonzero function f and a point c with $f(c) \neq 0$ the measure of the single point c is zero under ν but not under μ_f.

We do not need finiteness restrictions on ν when it is a positive measure, and we can take it to be perfectly general. In this case, the Radon–Nikodym derivative will be measurable but not necessarily in $L^1[\mu]$. We will not provide further discussion as this case is not needed for our applications of this theorem to the representations of dual spaces.

By looking at positive and negative parts, we can extend the theorem to the case for which ν is a finite signed measure, as we did in Chapter 6. Then, by looking at real and imaginary parts, we can extend to the case for which ν is a complex measure, which we will need for Section D.12.

There are further generalizations to the case where μ is a signed or complex measure. None of these are needed for the applications in the book.

D.11 PRODUCT MEASURES

In this section, we will sketch the main ideas but not give detailed proofs. These can be found in Rudin (1988) or Halmos (1974).

Suppose we are given two measure spaces $(X, \mathcal{S}, \mu)$ and $(Y, \mathcal{T}, \nu)$. For convenience in this exposition, we will take both to be σ-finite. There is a natural way of constructing a σ-algebra of subsets of the Cartesian product $X \times Y$. We define $\mathcal{S} \times \mathcal{T}$ to be the σ-algebra generated by sets of the form $A \times B$ for $A \in \mathcal{S}$ and $B \in \mathcal{T}$. By the word "generated," we mean that $\mathcal{S} \times \mathcal{T}$ is the intersection of all σ-algebras containing those sets. It can then be shown that that there is a unique measure on $\mathcal{S} \times \mathcal{T}$, denoted by $\mu \times \nu$, that satisfies the following property:

$$(\mu \times \nu)(A \times B) = \mu(A)\, \nu(B)$$

for all $A \in \mathcal{S}$ and $B \in \mathcal{T}$.

As a standard example, having constructed Lebesgue measure on the real line, which extends the concept of length, we would like to similarly extend the concept of area in the plane. This is done by taking the product of the real-line Lebesgue measure with itself. This procedure can be iterated to obtain an n-dimensional Lebesgue measure. To be more precise, a Lebesgue measure in dimensions greater than 1 involves a modification of our definition, whereby we complete the product measure. The product measure as given need not be complete even if both measures are. (Some authors in fact define the product measure to be the *completion* of that given above.)

The main result about product measures is known as *Fubini's theorem*. It essentially says that integrals with respect to product measures can be obtained by the same type of iterated procedure that one uses for continuous functions in basic multivariate calculus. Before stating this result, we need first some definitions and terminology.

Suppose that f is a function $X \times Y$. Then, given any $x \in X$, we define a function f_x on Y by $f_x(y) = f(x, y)$. Such a function is known as an *x-section* of f. Similarly, for any $y \in Y$, we define a function on X by $f^y(x) = f(x, y)$, which is known as a *y-section* of f. With our original definition of the σ-algebra $\mathcal{S} \times \mathcal{T}$, it turns out that if f is measurable, then all sections f_x and f^y are measurable as well.

The above statement is *not* true if we adopt the alternative definition for the product measure space, in which we enlarge the σ-algebra in the completion process and thereby allow more measurable functions. Indeed, we know from Proposition D.28 that in the presence of a complete measure we can arbitrarily change the values of a measurable function on a null set and preserve measurability. Therefore, take a measurable function on the unit square under a two-dimensional Lebesgue measure. For any fixed $x_0 \in [0, 1]$, the line segment $\{(x_0, y) : 0 \leq y \leq 1\}$ has a two-dimensional measure equal to 0. A measurable function can have arbitrary values on this segment; therefore, the section f_{x_0} need not be measurable. It can be shown in general, however, that the set of all points $x \in X$ for which the section f_x is not measurable is a set of measure 0 under μ. A similar statement holds for y-sections.

What happens if f is integrable? In both definitions of the product measure, we cannot say that all sections are integrable, but we can show that *almost* all of them are. This is stated precisely in Theorem D.34 below.

We now state the main theorem. For clarity, we put subscripts on the integral signs to indicate over which space we are integrating.

Theorem D.34 (Fubini's Theorem). *Suppose f is a measurable function on $X \times Y$ such that $\int_{X \times Y} f \, d(\mu \times \nu)$ exists and is finite.* [*In other words $f \in L^1(X \times Y)$.*] *Then*:

(*a*) *There exists $N \subseteq X$ with $\mu(N) = 0$ such that for $x \notin N$ the function $f_x \in L^1(Y)$.*

(*b*) *For the function h on X defined by $h(x) = \int_Y f_x \, d\nu$ for $x \notin N$ (and arbitrarily on N), we have*

$$\int_{X \times Y} f \, d(\mu \times \nu) = \int_X h \, d\mu. \qquad \square$$

The conclusion is sometimes written in the more familiar form

$$\int_{X\times Y} f\, d(\mu \times \nu) = \int_X \int_Y f(x,y)\, d\nu\, d\mu.$$

We omit the details of the proof. It follows the usual three-stage approach mentioned in Section D.4, in which we go from sets to simple functions to general nonnegative measurable functions, with a final fourth stage to go from nonnegative measurable functions to general measurable functions.

A frequently invoked corollary of Fubini's theorem is as follows.

Theorem D.35. *If $f \in L^1(X \times Y)$, then*

$$\int_X \int_Y f(x,y)\, d\nu\, d\mu = \int_Y \int_X f(x,y)\, d\mu\, d\nu.$$

Proof. By symmetry and Fubini's theorem, both quantities are equal to $\int_{X\times Y} f\, d(\mu \times \nu)$. □

EXAMPLE D.36. If f does not belong to $L^1(\mu \times \nu)$, it can happen that both sides of the equation of Theorem D.35 exist, but they are not equal. For an easy example, take both X and Y to be the positive integers with counting measure. Let f take the value of 1 on all elements of $X \times Y$ of the form (n, n), the value of -1 on all elements of the form $(n, n+1)$, and the value 0 at all other points. Then,

$$\int_Y f(x,y)\, d\nu = 0 \quad \text{for all } x = 1, 2, \ldots$$

and

$$\int_X f(x,y)\, d\mu = \begin{cases} 1, \text{for} & y = 1, \\ 0, \text{for} & y = 2, 3, \ldots . \end{cases}$$

So, in the expression of Theorem D.35, the left-hand side is 0, while the right-hand side is 1.

D.12 RIESZ REPRESENTATION THEOREM

The purpose of this section is to prove Theorem 6.4(b), given Theorem 6.4(a).

D.12.1 Variation

Let μ be a complex-valued measure. We will develop the main properties of the set function $|\mu|$, the *variation* of μ, defined in Section 6.1.4. (We defined $|\mu|$ for signed measures there, but the definition is the same in the complex-valued case.)

Theorem D.37. *$|\mu|$ is a measure.*

Proof. We prove this statement by invoking Proposition D.2. Given disjoint C and D in $\mathcal{S}$ and any $\epsilon > 0$, choose partitions $\{C_1, C_2, \ldots, C_k\}$ and $\{D_1, D_2, \ldots, D_n\}$ of C and D, respectively, such that

$$|\mu|(C) \le \sum_{i=1}^{k} |\mu(C_i)| + \frac{\epsilon}{2}, \qquad |\mu|(D) \le \sum_{j=1}^{n} |\mu(D_j)| + \frac{\epsilon}{2}.$$

Then, the C_i's and D_j's together form a partion of $C \cup D$, and we can say that

$$|\mu|(C \cup D) \ge \sum_{i=1}^{k} |\mu(C_i)| + \sum_{j=1}^{n} |\mu(D_j)| \ge |\mu|(C) + |\mu|(D) - \epsilon.$$

Since ϵ is arbitrary, it follows that

$$|\mu|(C \cup D) \ge |\mu|(C) + |\mu|(D).$$

It remains to prove that $|\mu|$ is countably subadditive. Let $\{A_i\}$ be a countable collection of sets in $\mathcal{S}$ with union A. We want to show that

$$|\mu|(A) \le \sum_{i=1}^{\infty} |\mu|(A_i). \qquad (*)$$

The finite superadditivity of $|\mu|$ proved above is enough to show, arguing as in Proposition D.1(a), that $|\mu|$ is monotone. That is, $B \subseteq A$ implies that $|\mu|(B) \le |\mu|(A)$. Then, by following the procedure to "disjointify" as used in the proof of Proposition D.1(b), we can assume that $A_i \cap A_j = \emptyset$ for $i \ne j$.

Given any $\epsilon > 0$, let $\{E_1, E_2, \ldots, E_n\}$ be a partition of A such that

$$|\mu|(A) \le \sum_{j=1}^{n} |\mu(E_j)| + \epsilon.$$

Since

$$E_j = \bigcup_{i=1}^{\infty} (E_j \cap A_i),$$

the countable additivity of μ shows that $\mu(E_j) = \sum_{i=1}^{\infty} \mu(E_j \cap A_i)$. Then,

$$|\mu(E_j)| \le \sum_{i=1}^{\infty} |\mu(E_j \cap A_i)| \qquad (\dagger),$$

for all j. Moreover, since $\{E_1 \cap A_i, \ldots, E_n \cap A_i\}$ is a partition of A_i, we have

$$|\mu|(A_i) \geq \sum_{j=1}^{n} |\mu(E_j \cap A_i)| \tag{‡}$$

for all i, by the definition of $|\mu|$. It then follows that

$$|\mu|(A) - \epsilon \leq \sum_{j=1}^{n} |\mu(E_j)| \leq \sum_{j=1}^{n} \sum_{i=1}^{\infty} |\mu(E_j \cap A_i)| = \sum_{i=1}^{\infty} \sum_{j=1}^{n} |\mu(E_j \cap A_i)| \leq \sum_{i=1}^{\infty} |\mu|(A_i).$$

Here, we use (†) for the second inequality and (‡) for the third inequality. Since all quantities are nonnegative, it is a standard theorem of real analysis that the interchange of summation is justified. (It also follows from Fubini's theorem applied to counting measure on the positive integers.) Since ϵ is arbitrary, we have verified $(*)$, which completes the proof. □

We now provide an alternative form of the variations.

Proposition D.38. *For any signed measure* μ,

$$|\mu| = \mu^+ + \mu^-.$$

Proof. Given any set E and a partition $\{E_1, E_2, \ldots, E_n\}$ of E, we have

$$\sum_{i=1}^{n} |\mu(E_i)| = \sum_{i=1}^{n} |\mu^+(E_i) - \mu^-(E_i)| \leq \sum_{i=1}^{n} \mu^+(E_i) + \sum_{i=1}^{n} \mu^-(E_i) = \mu^+(E) + \mu^-(E),$$

thereby showing that

$$|\mu|(E) \leq \mu^+(E) + \mu^-(E).$$

Conversely, choose a Hahn decomposition (P, N). By definition,

$$|\mu|(E) \geq |\mu(P)| + |\mu(N)| = \mu^+(E) + \mu^-(E)$$

since $\mu^+(N) = \mu^-(P) = 0$. □

We now investigate the Radon–Nikodym derivative $h = d\mu/d|\mu|$, which exists since it is clear that μ is absolutely continuous with respect to $|\mu|$. In the case of a real-valued measure, it is easy to see what h is. Indeed, if (P, N) is a Hahn decomposition of the space, then h will take the value of 1 on P and -1 on N a.e. $[|\mu|]$. In the complex-valued case, the function may be much more complicated, but we can say the following.

Proposition D.39. *If* $h = d\mu/d|\mu|$, *then* $|h| = 1$ *a.e.* $[\,|\mu|\,]$.

Proof. Let $A = \{\omega : |h(\omega)| < 1\}$, and let $B = \{\omega : |h(\omega)| > 1\}$. We must show that $|\mu|(A) = |\mu|(B) = 0$.

Consider B first. Fix any complex number z with $|z| > 1$, and fix any real number r such that $0 < r < |z| - 1$. Let $C = \{\omega : |h(\omega) - z| < r\}$. Since B can be written as a countable union of sets such as C, it is sufficient to show that $|\mu|(C) = 0$. We have

$$\left|\int_C (h - z)\, d|\mu|\right| = \left|\int_C h\, d|\mu| - z\,|\mu|(C)\right| \geq |z|\;|\mu|(C) - \left|\int_C h\, d|\mu|\right| \geq (|z| - 1)\;|\mu|(C)$$

by using the fact that $\left|\int_C h\,d|\mu|\right| = |\mu(C)| \leq |\mu|(C)$ from the definition of h. On the other hand,

$$\left|\int_C (h - z)\, d|\mu|\right| \leq \int_C |h - z|\, d|\mu| \leq r\,|\mu|(C) < (|z| - 1)\,|\mu|(C).$$

These two inequalities imply that $|\mu|(C) = 0$.

Now consider A. For a positive integer n, let $A_n = \{\omega : |h(\omega)| \leq 1 - 1/n\}$. Since $A = \bigcup_n A_n$, we complete the proof by showing that $|\mu|(A_n) = 0$ for all n. For any $E \subseteq A$, we invoke Proposition D.15 to see that

$$|\mu(E)| = \left|\int_E h\, d|\mu|\right| \leq \int_E |h|\, d|\mu| \leq \left(1 - \frac{1}{n}\right) |\mu|(E).$$

It follows that for any partition $\{E_1, E_2, \ldots, E_k\}$ of A_n

$$\sum_{i=1}^{k} |\mu(E_i)| \leq \left(1 - \frac{1}{n}\right) \sum_{i=1}^{k} |\mu|(E_i) = \left(1 - \frac{1}{n}\right) |\mu|(A_n).$$

By taking a supremum on the left-hand side of this inequality over such partitions, we obtain

$$|\mu|(A_n) \leq \left(1 - \frac{1}{n}\right) |\mu|(A_n),$$

implying that $|\mu|(A_n) = 0$. □

We can now state a further generalization of the triangle inequality.

Proposition D.40. *For all $f \in L^1(\mu)$,*

$$\left|\int f\,d\mu\right| \leq \int |f|\,d|\mu|.$$

Proof. The left-hand side is just equal to $|\int f\,h\,d|\mu||$, and we apply Propositions D.15 and D.39. □

D.12.2 Space *M*

We now return to the setting in which we are interested. We take $\Omega = [a, b]$, $\mathcal{S}$ the σ-algebra of Borel sets, and M the linear space of all complex-valued measures defined on $\mathcal{S}$. For all $\mu \in M$, we define

$$\| \mu \| = |\mu|([a, b]).$$

Theorem D.41. $\| \cdot \|$ *is a norm on* M.

Proof. The main point to show is that $\| \mu \| < \infty$, which is not obvious in view of the supremum in the definition of $|\mu|$. Suppose that $\mu = \nu + i\rho$, in which ν and ρ are signed measures. Then, for any measurable E, we have

$$|\mu(E)| = |\nu^+(E) - \nu^-(E) + i\rho^+(E) - i\rho^-(E)| \leq \nu^+(E) + \nu^-(E) + \rho^+(E) + \rho^-(E).$$

If $\{E_1, E_2, \ldots, E_n\}$ is a partition of $[a, b]$, we apply the above expression to each E_i and argue as in Proposition D.38 to see that

$$|\mu|([a, b]) \leq \nu^+([a, b]) + \nu^-([a, b]) + \rho^+([a, b]) + \rho^-([a, b]) < \infty.$$

The verification of the norm axioms is straightforward and is left to the reader. □

The result above on the Radon–Nikodym derivative has some important consequences. In particular, we will now show that M contains $L^1[a, b]$ as a subspace. In the following, let m denote Lebesgue measure on $[a, b]$.

Proposition D.42. *The mapping from $L^1[a, b]$ to M that takes f to m_f is an isometry.*

Proof. Fix any f, and let $h = dm_f/d|m_f|$. Then, from Proposition D.39, $h^{-1} = \bar{h}$, so that

$$\int g\,d|m_f| = \int g\,\bar{h}\,dm_f$$

for all $g \in M$. Then, for all measurable E,

$$|m_f|(E) = \int_E d|m_f| = \int_E \bar{h}\,dm_f = \int_E f\bar{h}\,dm.$$

Since the last integral is nonnegative for all measurable E, it must be that $f\bar{h} \geq 0$ a.e. $[m]$. (To verify this last statement, we can assume that the integrand is real by looking at its real and imaginary parts and then use a variation of the argument in the proof of Proposition D.8, in which we consider the measure of the set where the function takes a value less than $-1/n$.) This means that, a.e. $[m]$,

$$f\bar{h} = |f\bar{h}| = |f|,$$

and from the above we get

$$\| m_f \| = \int |f| \, dm = \| f \|_1 . \qquad \square$$

We now need a basic result of measure theory, which we state without proof. Some insight was provided in the remark following Theorem 6.4. For a complete proof in a general setting, see Hewitt and Stromberg (1975, Theorem 13.21).

Proposition D.43. *Let μ be a nonnegative measure in M, and let $h \in L^1(\mu)$. Then, given any $\epsilon > 0$, there exists $g \in C[a,b]$ such that $\| g - h \|_1 < \epsilon$. Moreover, if $|h| \leq K$ a.e., then we can choose g so that $|g| \leq K$.* $\square$

With this, we can deduce a key result.

Theorem D.44. *The mapping from M to $(C[a,b])^*$ that sends μ to the linear functional ψ_μ defined by*

$$\psi_\mu(f) = \int f \, d\mu$$

is an isometry.

Proof. From Proposition D.40, we know that

$$|\psi_\mu(f)| \leq \int |f| \, d|\mu| \leq \| f \|_{\sup} \| \mu \|,$$

thereby showing that

$$\| \psi_\mu \| \leq \| \mu \| .$$

For the other direction, let $h = d\mu/d|\mu|$. Take any $\epsilon > 0$, and choose, by Proposition D.43, $g \in C[a,b]$ such that $\| g - \bar{h} \|_1 < \epsilon$ and $\| g \|_{\sup} \leq 1$. Then,

$$\| \mu \| = |\mu|([a,b]) = \int h \bar{h} \, d|\mu| = \int \bar{h} \, d\mu = \int (\bar{h} - g) \, d\mu + \int g \, d\mu,$$

and the right-hand side, being positive, is less than $\epsilon + |\psi_\mu(g)|$. It follows that

$$|\psi_\mu(g)| \geq \| \mu \| - \epsilon.$$

Since ϵ is arbitrary and $\| g \|_{\sup} \leq 1$, we conclude that

$$\| \psi_\mu \| \geq \| \mu \| .$$

□

We now complete the proof of the Riesz representation theorem for the complex case. Given Theorem D.44, it remains to show the surjectivity of the map given in that theorem. As we saw with previous representation theorems, this is the most difficult part.

In preparation for this, we need the following technical lemma regarding elements of $C[a,b]$.

Lemma D.45. *Given nonnegative f_1 and f_2 and any g, all in $C[a,b]$, such that $|g| \leq f_1 + f_2$, we can find g_1 and g_2 in $C[a,b]$ such that*

$$|g_1| \leq f_1, \qquad |g_2| \leq f_2, \qquad g = g_1 + g_2.$$

Proof. Let

$$h_1 = \min(f_1, |g|), \qquad h_2 = |g| - h_1.$$

Then, $h_1 \leq f_1$, and $h_2 \leq f_2$. The latter follows since either $h_2(t)$ equals $|g(t)| - f_1(t)$ or 0, according as $f_1(t)$ is less than or greater than $|g(t)|$, respectively. Let

$$g_1 = h_1 \frac{g}{|g|} \qquad \text{and} \qquad g_2 = h_2 \frac{g}{|g|},$$

in which we set $g(t)/|g(t)|$ equal to 1 if $g(t) = 0$. The proof is completed by noting that

$$|g_1| = h_1, \qquad |g_2| = h_2, \qquad g_1 + g_2 = \frac{(h_1 + h_2)g}{|g|} = \frac{|g|\,g}{|g|} = g.$$

□

Completion of Proof of Theorem 6.4(*b*). Given a bounded linear functional ϕ on $C[a,b]$, we must produce a complex measure μ so that $\phi = \psi_\mu$, as defined in Theorem D.44. We do this in two stages. First, we find $|\mu|$ by utilizing the positive version of the theorem, namely Theorem 6.14(a). We then use the representation theorem we already have for $(L^1)^*$ (Theorem 6.3) to find the Radon–Nikodym derivative.

In fact, all we need is to produce a positive Borel measure $|\mu|$ on $[a, b]$ so that ϕ is bounded with respect to the $L^1(|\mu|)$ norm on $C[a, b]$.Once we have this $|\mu|$, we can then extend ϕ to a bounded linear functional $\hat{\phi}$ on the bigger space $L^1[a, b]$. We know that there exists $h \in L^\infty$ such that for all $f \in (L^1[a, b])^*$

$$\hat{\phi}(f) = \int f\, h\, d|\mu| = \int f\, d|\mu|_h,$$

and we can take $\mu = |\mu|_h$.

Of course, it is not at all obvious that this is possible, given that the $L^1(|\mu|)$ norm is weaker than the sup norm. We note first that to produce the measure $|\mu|$ with the required condition, it is sufficient to find a positive linear functional $\rho \in (C[a, b])^*$ such that for all $f \in C[a, b]$

$$|\phi(f)| \leq \rho(|f|). \tag{$*$}$$

Then, if $|\mu|$ is the measure corresponding to ρ by Theorem 6.4(a), the positive version of the Riesz respresentation theorem, we know that

$$|\phi(f)| \leq \rho(|f|) = \int |f|\, d|\mu| = \| f \|_1,$$

thereby showing that ϕ is indeed bounded with respect to $\| \cdot \|_1$.

We now proceed to find ρ satisfying $(*)$. Define ρ first on the nonnegative functions $g \in C[a, b]$ by

$$\rho(g) = \sup\{|\phi(f)| : |f| \leq g, f \in C[a, b]\}.$$

It is clear that $(*)$ holds for all $f \in C[a, b]$.

We then extend ρ to all of $C[a, b]$ by defining for any real-valued $g \in C[a, b]$

$$\rho(g) = \rho(g^+) - \rho(g^-),$$

and then for general $g \in C[a, b]$, by defining

$$\rho(g) = \rho(\text{Re } g) + i\rho(\text{Im } g).$$

The proof is completed by showing that ρ is linear. By using the same argument as the one in the proof of Proposition D.17, in which the linear functional was the integral, it is sufficient to show that ρ is additive on the nonnegative functions and has the scalar multiplication property for nonnegative scalars. The latter is clear. We will show the additivity. That is, for all nonnegative f_1 and f_2 in $C[a, b]$, we want to show that

$$\rho(f_1 + f_2) = \rho(f_1) + \rho(f_2). \tag{$\dagger$}$$

Let $\epsilon > 0$ be given. We will show that each side of (†) is less than the other side plus ϵ. Choose h_1 and h_2 in $C[a, b]$ such that

$$|h_1| \leq f_1 \qquad \text{and} \qquad |h_2| \leq f_2$$

and such that

$$\rho(f_1) \leq |\phi(h_1)| + \frac{\epsilon}{2} \qquad \text{and} \qquad \rho(f_2) \leq |\phi(h_2)| + \frac{\epsilon}{2}.$$

Now, consider the polar forms

$$\phi(h_1) = e^{\Theta_1}|\phi(h_1)| \qquad \text{and} \qquad \phi(h_2) = e^{\Theta_2}|\phi(h_2)|,$$

and note that

$$|e^{-\Theta_1}h_1 + e^{-\Theta_2}h_2| \leq |h_1| + |h_2| \leq f_1 + f_2,$$

from which it follows that

$$\rho(f_1 + f_2) \geq |\phi(e^{-\Theta_1}h_1 + e^{-\Theta_2}h_2)| = |\phi(h_1)| + |\phi(h_2)| \geq \rho(f_1) + \rho(f_2) - \epsilon.$$

For the other direction, pick $g \in C[a, b]$ with $|g| \leq f_1 + f_2$ and $\rho(f_1 + f_2) \leq |\phi(g)| + \epsilon$. Then, pick g_1 and g_2 satisfying the conditions in Lemma D.45. It follows that

$$\rho(f_1 + f_2) - \epsilon \leq |\phi(g)| = |\phi(g_1) + \phi(g_2)| \leq |\phi(g_1)| + |\phi(g_2)| \leq \rho(f_1) + \rho(f_2). \square$$

Remark. There are generalizations of the Riesz representation theorem. First, we can replace $[a, b]$ by any compact Hausdorff topological space X. The theorem goes through as stated, except for the following modification. It can happen, in general, that there are distinct Borel measures that induce the same linear functional. Then, the conclusion of Theorem D.44 cannot hold because the difference of the two is a measure that induces the 0 functional. What goes wrong in the proof is that we cannot invoke the density result, Proposition D.43. We need to restrict attention to what are called *regular* Borel measures. These are measures that are in some sense compatible with the topological structure; see Rudin (1974, Sections 2.15 and 6.18) for precise definitions. On $[a, b]$, indeed on any metric space that is a countable union of compact spaces, regularity is automatic for a Borel measure. If we define the space M to consist of regular Borel measures, then all arguments go through as above.

Furthermore, the theorem can be generalized to the case of a locally compact space and the space of continuous functions vanishing at infinity.

APPENDIX E

TYCHONOFF'S THEOREM

In this appendix, we prove the following theorem, which is due to Tychonoff.

Theorem E.1. *A product of compact spaces is compact.*

Before proving this theorem, we review the finite intersection property criterion for compactness. A collection of subsets of a given set X is said to have the *finite intersection property*, henceforth abbreviated as f.i.p., if any finite subcollection has nonempy intersection. A topological space X is compact if and only if for any collection of closed sets with the f.i.p. the intersection of the entire collection is nonempty. There is an equivalent formulation that allows us to deal with arbitrary sets, not necessarily closed. X is compact if and only if given any collection of sets with the f.i.p. there is a point in the closure of all the sets. In fact, we are going to consider only *maximal* collections of sets with the f.i.p. This is a collection of sets with the f.i.p. such that adding any set not already in the collection will yield a collection that does *not* have the f.i.p.

An obvious example of such a maximal f.i.p. collection is formed by choosing any point x and taking the collection of all subsets containing x. This trivially has the f.i.p. since any intersection contains the point x. It is maximal because adding any set B not in the collection gives a collection without the f.i.p. since $B \cap \{x\}$ is empty. There is no direct way to give other examples for an infinite set, but we can appeal to Zorn's lemma to show their existence, as we demonstrate.

In the following, we let $\mathbf{F}$ denote the set of all those collections of subsets of X that have the f.i.p. $\mathbf{F}$ becomes a poset under the partial order given by $\mathcal{A} \leq \mathcal{B}$ if every component set in $\mathcal{A}$ is also in $\mathcal{B}$. It is important to keep straight the fact that an element of $\mathbf{F}$ consists of several subsets of X, which, as above, we refer to as the *component sets* of $\mathcal{A}$. So, if $X = \{a, b, c\}$, a typical element of $\mathbf{F}$ would be something like $\mathcal{A} = \{\{a, b\}, \{a, c\}\}$. If $\mathcal{D} = \{\{a\}, \{a, b\}, \{a, c\}, \{a, b, c\}$, then $\mathcal{A} \leq \mathcal{D}$. In fact, $\mathcal{D}$ is a maximal element of $\mathbf{F}$ greater than $\mathcal{A}$.

The following shows that we can alway find such a maximal f.i.p collection containing any given f.i.p. collection.

A First Course in Functional Analysis. By S. David Promislow

Lemma E.2. *Given any $\mathcal{A} \in \mathbf{F}$, there exists a maximal element $\mathcal{D}$ in $\mathbf{F}$ such that $\mathcal{A} \leq \mathcal{D}$.*

Proof. Consider the subposet $\mathbf{H}$ consisting of elements of $\mathbf{F}$ larger than or equal to $\mathcal{A}$, which is not empty as it trivially contains $\mathcal{A}$. Given any linearly ordered subset $\mathbf{G}$ of $\mathbf{H}$, let $\mathcal{B}$ be the collection consisting of all subsets of X that are component sets of some $\mathcal{G} \in \mathbf{G}$. Clearly, $\mathcal{A} \leq \mathcal{B}$. We have to show that $\mathcal{B}$ has the f.i.p., which means it is an upper bound for $\mathbf{G}$. Zorn's lemma is then invoked to show the existence of a maximal element $\mathcal{D} \in \mathbf{H}$, which is necessarily a maximal element in $\mathbf{F}$ such that $\mathcal{A} \leq \mathcal{D}$.

Suppose we take any finite number of the component sets $Y_1, Y_2, \ldots, Y_n$ of $\mathcal{B}$. For each i, Y_i is a component set of $\mathcal{G}_i$, in which $\mathcal{G}_i \in \mathbf{G}$. From the linear ordering, there is some k such that $\mathcal{G}_i \leq \mathcal{G}_k$ for $i = 1, 2, \ldots, n$. Then, each of the sets Y_i is a component set in $\mathcal{G}_k$, and since $\mathcal{G}_k$ has the f.i.p, the intersection of the Y_i's is not empty. □

Lemma E.2 and the discussion at the beginning of this appendix yield the following corollary, which will be the basis of our proof of Theorem C.1.

Corollary E.3. *A set X is compact if and only if given any maximal subset $\mathcal{B} \in \mathbf{F}$ there is a point $x \in X$ that is in the closure of each set in $\mathcal{B}$.* □

We note that a maximal subset $\mathcal{B}$ of $\mathbf{F}$ has the following useful properties:

1. A finite intersection of sets in $\mathcal{B}$ is a component set of $\mathcal{B}$.
2. If a set A has nonempty intersection with each set in $\mathcal{B}$, then A is a component set of $\mathcal{B}$.

To verify property 1, suppose by contradiction that $D = \bigcap_{i=1}^{n} A_i$ is not in $\mathcal{B}$, in which each $A_i \in \mathcal{B}$. We claim that the collection $\mathcal{B}'$ formed by adding D to $\mathcal{B}$ has the f.i.p., thereby contradicting maximality of $\mathcal{B}$. Indeed, if we are given component sets $C_1, C_2, \ldots, C_m$ in $\mathcal{B}$, choose any

$$x \in C_1 \cap C_2 \cap \cdots \cap C_m \cap A_1 \cap A_2 \cap \cdots \cap A_n.$$

Then,

$$x \in C_1 \cap C_2 \cap \cdots \cap C_m \cap D.$$

The same type of argument shows that property 2 follows from property 1.

We can now complete our proof of Theorem E.1. The strategy is a familiar one when dealing with product spaces. Specifically, we use the projection mappings to infer properties of the whole space from those of the components.

Proof of Theorem E.1. We are given the set X consisting of all functions f on a set S, in which $f(s)$ is in the compact space X_s for all $s \in S$, and X is equipped with the product topology as defined in Section 7.1.4. For any $s \in S$, let $p_s : X \to X_s$ denote the projection mapping given by $p_s(f) = f(s)$.

Suppose we are given a maximal $\mathcal{B} \in \mathbf{F}$. By Corollary E.3, to show that X is compact, it is enough to find a function f in X such that every basic neighborhood V of f intersects every set in $\mathcal{B}$. Since $\mathcal{B}$ has the f.i.p., this will certainly be true if we show that every basic neighborhood, as defined in Section 7.1.4, is one of the component sets of $\mathcal{B}$.

We construct f by defining it at each point of S. Fix any $s \in S$. Since $\mathcal{B}$ has the f.i.p., it is not hard to see that $p_s(\mathcal{B})$ has the f.i.p. in X_s. By the compactness of X_s, there is a point, which we denote by $f(s)$, that is in $p_s(A)$ for each $A \in \mathcal{B}$. This gives us our candidate function f.

Next, take any basic neighborhood V of f. By the definition of the product topology, there exists a finite subset E of S and a neighborhood U_s of $f(s)$ in X_s for each $s \in E$ such that $V = \bigcap_{s \in E} V_s$, in which for $s \in E$, V_s is defined by

$$V_s = \{g \in X : g(s) \in U_s\}.$$

To conclude the proof, observe that for any $A \in \mathcal{B}$ the defining condition of f means we can find $h \in A$ such that $f(s) = p_s(h) = h(s)$. Therefore, $h \in V_s \cap A$. By property 2, each V_s is a component set in $\mathcal{B}$, and by property 1, V is a component set in $\mathcal{B}$. □

SYMBOLS

Symbol	Definition	Page Number
1_A	Function which is 1 on A, 0 elsewhere	5
$\downarrow$	Decreasing sequence of numbers, functions, or sets	257
$\uparrow$	Increasing sequence of numbers, functions, or sets	257
$A - B$	$A + (-1)B$	3
$A + B$	Sum of two subsets of linear space	3
A^c	Complement of A	15
a.e. $[\mu]$	Almost everywhere with respect to μ	42
A^o	Internal points of A	110
$A \subseteq^\epsilon B$	Approximate containment	188
$\bar{A}$	Closure of A	15
$B(S)$	Bounded functions on S	18
$B(x; r)$	Open ball	16
$\bar{B}(x; r)$	Closed ball	16
$\mathcal{B}(X)$	Space $\mathcal{B}(X, X)$	20
$\mathcal{B}(X, Y)$	Space of bounded linear operators from X to Y	20
c_0	Sequences converging to 0	26
c_{00}	Sequences with entries eventually equal to 0	8
co (A)	Convex hull of A	160
$C_0(\mathcal{R})$	Continuous functions vanishing at ∞	26
$C_{00}(\mathcal{R})$	Continuous functions with compact support	26
$C^2[a, b]$	Twice continuously differentiable functions	214
$C[a, b]$	Continuous functions on $[a, b]$	26
$\mathcal{C}^n$	The n-dimensional complex vectors	4
$\mathcal{C}$	Complex numbers	2
$d\nu/d\mu$	Radon–Nikodym derivative	125
$d(x, A)$	Distance from x to the set A	11

A First Course in Functional Analysis. By S. David Promislow

Symbol	Definition	Page Number
$d(x, y)$	Distance from x to y	11
diam (A)	Diameter of set A	12
$\|f\|_\infty, \|f\|_{\text{ess sup}}$	Essential sup of $\lvert f\rvert$	47
$\|f\|_p$	The p-norm of f	44
$\|f\|_{\text{sup}}$	Sup norm on $B(S)$	19
f^+, f^-	Positive and negative parts of function f	43
$\{F_\lambda\}$	Decomposition of identity	231
$F(S)$	Space of functions from S to $\mathcal{F}$	4
$\mathcal{F}$	Real or complex numbers	2
$f(T)$	Function of an operator	180
$f \vee g$	Maximum of two functions	249
$f \wedge g$	Minimum of two functions	249
Im z	Imaginary part of complex number	2
J_H	Natural map from Hilbert space to its dual	87
$k(s, t)$	Kernel of integral operator	206
L_0^+	Simple functions	257
$\lim x_n$	Limit of sequence	12
lim inf	Lower limit point of sequence	42
lim sup	Upper limit point of sequence	254
L^∞	Equivalence classes of essentially bounded measurable functions	46
L^p	Equivalence classes of measurable functions with finite p-norm	46
$\mathcal{L}^p$	Measurable function with finite p-norm	44
L	Measurable functions	257
L^+	Nonnegative measurable functions	257
ℓ^∞	Bounded sequences	50
ℓ^p	Sequences with finite p-norm	49
$\mathcal{L}(X)$	$\mathcal{L}(X, X)$	7
$\mathcal{L}(X, Z)$	Space of linear operators from X to Z	7
M	Space of measures	128
M_y	Multiplication operator	21
$\mathcal{M}$	Lebesgue measurable sets	270
$\mathcal{N}$	Positive integers	5
$N(T)$	Null space of T	5
p_K	Minkowski functional of K	113
P_M	Projection onto subspace	79
Re z	Real part of complex number	2
Re f, Im f	Real and imaginary parts of function f	43
$\mathcal{R}$	Real numbers	2
$\hat{\mathcal{R}}$	Extended real numbers	253
$\mathcal{R}^n$	The n-dimensional real vectors	4
$R(T)$	Range of T	6

Symbol	Defination	Page Number
$r(x)$	Spectral radius of x	173
span (A)	Smallest subspace containing A	4
$S^{\perp}$	Elements orthogonal to S	80
$S(T)$	Support of operator	91
$\\| T \\|$	Norm of operator	19
T^{-1}	Inverse of T	6
$T^{1/2}$	Positive square root of positive operator	94
T^*	Adjoint operator	131
$T : X \to Z$	Linear map from X to Z	5
$V(F, \epsilon)$	Weak neighborhood of 0	153
$V^*(F, \epsilon)$	Weak*neighborhood of 0	156
$V_{\text{op}}(F, \epsilon)$	Strong operator neighborhood of 0	164
$V^*_{\text{op}}(F, G\epsilon)$	Weak operator neighborhood of 0	164
$\mathcal{V}_x$	Neighborood base	144
$w(T)$	Numerical radius of T	177
$\\| x \\|$	Norm of x	12
$x \perp y$	Orthogonality relation	80
$x_n \to x$	Limit of sequence	12
X^{**}	Double dual	133
(X, d)	Metric space	11
X/N	Quotient space	31
X^*	Dual space	86
$Y \oplus Z$	Direct sum of two spaces	52
$Y \times Z$	Direct product of two spaces	52
$< x, y >$	Inner product	72
$\|z\|e^{i\Theta}$	Polar form of complex number	2
$\|z\|$	Absolute value	2
$\bar{z}$	Complex conjugate	2
αA	Scalar multiplication of subset of linear space	3
δ_i	Standard basis elements of $F(S)$	5
v^+, v^-	Positive and negative parts of signed measure	125
μ_f	Measure with $d\mu_f/du = f$	125
$\|\mu\|$	Variation of a measure	128
$\mu_{x,y}$	Measure associated with $\phi_{x,y}$	236
$\phi_{x,y}$	Sesquilinear form	236
π_N	Canonical projection onto X/N	32
π_s	Natural projection from direct product	52
$\prod_{s \in X} X_s$	Direct product	51
$\sigma(x)$	Spectrum of x	170
$(\Omega, \mathcal{S}, \mu)$	Measure space	40

REFERENCES

Apostol, T. M. (1974), *Mathematical Analysis*, 2nd ed., Addison-Wesley, Reading, MA.

Aronszajn, N., and Smith, K. T. (1954), Invariant subspaces of completely continuous operators, *Ann. Math*, 60, 345–350.

Banach, S.(1932), *Theorie des Operations Lineaires*, Chelea, New York.

Banach, S. (1988), *Theory of Linear Operators*, North Holland, Amersterdam (translated by F. Jellett).

Bobrowski, A. (2005), *Functional Analysis for Probability Theory and Stochastic Processes*, Cambridge University Press, Cambridge.

Bollobas, B. (1998), *Modern Graph Theory*, Springer, New York.

Brockwell, P. J. and Davis R. A. (1991), *Time Series: Theory and Methods*, Springer Verlag, New York.

Carothers, N. L. (2005), *A Short Course on Banach Spaces*, Cambridge University Press, Cambridge.

Dunford, N., and Schwartz, J. T. (1957), *Linear Operators*, Part I, Interscience, New York.

Enflo, P. (1973), A counterexample to the aproximation problem in Banach Spaces, *Acta Math.* 130, 309–317.

Enflo, P. (1987), On the invariant subspace problem in Banach Spaces, *Acta Math.* 158, 213–313.

Etheridge, A. (2002), *A Course in Financial Calculus*, Cambridge University Press, Cambridge.

Feintuch, A., and Saeks, R. (1982), *Systems Theory: A Hilbert Space Approach*, Academic, New York.

Giles, J. R. (2000), *Introduction to the Analysis of Normed Linear Spaces*, Cambridge University Press, Cambridge.

Halmos, P. R. (1974), *Measure Theory*, Springer-Verlag, New York.

Hewitt, E., and Stromberg, K. (1975), *Real and Abstract Analysis*, Springer-Verlag, New York.

Kadison, R. V., and Ringrose, J. R. (1983), *Fundamentals of the Theory of Operator Algebras*, Vol. I, Academic, New York.

Kelley, J. L. (1975), *General Topology*, Springer-Verlag, New York.

Lomonosov, V. I. (1973), On invariant subspaces of familes of operators, commuting with a compact operator, *Funk. Analiz i ego Prilozh*, 7, 55–56.

Read, C. J. (1984), A solution to the invariant subspace problem, *Bull. Lond. Math Soc.* 16, 337–401.

Read, C. J. (1985), A solution to the invariant subspace problem on ℓ^1, *Bull. Lond. Math Soc.* 17, 305–317.

Reed, S., and Simon, B. (1972), *Methods of Modern Mathematical Physics*, Vol I: *Functional Analysis*, Academic, New York.

Royden, H. L. (1988), *Real Analysis*, 3rd ed., Macmillan, New York.

Rudin, W. (1973), *Functional Analysis* McGraw-Hill, New York.

Rudin, W. (1974), *Real and Complex Analysis*, McGraw-Hill, New York.

Rudin, W. (1976), *Principles of Mathematical Analysis*, 3rd ed., McGraw-Hill, New York.

Schauder, J. (1927), Zur theorie stetiger abbildungen in funtionalraumen, *Mat. Zeit.*, 26, 47–65.

Scachermeyer, W. (1992), A Hilbert space proof of the fundamental theorem of asset pricing in discrete time, *Insurance: Mathematics and Economics*, 11, 249–257.

Schechter, M. (1971), *Principles of Functional Analysis*, Academic, New York.

Young, N. (1988), *An Introduction to Hilbert Space*, Cambridge University Press, Cambridge.

INDEX

Note: page references followed by *f* refer to figures.

A First Course in Functional Analysis. By S. David Promislow

PURE AND APPLIED MATHEMATICS

A Wiley-Interscience Series of Texts, Monographs, and Tracts

ADÁMEK, HERRLICH, and STRECKER—Abstract and Concrete Catetories
ADAMOWICZ and ZBIERSKI—Logic of Mathematics
AINSWORTH and ODEN—A Posteriori Error Estimation in Finite Element Analysis
AKIVIS and GOLDBERG—Conformal Differential Geometry and Its Generalizations
ALLEN and ISAACSON—Numerical Analysis for Applied Science
*ARTIN—Geometric Algebra
AUBIN—Applied Functional Analysis, Second Edition
AZIZOV and IOKHVIDOV—Linear Operators in Spaces with an Indefinite Metric
BASENER—Topology and Its Applications
BERG—The Fourier-Analytic Proof of Quadratic Reciprocity
BERMAN, NEUMANN, and STERN—Nonnegative Matrices in Dynamic Systems
BERKOVITZ—Convexity and Optimization in $\mathbb{R}^n$
BOYARINTSEV—Methods of Solving Singular Systems of Ordinary Differential Equations
BRIDGER—Real Analysis: A Constructive Approach
BURK—Lebesgue Measure and Integration: An Introduction
*CARTER—Finite Groups of Lie Type
CASTILLO, COBO, JUBETE, and PRUNEDA—Orthogonal Sets and Polar Methods in Linear Algebra: Applications to Matrix Calculations, Systems of Equations, Inequalities, and Linear Programming
CASTILLO, CONEJO, PEDREGAL, GARCIÁ, and ALGUACIL—Building and Solving Mathematical Programming Models in Engineering and Science
CHATELIN—Eigenvalues of Matrices
CLARK—Mathematical Bioeconomics: The Optimal Management of Renewable Resources, Second Edition
COX—Galois Theory
†COX—Primes of the Form $x^2 + ny^2$: Fermat, Class Field Theory, and Complex Multiplication
*CURTIS and REINER—Representation Theory of Finite Groups and Associative Algebras
*CURTIS and REINER—Methods of Representation Theory: With Applications to Finite Groups and Orders, Volume I
CURTIS and REINER—Methods of Representation Theory: With Applications to Finite Groups and Orders, Volume II
DINCULEANU—Vector Integration and Stochastic Integration in Banach Spaces
*DUNFORD and SCHWARTZ—Linear Operators
Part 1—General Theory
Part 2—Spectral Theory, Self Adjoint Operators in Hilbert Space
Part 3—Spectral Operators
FARINA and RINALDI—Positive Linear Systems: Theory and Applications

*Now available in a lower priced paperback edition in the Wiley Classics Library.
†Now available in paperback.

FATICONI—The Mathematics of Infinity: A Guide to Great Ideas
FOLLAND—Real Analysis: Modern Techniques and Their Applications
FRÖLICHER and KRIEGL—Linear Spaces and Differentiation Theory
GARDINER—Teichmüller Theory and Quadratic Differentials
GILBERT and NICHOLSON—Modern Algebra with Applications, Second Edition
*GRIFFITHS and HARRIS—Principles of Algebraic Geometry
GRILLET—Algebra
GROVE—Groups and Characters
GUSTAFSSON, KREISS and OLIGER—Time Dependent Problems and Difference Methods
HANNA and ROWLAND—Fourier Series, Transforms, and Boundary Value Problems, Second Edition
*HENRICI—Applied and Computational Complex Analysis
Volume 1, Power Series—Integration—Conformal Mapping—Location of Zeros
Volume 2, Special Functions—Integral Transforms—Asymptotics—Continued Fractions
Volume 3, Discrete Fourier Analysis, Cauchy Integrals, Construction of Conformal Maps, Univalent Functions
*HILTON and WU—A Course in Modern Algebra
*HOCHSTADT—Integral Equations
JOST—Two-Dimensional Geometric Variational Procedures
KHAMSI and KIRK—An Introduction to Metric Spaces and Fixed Point Theory
*KOBAYASHI and NOMIZU—Foundations of Differential Geometry, Volume I
*KOBAYASHI and NOMIZU—Foundations of Differential Geometry, Volume II
KOSHY—Fibonacci and Lucas Numbers with Applications
LAX—Functional Analysis
LAX—Linear Algebra and Its Applications, Second Edition
LOGAN—An Introduction to Nonlinear Partial Differential Equations, Second Edition
MARKLEY—Principles of Differential Equations
MORRISON—Functional Analysis: An Introduction to Banach Space Theory
NAYFEH—Perturbation Methods
NAYFEH and MOOK—Nonlinear Oscillations
O'NEIL—Beginning Partial Differential Equations, Second Edition
PANDEY—The Hilbert Transform of Schwartz Distributions and Applications
PETKOV—Geometry of Reflecting Rays and Inverse Spectral Problems
*PRENTER—Splines and Variational Methods
PROMISLOW—A First Course in Functional Analysis
RAO—Measure Theory and Integration
RASSIAS and SIMSA—Finite Sums Decompositions in Mathematical Analysis
RENELT—Elliptic Systems and Quasiconformal Mappings
RIVLIN—Chebyshev Polynomials: From Approximation Theory to Algebra and Number Theory, Second Edition
ROCKAFELLAR—Network Flows and Monotropic Optimization
ROITMAN—Introduction to Modern Set Theory
ROSSI—Theorems, Corollaries, Lemmas, and Methods of Proof
*RUDIN—Fourier Analysis on Groups
SENDOV—The Averaged Moduli of Smoothness: Applications in Numerical Methods and Approximations
SENDOV and POPOV—The Averaged Moduli of Smoothness
SEWELL—The Numerical Solution of Ordinary and Partial Differential Equations, Second Edition

*Now available in a lower priced paperback edition in the Wiley Classics Library.
†Now available in paperback.

SEWELL—Computational Methods of Linear Algebra, Second Edition
SHICK—Topology: Point-Set and Geometric
*SIEGEL—Topics in Complex Function Theory
Volume 1—Elliptic Functions and Uniformization Theory
Volume 2—Automorphic Functions and Abelian Integrals
Volume 3—Abelian Functions and Modular Functions of Several Variables
SMITH and ROMANOWSKA—Post-Modern Algebra
ŠOLÍN–Partial Differential Equations and the Finite Element Method
STADE—Fourier Analysis
STAKGOLD—Green's Functions and Boundary Value Problems, Second Editon
STAHL—Introduction to Topology and Geometry
STANOYEVITCH—Introduction to Numerical Ordinary and Partial Differential Equations Using MATLAB®
*STOKER—Differential Geometry
*STOKER—Nonlinear Vibrations in Mechanical and Electrical Systems
*STOKER—Water Waves: The Mathematical Theory with Applications
WATKINS—Fundamentals of Matrix Computations, Second Edition
WESSELING—An Introduction to Multigrid Methods
†WHITHAM—Linear and Nonlinear Waves
ZAUDERER—Partial Differential Equations of Applied Mathematics, Third Edition

*Now available in a lower priced paperback edition in the Wiley Classics Library.
†Now available in paperback.

Printed and bound by CPI Group (UK) Ltd, Croydon, CR0 4YY
16/09/2026
14963459-0004